Fuel, Energy and Net Zero

Richard Skiba

AFTER MIDNIGHT
PUBLISHING

Skiba, Richard (author)

Fuel, Energy and Net Zero

ISBN 978-1-7641699-0-5 (Paperback) 978-1-7641699-1-2 (eBook)

978-1-7641699-2-9 (Hardcover)

Non-fiction

Contents

Introduction

Fuel, Energy and Net Zero provides an exploration of the evolution, science, economics, and future of fuels in human society. Spanning ancient discoveries to cutting-edge innovations, this book guides readers through the critical role that fuels have played—and will continue to play—in shaping economies, technologies, and geopolitics. From fossil fuels such as coal, oil, and natural gas, to emerging alternatives like biofuels, green hydrogen, and synthetic fuels, the book examines the complex systems that sustain modern life and the urgent transition needed to build a sustainable energy future.

This work is especially relevant in today's context of climate change, energy security concerns, and accelerating technological transformation. It bridges scientific understanding with policy debates and market dynamics, making it suitable for students, professionals, and general readers alike who are interested in energy systems, environmental sustainability, or global development.

Readers will gain insights into the physical and chemical properties of various fuels, the history of fuel use in industrial societies, and the geopolitical and economic factors that shape global energy markets. They will also learn about renewable fuel technologies, decarbonisation strategies, carbon pricing, energy storage, and the post-fossil fuel transition—all framed within the global effort to achieve net-zero emissions by 2050.

Whether you are an environmental science student, a policymaker, an industry professional, or simply a curious reader, this book offers a clear and engaging roadmap to understanding one of the most critical topics of our time: how we power our world, and how we must change it.

Part 1

Understanding Fuel

Chapter 1

What Is Fuel?

In a world racing toward decarbonisation, the term *fuel* is no longer confined to petrol pumps or coal-fired power plants—it is central to global discussions on climate policy, energy transition, and technological innovation. *Fuel, Energy and Net Zero* sets the stage for understanding how society's evolving relationship with energy is driving transformation across every sector, from transportation and industry to homes and cities. Before we can chart a path to net zero emissions, we must first return to basics: What exactly is fuel?

Chapter 1, *What Is Fuel?*, opens this conversation by unpacking the fundamental concept of fuel as a source of stored energy. It examines the physical and chemical principles that make substances useful for powering machines, generating electricity, and sustaining modern life. From ancient wood fires to advanced synthetic fuels, this chapter explores how fuel is defined, categorised, and measured. It also introduces the underlying physics—such as energy content and combustion—that determine fuel performance.

Understanding what fuel is provides the foundation for exploring its history, forms, applications, and future directions in the chapters that follow. As you begin this chapter, consider how something as seemingly simple as "fuel" has shaped economies, ecosystems, and geopolitical landscapes—and how redefining it will shape the future.

Definition and Purpose

We need fuel because it is one of the most important sources of energy that powers our modern world. Fuel provides the stored energy necessary to perform work, drive machines, generate electricity, and support basic human activities like cooking, heating, and transportation. Whether in the form of fossil fuels, biofuels, or emerging alternatives, fuel is essential to the functioning of societies, economies, and daily life.

At its core, fuel enables the conversion of stored chemical energy into usable energy—such as heat, motion, or electricity. In homes, fuel allows us to cook food, heat water, and stay warm during cold seasons. In industry, it powers the engines, machines, and furnaces that produce everything from metals and chemicals to textiles and electronics. Fuel is also fundamental to agriculture, powering equipment like tractors and irrigation systems that help feed growing populations.

One of the most visible roles of fuel is in transportation. Fuels like gasoline, diesel, aviation fuel, and natural gas power the cars, trucks, trains, ships, and planes that move people and goods across the globe. Without fuel, supply chains would break down, travel would grind to a halt, and economic activity would be severely disrupted.

Fuel is also essential in electricity generation, especially in regions that rely heavily on fossil fuels like coal, oil, or natural gas to produce power. Even as renewable energy sources grow, many countries still depend on fuel-based power plants to maintain a stable energy supply, especially during peak demand or when solar and wind sources are unavailable.

In emergency situations, backup generators that run on fuel ensure continuity in hospitals, communication systems, and disaster response operations. Fuel even plays a principal role in space exploration, where high-energy fuels are needed to launch rockets and power spacecraft.

We need fuel because it is a reliable, portable, and versatile energy source that supports almost every aspect of human life. From powering homes and factories to driving vehicles and supporting global infrastructure, fuel continues to be an essential foundation of modern civilization. As we face environmental challenges, the goal is not to eliminate fuel use entirely but to shift toward cleaner, more sustainable fuels that meet our needs while protecting the planet.

The concept of fuel encompasses any material capable of being burned or otherwise consumed in reactions that release stored chemical potential energy. Fundamentally, this energy is liberated through combustion—a process in which fuel reacts, typically with oxygen, to produce heat. Additionally, energy release can occur through other chemical and physical reactions [1, 2]. The calorific value of a fuel, which quantifies the total energy it can deliver, is a core measure used in both scientific research and industrial applications to ensure stability and predictability in energy conversion processes [1].

This released energy can be directly converted into several useful forms. First, heat energy, often harnessed for domestic purposes such as cooking and heating, results from the exothermic nature of combustion reactions [2]. Second, mechanical energy is generated when the chemical energy stored within the fuel is converted into work, for example, in the internal combustion engines of vehicles and various industrial machines [2]. Lastly, electrical energy, critical for powering homes and cities, can be derived from generators linked to combustion processes, or through advanced methods like fuel cells, where the conversion from chemical to electrical energy occurs without intermediate mechanical work [3]. Research into novel platforms—such as thermopower waves in nanostructured materials—illustrates emerging methods to directly convert combustion energy into electrical power, thus expanding the spectrum of energy conversion (Lee et al., 2014).

The cascade utilization of energy highlights the potential of fuel to be deployed in multiple successive applications, where the primary conversion process may yield heat that is then used to generate mechanical or electrical energy. Such integrated energy systems provide theoretical frameworks and practical designs for optimally harnessing fuels' inherent chemical potential in multi-step processes [4]. In this context, fuels are essential not only to conventional combustion-based applications but also to emerging, more sophisticated energy technologies aiming to improve efficiency and reduce environmental impact [3].

Cascade utilization of energy refers to the process of using energy from fuel in multiple steps rather than in a single-use application. Instead of letting heat go to waste after the initial conversion, it is captured and used again in another form. For instance, fuel might first be burned to produce heat, and then that heat is reused to generate electricity or power machinery.

This approach allows us to extract more useful energy from the same amount of fuel, which significantly increases overall efficiency and reduces waste. By doing so, we maximize the energy potential locked within the fuel.

Cascade utilization is important because it offers a smarter and more sustainable way to use fuel. By making each step of the energy conversion process more efficient, we reduce the amount of fuel needed to achieve the same outcomes. This leads to lower fuel consumption and less environmental harm due to decreased emissions and waste.

This principle is already being applied in advanced energy systems such as combined heat and power (CHP) plants, cogeneration systems, and some renewable energy technologies. These systems are designed to capture and reuse heat that would otherwise be lost, making them far more efficient than traditional single-use methods.

Fuel doesn't have to be used just once and then discarded. Through cascade utilization, it can be part of a chain of processes where the energy is reused at different stages. This not only improves efficiency but also supports more sustainable and environmentally friendly energy systems.

Overall, the definition of fuel as a material that can be burned to release energy, and its subsequent conversion to heat, mechanical work, and electricity, forms the cornerstone of a broad spectrum of energy conversion technologies. The literature consistently reflects that the principles of combustion and energy conversion remain critical both in traditional settings (e.g., heating and transportation) and in advanced applications (e.g., fuel cell technologies and thermopower wave systems) [1-4].

Fuel is central to powering nearly every aspect of modern society, serving as the primary energy source for multiple sectors including transportation, electricity generation, heating and cooking, industry and manufacturing, agriculture, and military and aviation applications. Foundational texts on energy systems affirm that fossil fuels and other combustible materials are integral to societal frameworks, driving innovation and economic growth across sectors [5, 6].

In transportation, fuels such as petrol (gasoline) and diesel are extensively used in cars, trucks, buses, and motorcycles. Jet fuel is formulated specifically to power airplanes, while marine fuels are tailored for ships and tankers. Additionally, biofuels have emerged as sustainable alternatives for both road vehicles and aircraft. Detailed analyses of modern vehicular systems show that diesel engines and gas turbines remain integral for vehicle propulsion, evidencing the critical role of fuel in ensuring efficient transportation [7]. Research on urban energy demand models further emphasizes the importance of conventional fuels in supporting mass transit and aviation, while also noting the increasing role of biofuels in reducing environmental footprints [8].

Electricity generation is another principal application of fuels. In power plants, coal, natural gas, oil, and biomass are combusted to produce steam, which drives turbines to generate electricity. This conventional method remains indispensable not only in large-scale power generation but also in backup generators that ensure grid stability during outages [6]. The reliance on fuel-derived energy for electricity production is well documented in studies that analyse the structure and sustainability of modern energy systems [5].

Domestic heating and cooking heavily depend on various fuels. Fuels such as natural gas, wood, charcoal, and kerosene are commonly utilized in residential and commercial settings to provide heat for buildings and to cook food. Investigations into household energy consumption have revealed that fuel stacking—a practice of using a mix of traditional and modern fuels—is widespread, often driven by socioeconomic factors and efficiency needs [9]. This multifaceted fuel usage in the household sector is crucial for energy consumption dynamics and the transition towards cleaner energy sources [9].

The industrial and manufacturing sectors depend significantly on fuel for running machines and providing the necessary thermal energy for processing and refining materials. In sectors such as metal and glass production and in oil refining, fuels are combusted to achieve the high temperatures required for these processes. Integrated energy system analyses indicate that industries are heavily reliant on a steady supply of energy from fuels to maintain operational efficiency and competitive production [5].

In agriculture, fuel is essential for powering tractors and other farm equipment through diesel and petrol engines. Fuels are also used to heat greenhouses and facilitate the drying of crops, critical to modern agricultural practices. Recent studies point to the evolving nature of agricultural energy solutions, including the adoption of decarbonized fuels in mobile energy vehicles used for agricultural production, underscoring the sector's transition toward more sustainable energy use [10].

Military and aviation applications exemplify the specialization in fuel formulations. While conventional fuels such as jet fuel dominate commercial aviation, dedicated formulations and specialized fuels have been developed particularly for military vehicles and advanced weapons systems, enhancing performance and reliability under extreme conditions. Reviews of energy systems confirm that these specialized fuels are crucial to national defence and advanced aeronautical applications, reflecting stringent quality and performance requirements [5].

Fuel functions as the backbone of modern energy usage, supporting diverse and critical applications from transportation and electricity generation to domestic, industrial, agricultural, and military needs. The extensive literature on energy systems and fuel technology consistently highlights how fuel-based energy conversion remains indispensable while also evolving toward cleaner alternatives in response to environmental and economic challenges [5].

The following provides examples of different types of fuel, grouped by category:

Solid Fuels: These are typically burned to produce heat and energy.

- **Coal** – Used in power plants and industrial furnaces.
- **Wood** – Used for heating and cooking (especially in rural areas).
- **Charcoal** – Used for cooking and in barbecues.
- **Peat** – Partially decayed plant matter, used as a traditional fuel in some regions.
- **Pellets** – Compressed biomass (like wood waste), used in pellet stoves.

Liquid Fuels: These are commonly used in transportation and heating.

- **Petrol (Gasoline)** – Fuels cars, motorcycles, and light trucks.
- **Diesel** – Used in trucks, buses, trains, and some cars.
- **Kerosene** – Used in jet engines and as heating fuel.
- **Heavy Fuel Oil** – Used in ships and industrial engines.
- **Biodiesel** – A renewable fuel made from vegetable oils or animal fats.
- **Ethanol** – Often blended with petrol to reduce emissions.

Gaseous Fuels: Gases used as fuel are often more efficient and cleaner burning.

- **Natural Gas (Methane)** – Used for heating, cooking, and electricity generation.
- **LPG (Liquefied Petroleum Gas)** – Used in home heating and cooking; also powers some vehicles.
- **Hydrogen** – A clean fuel for fuel cell vehicles and emerging technologies.
- **Biogas** – Produced from the decomposition of organic matter (e.g., in landfills or digesters).

Each type of fuel has its own advantages, disadvantages, and uses, depending on the application, location, and available technology.

Modern fuels are widely used today and benefit from established infrastructure and technology. These fuels support current global transportation, industry, and energy systems while offering improved efficiency and reduced emissions compared to traditional options.

Compressed Natural Gas (CNG) is commonly used in buses, taxis, and some private vehicles. It burns cleaner than petrol or diesel, resulting in lower emissions and less air pollution in urban areas.

Liquefied Natural Gas (LNG) is used in heavy-duty trucks, ships, and power plants. It has a higher energy density than CNG and is easier to transport in large volumes, making it suitable for long-distance and industrial applications.

Biodiesel is a renewable fuel made from vegetable oils or animal fats. It can be used in most diesel engines with little or no modification, offering a more sustainable alternative to petroleum-based diesel.

Ethanol is an alcohol-based fuel derived from crops such as corn or sugarcane. It is often blended with petrol—such as in E10 or E85 fuel mixes—to reduce greenhouse gas emissions and dependence on fossil fuels.

Electricity (Battery Storage), although not a traditional fuel, powers electric vehicles (EVs) and is an increasingly important part of modern transportation. Electricity can be generated from both renewable sources like solar and wind, and from fossil fuels, making its environmental impact dependent on the source of generation.

Emerging fuels are still in development or early adoption stages, with the goal of providing low-carbon, renewable alternatives to traditional fossil fuels. These innovative fuels aim to improve sustainability and reduce environmental impact.

Green Hydrogen is produced by using renewable electricity to split water molecules through a process called electrolysis. It emits no greenhouse gases when used in fuel cells or burned and is ideal for heavy transport, industrial processes, and long-term energy storage.

Ammonia (NH_3) is being explored as a fuel for shipping and power generation. It can be used directly or serve as a hydrogen carrier, offering a potential zero-carbon solution when produced from renewable sources.

Synthetic Fuels (E-fuels) are created by combining captured carbon dioxide (CO_2) with green hydrogen. These fuels can directly replace petrol, diesel, or jet fuel in existing engines and are considered carbon-neutral if produced using renewable energy.

Algae-Based Biofuels are derived from microalgae that naturally produce oil. These oils can be refined into biodiesel or jet fuel. Algae grow rapidly and don't compete with food crops, making this a promising and sustainable biofuel option.

Fuel, Energy and Net Zero

Dimethyl Ether (DME) is a clean-burning alternative to diesel and LPG. It can be produced from biomass or natural gas and is compatible with modified diesel engines.

Methanol, made from natural gas, biomass, or captured CO_2, is used as fuel in some race cars and ships. It is also a key ingredient in the production of other chemicals and synthetic fuels, making it a versatile and potentially greener fuel option.

As further examples, the following describes fuels used for rocket propulsion and space travel.

One of the most commonly used fuel types in modern space travel is liquid propellant, which typically consists of a liquid fuel and a liquid oxidizer. A classic example is the combination of liquid hydrogen (LH_2) and liquid oxygen (LOX), used in the main engines of NASA's Space Shuttle and in the upper stages of many launch vehicles. When ignited, this mixture produces extremely high thrust and clean byproducts—mainly water vapor. Another well-known liquid propellant is RP-1, a highly refined kerosene, often used with LOX. This combination powers rockets such as the Falcon 9 from SpaceX and the historical Saturn V first stage. RP-1 is denser and easier to handle than hydrogen, though it produces more carbon emissions.

Solid rocket fuels are pre-mixed combinations of fuel and oxidizer in a solid state. They are known for their simplicity, reliability, and ability to provide instant thrust. These fuels are used in booster rockets like those on the Space Shuttle's solid rocket boosters (SRBs) and intercontinental ballistic missiles (ICBMs). A typical solid propellant might include ammonium perchlorate as the oxidizer and powdered aluminium as the fuel, mixed into a rubbery binder that also acts as a fuel. Once ignited, solid fuel burns rapidly and cannot be stopped or throttled, making it ideal for quick, powerful liftoffs.

Hypergolic fuels ignite instantly upon contact with an oxidizer, making them valuable for spacecraft that need reliable engine restarts, such as for orbital manoeuvring or descent stages. A common hypergolic fuel is unsymmetrical dimethylhydrazine (UDMH) paired with nitrogen tetroxide (N_2O_4). These were used in the Apollo Lunar Module, Soviet Soyuz spacecraft, and many interplanetary missions. Hypergolic propellants are highly toxic and corrosive, but they provide simplicity in ignition systems and long-term storability, which is critical for deep space missions.

Future and experimental propulsion systems are exploring advanced fuels and technologies. Ion propulsion systems, for example, use gases like xenon, which are electrically charged (ionized) and expelled at high speeds to produce thrust. These systems are extremely fuel-efficient but provide very low thrust, suitable for long-duration deep space missions such as NASA's Dawn spacecraft. Nuclear thermal propulsion is another concept under development, where a nuclear reactor heats a propellant like hydrogen, which then expands through a nozzle to create thrust. This technology promises faster travel to distant planets like Mars.

Energy and Combustion

The concepts of energy and combustion are central to any discussion about fuel because they explain how fuels work, why they are valuable, and how they are applied in real-world systems.

Understanding these concepts allows us to measure, compare, and improve the performance and sustainability of various fuel types.

At the core of fuel usage is energy—the capacity to do work. Fuels are essentially carriers of stored energy, and the reason we use them is to release that energy in useful forms, such as heat, motion, or electricity. Whether we're driving a car, powering a factory, flying a plane, or heating a home, the energy stored in fuels is what makes it all possible. Without an understanding of how energy is stored, transferred, and used, we can't evaluate the true effectiveness of one fuel over another or design systems that use fuel efficiently.

Combustion is the main method through which fuel's stored chemical energy is released. It is a chemical reaction—usually between a fuel and oxygen—that produces heat and light. This heat is then converted into mechanical or electrical energy through engines or turbines. Understanding combustion helps us identify the conditions needed for complete fuel burning, which leads to maximum energy output and minimal emissions. Concepts like stoichiometric ratios, efficiency, and emission control all depend on a deep understanding of combustion.

In practical terms, knowledge of energy and combustion enables engineers and scientists to:

- Select the most appropriate fuel for a specific task, considering factors like energy density, environmental impact, and availability.

- Design efficient engines and power systems that use fuel effectively and safely.

- Improve combustion technologies to reduce fuel consumption and harmful emissions.

- Explore alternatives to combustion, such as fuel cells or electric power, especially for cleaner energy solutions.

These concepts are also essential when addressing modern challenges like climate change, fuel sustainability, and energy transition. To reduce the global reliance on fossil fuels and move toward renewable energy, we must understand how fuels work at a chemical and energetic level so we can develop cleaner alternatives that still meet society's power demands.

Energy is the capacity to do work or produce change. In the context of fuel and power systems, energy is stored in the chemical bonds of substances and can be released through processes like combustion. Combustion is a chemical reaction, usually between a fuel and an oxidizer (most commonly oxygen from the air), that produces heat and light. This reaction transforms the chemical energy stored in the fuel into thermal energy, which can then be used to generate motion, electricity, or other forms of useful energy.

At its core, combustion is an exothermic reaction, meaning it releases energy. When a fuel like wood, coal, petrol, or natural gas burns, its carbon (C) and hydrogen (H) atoms react with oxygen (O_2). The carbon forms carbon dioxide (CO_2), and the hydrogen forms water (H_2O), releasing heat in the process. The energy output of this reaction depends on the type of fuel and its composition. For instance, hydrogen-rich fuels like natural gas tend to burn hotter and cleaner than carbon-heavy fuels like coal.

Fuel, Energy and Net Zero

The energy released from combustion is often harnessed in mechanical systems such as car engines, turbines, and power plants. In a car engine, for example, fuel combustion pushes pistons, which turns the crankshaft and drives the vehicle. In power stations, the heat from burning fuel turns water into steam, which spins turbines connected to electricity generators.

Combustion can occur in open systems, like a fireplace, or in controlled environments, such as engines or furnaces. The efficiency of combustion—the percentage of fuel's energy that is actually converted into useful energy—can vary. Incomplete combustion, for example, happens when there is not enough oxygen, producing carbon monoxide (CO) and soot instead of CO_2 and water. This not only reduces efficiency but also increases pollution.

Modern technology focuses on improving combustion efficiency while minimizing harmful emissions. This includes better engine design, cleaner-burning fuels, and emission control systems. Additionally, alternative energy systems like fuel cells and electric power aim to reduce or replace combustion entirely to reduce environmental impact.

The amount of energy that can be extracted from a fuel depends on two main factors: its specific energy and its stoichiometric air–fuel ratio. These factors together determine how efficiently and completely a fuel can be burned in a controlled environment, such as an engine or a combustion chamber.

Specific energy refers to the amount of energy released per unit mass of fuel, typically measured in megajoules per kilogram (MJ/kg). This tells us how much energy is stored in each kilogram of the fuel. For instance, hydrogen has a very high specific energy of 142 MJ/kg, while methanol is much lower at 19.7 MJ/kg. This makes hydrogen an extremely energy-dense fuel by weight.

Specific energy is a fundamental concept in energy science and engineering that refers to the amount of energy stored in a given mass of a substance. It is typically expressed in units such as megajoules per kilogram (MJ/kg) or sometimes kilowatt-hours per kilogram (kWh/kg). The term is used to describe how much energy a fuel or material can release per kilogram when it is fully consumed or converted. This measurement is especially important when comparing different types of fuels and energy storage systems, as it helps determine which fuels are more energy-dense and efficient by weight.

The higher the specific energy of a fuel, the more energy it can deliver for each kilogram used. For example, hydrogen has an exceptionally high specific energy of about 142 MJ/kg, which means it can release more energy per kilogram than most other fuels. In contrast, methanol has a much lower specific energy of around 19.7 MJ/kg, meaning more methanol would be required to produce the same amount of energy as a smaller quantity of hydrogen. This is why hydrogen is often considered a promising fuel for future applications, especially in aerospace and long-distance transportation, where weight is a critical factor.

Specific energy is not only important in combustion-based fuels like diesel, gasoline, and natural gas, but also in batteries and other energy storage systems. For instance, lithium-ion batteries, commonly used in electric vehicles and portable electronics, have specific energy values in the range of 100–250 Wh/kg (which converts to about 0.36–0.90 MJ/kg), significantly lower than conventional fuels. This comparison explains why, despite their environmental advantages, electric vehicles currently face limitations in driving range and require heavier battery packs compared to the lightweight energy storage of liquid fuels.

In practical terms, specific energy helps engineers and designers make informed decisions about which energy source to use for a particular application. For air travel, where every kilogram matters, fuels with high specific energy are ideal. For stationary energy storage, where space and weight are less important, fuels or systems with lower specific energy might still be acceptable if they offer other advantages like safety, availability, or environmental benefits.

However, simply knowing the energy per kilogram is not enough. Combustion requires oxygen, usually supplied through air, and the stoichiometric air–fuel ratio tells us the chemically correct amount of air needed to completely burn a specific fuel. This ratio ensures that both the fuel and the oxygen are fully consumed in the combustion process, with no leftover fuel (which causes pollution) or excess air (which cools the flame and reduces efficiency). For example, gasoline requires about 14.7 parts of air for every 1 part of fuel, while hydrogen requires much more— 34.3:1—due to its light weight and high reactivity.

The stoichiometric ratio, also known as the stoichiometric air–fuel ratio, is the chemically correct proportion of air to fuel needed for complete combustion. This means that all of the fuel and all of the oxygen in the reaction are entirely consumed, leaving no excess fuel or oxygen. When this balance is achieved, the reaction is most efficient, producing the maximum possible energy output and minimizing the creation of harmful byproducts such as carbon monoxide (CO) or unburned hydrocarbons.

During combustion, oxygen from the air reacts with the hydrocarbons present in fuel. The precise air–fuel ratio required for complete combustion depends on the chemical makeup of the fuel, since different fuels contain different amounts of carbon, hydrogen, and sometimes oxygen. Gasoline, for example, requires about 14.7 parts of air for every one part of fuel to achieve stoichiometric combustion. Ethanol, which already contains oxygen in its molecular structure, requires less air, with a stoichiometric ratio of approximately 9:1. Hydrogen, being a very light and reactive element, needs significantly more air to burn completely, with a stoichiometric ratio of around 34.3:1.

The air–fuel equivalence ratio, denoted by λ (lambda), is a way of describing whether the mixture is rich, lean, or perfectly balanced. A λ value of 1 indicates a stoichiometric mixture, where the air and fuel are in the exact proportions needed for complete combustion. This is the ideal condition assumed in most fuel efficiency and emission calculations.

While the stoichiometric ratio represents the ideal mixture, real-world engines often adjust the air–fuel balance depending on performance needs. Engines running a lean mixture have more air than the stoichiometric requirement (λ > 1). This can improve fuel economy and reduce some pollutants, but it may also lead to higher combustion temperatures and increased nitrogen oxide (NO_x) emissions. On the other hand, engines running a rich mixture have less air than needed (λ < 1), which can produce more power but also results in higher emissions and less efficient fuel use.

Fuel, Energy and Net Zero

Understanding the stoichiometric ratio is essential for designing and operating efficient combustion systems. It helps engineers develop engines and energy systems that deliver optimal performance while controlling emissions and minimizing fuel waste. In both automotive and industrial applications, maintaining or carefully adjusting the stoichiometric ratio is a key factor in achieving clean, efficient, and effective combustion.

When we combine the specific energy and the stoichiometric air–fuel ratio, we can calculate how much energy is released per kilogram of air, which is useful for understanding how much usable energy we get in practical combustion systems.

Table 1: Energy capacities of common types of fuel.

Fuel type	Specific energy (MJ/kg)	Air–fuel ratio (stoichiometric)	Energy @ λ=1 (MJ/kg$_{(Air)}$)
Diesel	48	14.5: 1	3.310
Ethanol	26.4	9: 1	2.933
Gasoline	46.4	14.7: 1	3.156
Hydrogen	142	34.3: 1	4.140
Kerosene	46	15.6: 1	2.949
LPG	46.4	17.2: 1	2.698
Methanol	19.7	6.47: 1	3.045
Methane	55.5	17.2: 1	3.219
Nitromethane	11.63	1.7: 1	6.841

This is shown in the column labelled "Energy @ λ=1 (MJ/kg(Air))". It gives a more realistic picture of the fuel's performance in air. For example:

- Nitromethane, although it has a low specific energy of 11.63 MJ/kg, has a very low air–fuel ratio of 1.7:1, meaning it needs very little air to burn. This gives it a surprisingly high energy per kilogram of air—6.841 MJ/kg(Air)—which makes it useful in drag racing engines where extreme short bursts of power are needed.

- Hydrogen, with the highest specific energy, also performs well in terms of energy per kilogram of air (4.140 MJ/kg(Air)), even though it needs a large volume of air to burn. This is why it's considered a top candidate for clean energy applications.

Other fuels such as diesel, gasoline, and methane offer a balance between energy density and air requirements. Diesel, for example, has a high specific energy of 48 MJ/kg and a typical air–fuel ratio of 14.5:1, yielding 3.310 MJ/kg(Air). Methane, used in natural gas systems, also performs well with 3.219 MJ/kg(Air).

In practice, these values help engineers and scientists design efficient combustion systems, such as engines and turbines, by selecting the right fuel for the application, balancing performance, fuel availability, and emissions. It also highlights why some fuels are more suited to specific uses—hydrogen for space and clean transport, diesel for heavy-duty engines, and LPG or ethanol for cleaner-burning alternatives.

Hydrocarbon fuels such as diesel, gasoline, and kerosene are known to possess high specific energy values, typically in the range of 46–48 MJ/kg. Their stoichiometric air–fuel ratios are generally around 14–15:1, which results in energy releases per kilogram of air (MJ/kg(Air)) in the vicinity of 3.1–3.3 MJ/kg(Air). Such data are reflective of the concentrated carbon–hydrogen bonds in these fuels and have been consistently reported in works analysing stoichiometric combustion through air oxidation equations [11, 12]. These values underscore the energy density of traditional hydrocarbon fuels and provide a benchmark for comparison with oxygenated and alternative fuels.

Oxygenated fuels, by contrast, tend to exhibit lower specific energy values due to the intrinsic presence of oxygen in their molecular structure. Ethanol and methanol, for example, have specific energies of approximately 26.4 MJ/kg and 19.7 MJ/kg, respectively. Their corresponding stoichiometric air–fuel ratios are lower than those of hydrocarbons (about 9:1 for ethanol and 6.47:1 for methanol), leading to energy outputs per kilogram of air that are slightly less than those for diesel or gasoline, despite maintaining reasonable combustion characteristics [11]. In some applications, these fuels are blended with hydrocarbon fuels to leverage their high octane numbers and improved emission profiles while attempting to sustain energy delivery [11].

Hydrogen presents a markedly different profile, with an exceptionally high specific energy of roughly 120 MJ/kg. However, its very lean stoichiometric air–fuel ratio (approximately 34.3:1) means that a larger mass of air is required to achieve complete combustion. Nevertheless, when normalized to the mass of air, hydrogen delivers an energy value of about 3.0 MJ/kg(Air), which is beneficial in contexts where weight is a critical factor [13]. Similarly, liquefied petroleum gas (LPG) and methane, which are commonly used in alternative fuel scenarios, have specific energy values of approximately 46.4 MJ/kg and 55.5 MJ/kg, respectively, with stoichiometric ratios near 17.2:1. These figures ensure that—with normalization to the amount of air—the energy release remains competitive with that of conventional fuels [12].

Nitromethane, though less common and primarily used in high-performance or specialized applications (such as drag racing), exhibits a unique behaviour with a low specific energy of about 11.63 MJ/kg; its stoichiometric air–fuel ratio is exceptionally low (around 1.7:1), resulting in a high energy-per-unit-air mass of approximately 6.84 MJ/kg(Air). This anomaly is attributed to its high oxygen content, which facilitates combustion with a relatively small volume of air [11]. Despite its lower overall energy content, the high combustion intensity of nitromethane makes it a point of interest in niche applications.

Fuel vs. Energy

The concepts of fuel and energy, although interconnected, represent fundamentally distinct ideas crucial for understanding energy systems and their applications. Fuel is defined as a material—be it solid, liquid, or gas—that possesses stored chemical energy, which can be released through various processes such as combustion or electrochemical reactions. Common examples include coal, gasoline, natural gas, and alternative sources like ethanol and hydrogen. Fuels, therefore, act as carriers of energy but do not constitute energy in and of themselves; rather they are the mediums through which energy is unlocked and made usable [14, 15].

Energy, conversely, refers to the capacity to perform work or instigate change and is not a tangible substance. It manifests in various forms, including thermal (heat), mechanical, electrical, chemical, and kinetic energy. The transition of energy from one form to another underpins the utility of fuels. For instance, when wood is combusted, the chemical energy contained within its structure is transformed into thermal energy, which can subsequently be utilized for heating or cooking [16, 17]. This illustrates the dynamic where fuel functions primarily as the source, and energy is the resultant product of its conversion [18].

Understanding the distinction between fuel and energy has significant implications, especially in the context of modern applications across environmental policy, engineering, and energy efficiency strategies. As nations transition towards more sustainable practices, the selection of fuels and the pursuit of renewable energy sources become pivotal in reducing dependency on fossil fuels and mitigating environmental impacts. The efficacy of energy systems hinges on the effective transformation of fuel into usable energy forms, necessitating a careful consideration of energy efficiency, emissions, and technological advancements in fuel production and utilization [19, 20].

The necessity for efficient energy use is underscored by the global commitment to sustainable energy transitions. As economic development progresses, the "energy ladder" theory suggests that improvements in income lead individuals to migrate from traditional biomass fuels to cleaner, more modern energy sources [21, 22]. This shift is indicative of a broader movement towards accessing not only cleaner but also more efficient energy solutions, facilitating growth without compromising environmental integrity or resource availability. Therefore, delineating the roles of fuel as a source and energy as its product is critical for informed decision-making in energy policy and management [23, 24].

Energy Density

In physics, energy density refers to the amount of energy stored in a given volume of a system or material. It is calculated by dividing the total energy by the volume in which it is contained, resulting in units such as joules per cubic meter (J/m^3). This measurement is crucial for evaluating how much usable energy a substance or system can deliver in relation to its physical space. Typically, when we talk about energy density in practical contexts, we are referring to the extractable or usable energy, not necessarily the total theoretical energy present.

It's important to distinguish energy density from specific energy. While energy density measures energy per unit volume, specific energy (also called gravimetric energy density) refers to energy per unit mass, typically measured in megajoules per kilogram (MJ/kg). These two properties help us evaluate and compare fuels or energy storage systems based on how much energy they contain and how compact or lightweight they are.

Different types of energy storage correspond to different types of reactions. These reactions vary in the amount of energy they release. At the highest end of the energy spectrum are nuclear reactions, which occur in stars and nuclear reactors. These release enormous amounts of energy due to changes in the nucleus of atoms. Next are chemical reactions, such as combustion and electrochemical processes, which power cars, machines, and biological organisms. Liquid hydrocarbons—such as gasoline, diesel, and kerosene—are among the densest and most practical chemical fuels available today. For instance, 1 kg of diesel burns with about 15 kg of air, demonstrating its high chemical energy density.

Other sources of energy include electrical energy, pressure energy, material deformation, and electromagnetic fields. In homes around the world, people rely on the combustion of local biomass fuels—like wood, charcoal, or kerosene—for heating, cooking, and lighting. In electronics, electrochemical energy stored in batteries powers devices like phones and laptops through controlled chemical reactions.

Interestingly, energy density has the same physical units as pressure, and in many systems, it behaves in similar ways. For example, the energy density of a magnetic field can be expressed as an equivalent pressure. In thermodynamics, compressing a gas involves work that can be measured by multiplying pressure differences by changes in volume. This demonstrates how energy density, pressure, and the ability to perform work are deeply linked. In fields like cosmology and general relativity, energy density is part of the stress–energy tensor, which includes both mass energy and the effects of pressure, influencing the curvature of spacetime and the dynamics of the universe.

In chemical systems, the amount of energy we consider depends on the intended use. One measure is exergy, which represents the theoretical amount of useful thermodynamic work a system can perform under certain conditions. Another is the Gibbs free energy, which is particularly useful when calculating the electrical energy that can be extracted from chemical reactions under standard conditions. However, when fuels are burned for heat or used in heat engines, the most relevant measurement is the enthalpy change, also known as the heat of combustion.

The heat of combustion can be expressed in two ways: the higher heating value (HHV) and the lower heating value (LHV). The HHV includes all the heat released when the combustion products cool to room temperature and the water vapor condenses. The LHV, on the other hand, excludes this condensation heat and sometimes the full cooling effect, making it a more conservative estimate of usable heat. These values are important when comparing fuels for heating and engine applications.

Fuel, Energy and Net Zero

When it comes to energy storage, energy density is used to assess how much energy can be stored in a given volume, such as a fuel tank or battery. This is especially important in transportation, where space and weight are limited. Fuels with high volumetric energy density, like gasoline, are ideal for vehicles because they allow for long travel ranges without occupying too much space. In contrast, batteries like lithium-ion have relatively low energy density, meaning they must be larger or heavier to deliver the same amount of energy as a small tank of fuel.

Some values used to compare fuels may not include the weight of the oxidizer, which is often drawn from the air in practical systems. This is why substances that contain their own oxidizer, like gunpowder or TNT, may appear to have lower energy densities. The added mass of the oxidizer, and the energy used to release oxygen during combustion, affect their overall energy performance. In some cases, this leads to surprising comparisons—for instance, a sandwich might technically have a higher energy density than a stick of dynamite when measured without the need for external oxygen.

Because of gasoline's high energy density, finding alternatives that can store similar amounts of energy in the same volume is a major challenge. Hydrogen fuel cells and batteries are promising technologies but often have far lower energy densities. For example, the same mass of lithium-ion batteries might only offer about 2% of the driving range of gasoline. To match that range, significantly more volume or advanced storage solutions would be needed. Research into supercapacitors, solid-state batteries, and other emerging technologies is underway to improve energy density and charging speed.

No single energy storage solution is perfect in terms of specific power, specific energy, and energy density. For example, Peukert's law explains that in a lead-acid battery, the faster energy is drawn, the less energy is actually available, showing how usage rate impacts performance [25-27].

While energy density tells us how much energy is stored, it doesn't tell us how efficiently that energy can be converted into useful work. Real-world engines and systems lose energy to friction, heat, and other inefficiencies. As a result, the specific fuel consumption—how much fuel is used to perform a task—is always higher than the actual kinetic energy output.

It's also important to distinguish energy density from energy conversion efficiency and embodied energy, which refers to the total energy required to produce, refine, transport, and manage the environmental impact of a fuel or storage device. High-energy-density systems may still be environmentally costly, so energy choices must consider not just performance, but also climate impact, resource use, and long-term sustainability.

Chapter 2

A Brief History of Fuel Use

Having explored the definition, types, and scientific foundations of fuel in Chapter 1, we now turn to how fuel has shaped human civilisation over millennia. To understand today's energy challenges—and the urgency of moving toward net zero—it is vital to look back at how our use of fuel evolved, from primitive biomass to the industrial-scale extraction of coal, oil, and gas.

Chapter 2 takes us on a journey through time, examining the critical role of fuels in economic development, technological advancement, and global power dynamics. From early wood-burning hearths and charcoal kilns to the transformative age of steam, oil booms, and nuclear ambition, this chapter traces how each energy era redefined society's capabilities and constraints.

By understanding the historical patterns of fuel discovery, adoption, and dominance, we gain perspective on the cultural, political, and environmental legacies still influencing energy decisions today. This context is essential as we examine the modern fuel landscape—and the shift it must now undergo—to meet the demands of a sustainable and carbon-neutral future.

Early Human Use of Biomass

Biomass as a Fuel

Biomass is organic material that originates from plants, animals, and microorganisms. It includes a wide range of natural substances such as wood, crop residues, agricultural waste, animal manure, algae, and even organic waste generated from households and industries. Biomass is regarded as a renewable energy source because its raw materials—plants and organic matter—can be regrown or replenished within a relatively short time frame, especially when compared to fossil fuels that take millions of years to form.

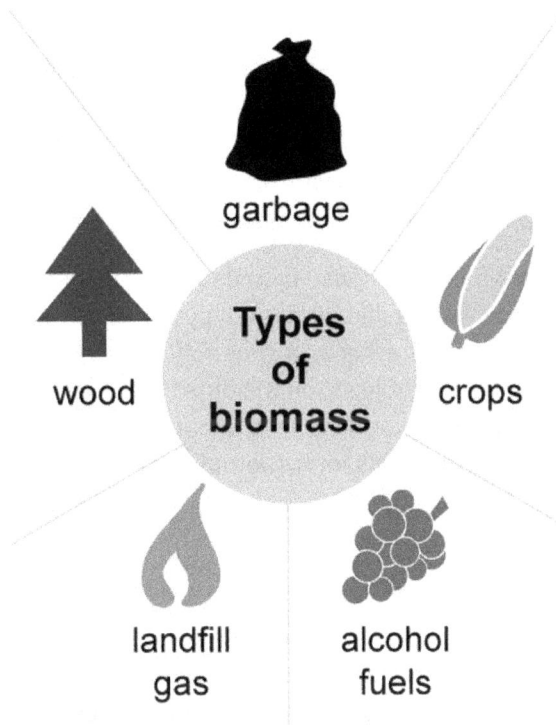

Figure 1: Types of biomass. Shunkehe, CC BY-SA 4.0, via Wikimedia Commons.

The value of biomass as a fuel lies in the chemical energy stored within its organic molecules, primarily carbon, hydrogen, and oxygen. This energy originally comes from the sun. Plants absorb solar energy during photosynthesis and store it in the form of carbohydrates, oils, and other compounds. When biomass is burned or otherwise processed, this stored energy is released, making it available for use in various forms of energy production.

One of the most traditional uses of biomass is direct combustion. This involves burning biomass materials to generate heat, often for cooking, space heating, or producing steam in industrial settings. Materials like wood logs, wood chips, and dried agricultural waste are commonly used in open fires or specifically designed stoves and furnaces, especially in rural and developing regions.

Biomass can also be used to generate electricity. In power plants, organic materials are burned to produce steam, which drives turbines that in turn generate electrical power. Some modern facilities use a method called co-firing, where biomass is burned alongside coal to lower greenhouse gas emissions while maintaining reliable power output.

Another important use of biomass is in biogas production. This process involves anaerobic digestion, where organic material decomposes in the absence of oxygen. The result is biogas—a mixture of methane and carbon dioxide—which can be used for cooking, heating, or producing electricity. Common feedstocks for biogas include animal manure, food waste, and sewage sludge, making it a valuable solution for both energy production and waste management.

Biomass is also converted into liquid fuels known as biofuels, which are used in the transportation sector. Ethanol, derived from crops like corn or sugarcane, is often blended with petrol and used in conventional petrol engines. Biodiesel, produced from vegetable oils or animal fats, is suitable for diesel engines. These fuels are renewable and compatible with existing vehicle technologies, offering a cleaner alternative to petroleum-based fuels.

In addition to these conventional methods, biomass can be processed through advanced conversion technologies such as gasification and pyrolysis. Gasification involves heating biomass in limited oxygen to create syngas, a fuel-rich gas that can be used for electricity generation or chemical production. Pyrolysis, on the other hand, involves heating biomass in the absence of oxygen to produce bio-oil, gases, and biochar. These technologies are aimed at producing more efficient and cleaner fuels for industrial use and emerging energy systems.

Historical Use of Biomass

The history of fuel use reflects a long evolutionary process that begins with early hominids and extends to the complex energy regimes of modern society. It is widely acknowledged that the combustion of wood and other biomass sources was employed by early human ancestors, with researchers suggesting that even Homo erectus may have used fire for warmth, cooking, and protection [28]. This early use of organic fuel, derived from plants or animal fat, remained the dominant energy source for the vast majority of human history until more concentrated fuels were harnessed.

Figure 2: Wood as fuel for combustion. Scott Marley from Falkirk, Scotland, CC BY 2.0, via Wikimedia Commons.

The production and extensive use of charcoal mark one of the earliest technological innovations in fuel usage. Charcoal has been documented to have been produced as early as 6000 BCE for applications such as metal melting, with studies linking charcoal kiln sites to historical forest management practices [29]. Research demonstrates that charcoal production played a significant role in various regions during early industrialization; however, its role gradually diminished as European furnaces began to rely more on fossil fuels like coke derived from coal [30]. The depletion of European forests in the 18th century, along with the subsequent transition from charcoal to coke, illustrates a critical shift in fuel sources driven by resource scarcity and technological change [29, 30].

In a parallel development, the distillation of crude oil was pioneered by Persian chemists, with medieval Arabic texts providing some of the earliest detailed accounts of transforming petroleum into kerosene and other hydrocarbon compounds. This tradition continued well into the medieval period, evidenced by the exploitation of natural petroleum deposits for urban infrastructure, as shown by examples such as tar paving in Baghdad and the oil fields of the Caucasus region noted by both Arab geographers and travellers like Marco Polo [28].

The advent of the steam engine in Britain circa 1769 initiated a transformative era in energy usage. The steam engine's reliance on coal, which provided a more energy-dense alternative to organic fuels, catalysed a widespread shift in energy consumption patterns. Detailed examinations of steam engine technology and its dissemination have been provided by various researchers [31, 32], who document the spatial and technological spread of steam power. The adoption of steam engines was influenced by coal prices and availability, which were critical factors in the transition from traditional energy sources [32]. This period also witnessed the extensive use of coal in diverse applications—from powering locomotives and ships to providing gas for street lighting in cities like London [33].

The extensive use of fossil fuels during the Industrial Revolution—driven by their superior concentration and flexibility compared to traditional biomass fuels—has profoundly shaped the energy landscape into the modern era. Industrial advancements have heavily relied on coal to power steam engines and generate electricity, accounting for a significant portion of global electrical production [33]. However, the concurrent environmental challenges, including global warming and air pollution, have led to a shift toward renewable fuels, such as biofuels and alcohol-based energy carriers. Contemporary studies emphasizing the integration of fossil energy with renewable systems highlight efforts to reduce CO_2 emissions while meeting energy demands [34].

The Discovery of Coal and Oil

The discovery and use of coal and oil span thousands of years and have played a vital role in shaping human civilization, technology, and the global economy. These two fossil fuels—formed from ancient organic matter through heat and pressure over millions of years—have provided the energy foundation for societies past and present. Understanding their historical journey helps to explain how modern energy systems evolved and why transitioning away from fossil fuels is a global priority today.

The discovery, utilization, and transformation of fossil fuels such as coal and oil have fundamentally shaped human civilization, technological progress, and the global economy over millennia. Fossil fuels, generated from ancient organic matter that, over millions of years, underwent chemical transformation under heat and pressure, provided the energy foundation for early human societies and later catalysed the industrial transformations that underpin contemporary economic systems [35]. Early evidence from various cultures, including ancient Chinese and Roman societies, indicates that coal was initially discovered and used as a basic source of heat—for cooking and warming—long before its systematic exploitation during the Industrial Revolution [35]. The gradual embrace of coal culminated in its pivotal role during the steam-engine era, which not only revolutionized manufacturing and transportation but also laid the groundwork for modern energy systems.

Coal is one of the earliest known fuels used by humans. Its origins as a heat source date back to prehistoric times, when ancient people discovered that certain black rocks could burn. Archaeological evidence indicates that coal was used in China and among the Romans for heating and cooking. Around 4,000 years ago in northern China, people began burning coal for warmth and other basic needs. The Romans later adopted coal, even transporting it from Britain to Rome for use in public bathhouses.

Figure 3: Bituminous coal is a type of sedimentary rock and a medium-grade coal that forms part of the geological progression of coal types. It sits between sub-bituminous coal and anthracite in terms of carbon content, moisture, and energy output. Amcyrus2012, CC BY 4.0, via Wikimedia Commons.

Coal's importance grew slowly but steadily. By the 1200s, it was becoming more common in places like London, where growing populations made it difficult to source enough firewood. During the Industrial Revolution in the 18th and 19th centuries, coal use surged dramatically. The invention of the steam engine, particularly the Newcomen engine in 1712, created massive demand for coal as a fuel for steam power. Factories, trains, and ships relied on it, turning coal into the powerhouse of industrial society.

In Australia, coal mining began in the late 1700s, coinciding with European colonization. It quickly became a cornerstone of the developing economy. Today, Australia is one of the world's leading producers and exporters of coal, particularly black coal used for power generation and steel production, and brown coal, often used in domestic energy markets.

Similarly, the early use of oil can be traced back thousands of years to civilizations such as China, where crude petroleum was used for lighting and heating purposes [35]. The emergence of the modern oil industry in the 19th century, marked by pivotal events such as Edwin Drake's pioneering commercial drilling, signalled a rapid evolution from localized use to global reliance on oil as a versatile energy source [35]. This transition was characterized by a diverse range of applications—from lubricants and kerosene for lighting to fuelling the internal combustion engines of automobiles—which in turn transformed economic structures and geopolitical dynamics [35]. The extensive economic and political influence of oil has not only shaped trade and industry but also established it as a crucial factor in international relations and energy security.

Like coal, oil has been used by humans for thousands of years. Chinese civilizations were among the first to discover and use petroleum, as early as the 4th century BC. They transported crude oil through bamboo pipelines and used it for lighting and heating. However, oil remained a limited and localized energy source for centuries.

The modern oil era began in the 19th century. In 1858, oil was discovered in Oil Springs, Ontario, Canada, sparking one of North America's first oil booms. Just a year later, in 1859, Edwin Drake successfully drilled the first commercial oil well in Titusville, Pennsylvania, setting the stage for the global oil industry. The discovery of vast reserves at Spindletop, Texas, in 1901 further transformed oil from a novelty into an energy giant.

Initially used as a lubricant and for lighting (in the form of kerosene), oil's role expanded with the invention of the internal combustion engine by Nikolaus Otto in the 1860s. This engine, which powered early automobiles, ran on fuels refined from crude oil. The rise of motor vehicles, along with increasing demand for electricity and industry, cemented oil's position as a dominant fuel throughout the 20th century.

By the early 1900s, products like gasoline, kerosene, and diesel were refined from crude oil and widely used in transportation, agriculture, and military applications. Bertha Benz's pioneering 66-mile car trip in 1888, where she bought benzene at a pharmacy to refuel, famously marked the first long-distance automobile journey and highlighted oil's role in shaping modern mobility.

The continuing legacy of fossil fuels is underscored by their deeply embedded infrastructure in modern economies. The sustained use of these energy sources has underpinned industrial development around the world, while simultaneously contributing to significant environmental impacts such as greenhouse gas emissions and climate change concerns [35]. As much as fossil fuels have driven progress, their formation over geological timescales and the rapid pace of their consumption highlight their non-renewable nature—a duality that presents enduring challenges in the transition toward more sustainable energy systems [36]. Consequently, understanding the extensive historical journey of coal and oil—from their early discovery and primitive applications to their central role in today's energy mix—not only illuminates the evolution of global energy systems but also clarifies the urgent need for a strategic shift to cleaner alternatives.

By 2022, fossil fuels accounted for over 80% of the world's primary energy consumption and more than 60% of global electricity generation [37]. Their widespread use has made them the backbone of industrial development and economic growth. However, the environmental consequences have been severe. Burning fossil fuels is the largest source of greenhouse gas emissions, particularly carbon dioxide (CO_2), contributing significantly to climate change, ocean acidification, and air pollution [38].

Fossil fuel combustion releases not only CO_2 but also harmful particulates and toxic gases, which are responsible for millions of premature deaths globally each year. In fact, air pollution related to fossil fuel use is estimated to cost the world more than 3% of its global GDP annually.

Due to these impacts, there has been growing momentum to transition to cleaner and renewable energy sources, such as solar, wind, and bioenergy. International agreements like the Paris

Climate Accord, and the United Nations Sustainable Development Goals, emphasize the importance of reducing dependence on fossil fuels. Agencies like the International Energy Agency (IEA) have stated that no new fossil fuel extraction projects can be justified if the world is to meet its climate targets.

Oil has not only powered industries but has also become a major geopolitical asset. Since oil fields are concentrated in specific regions of the world, many countries are oil-dependent on a few producing nations. This dependence has shaped international relations, trade policies, and even military strategies.

The rise of companies like the Pennsylvania Rock Oil Company, and later Standard Oil, founded by John D. Rockefeller, marked the birth of the global oil industry. The economic and strategic importance of oil increased significantly during the World Wars, and since then, oil has remained a key factor in global politics and economics.

The Rise of Industrial Fuel Consumption

The Industrial Revolution marked a significant transition in energy consumption, fundamentally altering economies from agrarian bases to industrial powerhouses reliant on fossil fuels. This shift commenced in the 18th century, primarily as societies began to abandon traditional biomass sources—such as wood—for more energy-dense alternatives like coal. Prior to the Industrial Revolution, energy for heating and cooking was predominantly derived from organic material, but the growing urban population necessitated a more robust and efficient energy source. Schuster et al. noted that the anthropogenic emissions of mercury are primarily linked to coal combustion, illuminating the escalating scale of industrial fuel usage during this era [39].

As industries expanded, the demand for energy surged. Coal emerged as the primary fuel source due to its abundance and significantly higher energy density compared to wood, which was insufficient for the increasing power needs of industrial processes [40]. The introduction of the steam engine, which utilized coal to produce mechanical energy, revolutionized manufacturing. Factories proliferated, railways expanded, and steam-powered vessels became prevalent, driving rapid industrial growth. This pivotal change in production methods not only met the demands of burgeoning industries but also facilitated urbanization, as noted by Wrigley, who highlighted that the Industrial Revolution vastly increased productivity and transformed societal structures [41].

Manufacturing capabilities were significantly enhanced by the efficiency and cost-effectiveness of coal. The UK, with its abundant coal reserves, rapidly advanced in industrial production, with coal output playing a crucial role in urban population growth during the period from 1750 to 1900, effectively driving the narrative of industrialization in Europe [42]. Concurrently, the exploitation of coal resources led to the establishment of vast transportation networks to supply industries, fundamentally changing energy infrastructures [43].

As the 19th century progressed into the 20th century, a new evolution in fuel consumption arose with the advent of petroleum. The discovery and commercialization of oil refined its influence on industrial activity, marking the transition from coal as the dominant energy source [44]. Oil's advantages in transportability and versatility became particularly integral with the development of the internal combustion engine, facilitating not only increased efficiency in industrial machinery but also revolutionizing transportation and military applications [45]. This shift further accelerated global industrialization and integration into the world economy.

Throughout the 20th century, industrial fuel consumption remained on an upward trajectory. The growth of heavy industries and the mass production of goods continued to drive energy demands [46]. However, this relentless pursuit of energy came at a cost, as extensive fossil fuel combustion led to significant environmental challenges, including heightened greenhouse gas emissions and pollution [47]. The impacts of this industrialization narrative have cemented fossil fuels as pivotal to economic development but have also necessitated discussions about sustainability and cleaner energy alternatives in contemporary times.

Chapter 3

Types of Fuel

Building on the foundational understanding of what fuel is and how it has evolved throughout history, Chapter 3 introduces the many types of fuel that power our world. This chapter forms a crucial bridge between the theoretical underpinnings of fuel and the practical energy sources used across different sectors today.

From the dense carbon richness of coal to the volatile gases of methane and the promise of renewable options like biodiesel and hydrogen, this chapter explores the wide spectrum of fuels—both conventional and alternative. Each type is examined for its physical characteristics, methods of extraction or production, energy content, and typical applications.

As the global energy system moves toward decarbonisation, understanding the distinct properties and uses of different fuels becomes vital. This chapter not only identifies and compares fuels by origin and state but also foreshadows the environmental and economic implications that will be discussed in later chapters. Whether you are a student, policymaker, or energy professional, this overview will provide you with the necessary framework to critically assess fuel choices in the journey to net zero.

Fuel Types and Classifications

Fuels come in many forms and are categorized based on their physical state, origin, and how they release energy.

Solid fuels are some of the oldest and most widely used energy sources in human history. These fuels are generally natural or processed materials that are burned to produce heat. One of the most common solid fuels is wood, which is still widely used for cooking and heating, especially in rural and developing regions. Coal, another key solid fuel, is a fossil fuel that plays a major role in electricity generation and industrial processes due to its high energy content. Charcoal, which is made by heating wood in the absence of oxygen, is used for cooking and in metalworking applications like blacksmithing. Peat, a partially decomposed plant material found in wetlands, has traditionally been used for heating in some parts of the world. Modern biomass pellets, made by compressing plant waste, offer a cleaner and more efficient form of solid fuel for use in stoves and boilers.

Liquid fuels are particularly important in transportation and industrial applications because of their high energy density and ease of storage and handling. Gasoline, or petrol, is widely used in cars and motorcycles with internal combustion engines. Diesel, a denser fuel, powers larger vehicles such as trucks, buses, and trains, as well as some power generators. Kerosene serves various purposes, from jet fuel to heating and lighting. Heavy fuel oil is commonly used in ships and large industrial boilers. Biodiesel, derived from vegetable oils or animal fats, is a renewable alternative that can often be used in existing diesel engines. Ethanol, an alcohol-based fuel made from crops like corn and sugarcane, is frequently blended with gasoline to reduce emissions.

Gaseous fuels are used for cooking, heating, electricity generation, and increasingly in transportation. Natural gas, primarily composed of methane, is one of the most widely used gaseous fuels for household and commercial heating and power generation. Liquefied petroleum gas (LPG), a blend of propane and butane, is used in heating systems, stoves, and vehicles. Biogas, which is produced from the breakdown of organic waste in the absence of oxygen, provides a renewable option for heating and electricity. Hydrogen, a clean-burning gas, is gaining attention for its use in fuel cells and emerging zero-emission transport technologies. Propane and butane, while part of LPG, are also separately used for portable heating, camping, and small-scale energy needs.

Biofuels and renewable fuels come from biological sources and are increasingly seen as sustainable alternatives to fossil fuels. Ethanol and methanol are alcohol fuels made from plant-based materials, often used as additives or substitutes for gasoline. Biodiesel, produced from renewable oils or fats, is used in diesel engines. Biogas, derived from the decomposition of organic waste, is used in homes and industrial settings. Algae-based fuels are an emerging technology, where oil-rich algae are cultivated and processed into biofuels. Although still under development, algae fuels hold promise due to their fast growth rates and minimal land use.

Nuclear fuel is distinct from other types of fuel in that it releases energy through nuclear fission rather than combustion. Uranium-235 is the most common nuclear fuel used in reactors to generate electricity. Plutonium-239, a man-made isotope, is used in some advanced reactors and in weapons-grade materials. Thorium, another naturally occurring element, is being researched as a safer, potentially more abundant alternative to uranium. Nuclear fuels power not only electric power plants but also military submarines and aircraft carriers.

Figure 4: A billet of highly enriched uranium that was recovered from scrap processed at the Y-12 National Security Complex Plant. United States Department of Energy, Public Domain, , via Wikimedia Commons.

Electrical energy, while not a fuel in the traditional sense, is increasingly used as a substitute for fuel, particularly in transportation and heating. Electric vehicles rely on batteries to store and supply energy for propulsion. Electric heating systems and public transport such as trams and trains also rely on electricity. However, electricity itself must be generated from a primary fuel source—whether fossil-based, nuclear, or renewable—so it serves as both a product and a carrier of fuel-derived energy.

Fuels can exist in solid, liquid, gas, biological, nuclear, or electrical forms. Each type has its own unique benefits and is chosen based on the needs of the application, the energy content, availability, and environmental impact. As the global energy landscape evolves, there's a growing shift toward renewable and sustainable fuels that aim to reduce our dependence on fossil fuels and minimize environmental harm.

Fuels can be classified based on their origin, which refers to where and how they are formed. This classification is important because it helps us better understand their renewability, environmental impact, availability, and overall suitability for different uses. Broadly speaking, fuels are divided into two main categories according to their origin: natural (also known as primary) fuels and man-made (or secondary) fuels.

Natural fuels are those that are found directly in nature and can be used with minimal processing. They are either extracted from the Earth's crust or produced through biological and geological processes. Within this category, fossil fuels and biomass fuels are the two primary subtypes. Fossil fuels are formed from the decomposition of ancient plants and animals buried deep underground for millions of years. Under intense heat and pressure, these remains transform into energy-rich substances. Fossil fuels are considered non-renewable because they form over geological time and are being consumed far faster than they can naturally replenish. Common examples include coal, which is primarily formed from plant material and widely used in power generation and industry; crude oil, a liquid hydrocarbon used to produce gasoline, diesel, and many other refined products; and natural gas, composed mostly of methane and used for heating, electricity generation, and cooking.

Biomass fuels, on the other hand, originate from recently living organisms such as plants, agricultural byproducts, and animal waste. These fuels are considered renewable because they can be grown or collected repeatedly through agricultural or natural cycles. Examples include wood, which is widely used for cooking and heating; crop residues like husks and stalks; animal dung, which is dried and used as fuel in many rural areas; and specially cultivated energy crops and algae, which are processed into usable energy forms.

Man-made or secondary fuels are created by refining, processing, or chemically altering natural fuels to improve their energy content, usability, and efficiency. These fuels do not exist naturally in usable form and require human intervention to be produced. One type of man-made fuel includes refined fossil fuels, which are produced by processing crude oil or natural gas. These include gasoline, commonly used in vehicles; diesel, used in trucks, buses, and machinery; kerosene, which is widely used in aviation and heating; and liquefied petroleum gas (LPG), a processed blend of propane and butane used in heating, cooking, and transport.

Another category of man-made fuels is biofuels, which are derived from organic materials through chemical, biological, or thermal processes. Examples include ethanol, an alcohol-based fuel made from sugarcane or corn; biodiesel, produced from vegetable oils or animal fats; and biogas, generated through the anaerobic digestion of organic waste like food scraps or manure. These fuels offer renewable alternatives to fossil fuels and are gaining popularity due to their lower environmental impact.

Synthetic and alternative fuels are also part of the man-made fuel category. These are created through more complex industrial processes, often aimed at reducing reliance on traditional fossil fuels. Hydrogen fuel, for example, can be produced from water through electrolysis or from natural gas, and is increasingly used in clean transport and fuel cell technologies. Synthetic fuels, or e-fuels, are made by combining captured carbon dioxide with hydrogen using renewable electricity. Other examples include coal gas and producer gas, which are formed by gasifying coal or biomass to create gaseous fuel for heat and power applications.

Classifying fuels based on their origin is essential for understanding their renewability, environmental footprint, and how they fit into energy systems. It also informs decisions about economic viability, infrastructure needs, and energy policy. Recognizing whether a fuel is natural or man-made allows energy planners and governments to prioritize sustainable sources and invest in technologies that support the transition to cleaner, more resilient energy systems.

Fuels can be categorised not only by their origin or physical state but also by how they release energy when used. This classification is important because it helps determine the application, efficiency, safety, and environmental impact of different fuels. Generally, fuels are grouped into three main categories based on the mechanism of energy release: combustion fuels, electrochemical fuels, and nuclear fuels.

The most common way fuels release energy is through combustion, a chemical reaction that occurs when a fuel reacts with an oxidizer—usually oxygen from the air—to produce heat and light. This process breaks the chemical bonds in the fuel and releases thermal energy, which can be converted into mechanical or electrical energy.

Combustion fuels include most solid, liquid, and gaseous fuels, such as coal, wood, gasoline, diesel, natural gas, kerosene, and biomass. These fuels are burned in engines, power plants, stoves, and furnaces to produce heat or power machinery. For example, internal combustion engines in cars burn petrol or diesel to create the energy that moves the vehicle. Steam turbines in coal-fired power plants burn coal to produce electricity.

While combustion is a powerful and efficient way to release energy, it often comes with significant downsides, including air pollution, carbon dioxide (CO_2) emissions, and health risks from toxic by-products.

Electrochemical fuels release energy through electrochemical reactions, rather than combustion. These reactions involve the movement of electrons from one chemical substance to another, which generates electricity directly without a flame or high heat. This process takes place in devices like batteries and fuel cells.

In batteries, stored chemical energy is converted into electrical energy when the battery is in use. Common battery types include lithium-ion, lead-acid, and nickel-metal hydride, all of which rely on electrochemical reactions.

In hydrogen fuel cells, hydrogen reacts with oxygen in an electrochemical cell to produce electricity, water, and heat. This process is highly efficient and emits no harmful gases, making it a clean alternative for powering vehicles and equipment.

Electrochemical fuels are ideal for portable electronics, electric vehicles, and backup power systems. They are also central to renewable energy storage technologies, helping to balance energy supply and demand in modern power grids.

Nuclear fuels release energy through nuclear reactions, particularly fission, where the nucleus of an atom is split into smaller parts. This process releases a massive amount of energy compared to chemical reactions. Nuclear fuels are typically uranium-235, plutonium-239, or thorium, used in nuclear reactors and some military applications.

In a nuclear reactor, fission produces heat, which is used to generate steam and drive turbines that produce electricity. Unlike combustion, nuclear reactions do not involve carbon emissions. However, they pose other challenges, including radioactive waste, high risk in case of accidents, and long-term environmental concerns.

Despite these risks, nuclear energy is a low-carbon and high-output option for countries seeking to reduce greenhouse gas emissions while maintaining a stable energy supply.

Solid Fuels: Wood, Coal, Peat

Solid fuels are materials that can be burned to release energy in the form of heat and are primarily used for cooking, heating, and electricity generation.

Solid fuel refers to a wide range of solid materials that can be burned to produce heat and light through the process of combustion. Unlike liquid and gaseous fuels, solid fuels exist in physical, often compact forms and have been used by humans for millennia, dating back to the earliest discovery of fire. Today, solid fuels remain essential for domestic heating, cooking, industrial processes, electricity generation, and even specialized fields like rocketry. Common examples of solid fuels include wood, charcoal, peat, coal, animal dung, agricultural residues, wood pellets, and processed grains like corn and wheat. In high-tech applications, specially designed solid fuels are used as propellants in rockets, due to their high energy density and stability.

Among solid fuels, those derived from biomass—plant and animal material—are considered renewable, especially when harvested or cultivated sustainably. In contrast, fossil-based solid fuels, like coal and peat, are non-renewable, as they form over millions of years and are being depleted far more rapidly than they are replaced.

Biomass and Renewable Solid Fuels

Biomass solid fuels originate from recently living plants and organic waste materials. These can be used in their raw form or processed into pellets, briquettes, or chips for improved combustion efficiency. The term "biofuel" often refers to liquid and gaseous fuels from organic matter, particularly for transportation, but solid biomass fuels play a crucial role in heating, cooking, and power generation.

Pellet fuels, a popular biomass product, are made by compressing organic materials such as wood waste, agricultural residues, food waste, and energy crops. Wood pellets are the most widely used, offering high energy output and low moisture content. In energy systems, wood and wood residues are still the dominant biomass sources. These include firewood, sawdust, wood chips, and charcoal. Wood can be burned directly in stoves and furnaces, or processed into other forms. Other plants like maize, miscanthus, switchgrass, and bamboo are also cultivated as energy crops. Upgrading raw biomass into more efficient solid fuels can be achieved through thermal, chemical, or biochemical processing.

Wood fuel is especially versatile. It can be used in simple forms like branches and logs, or processed into pellets, sheets, and chips. Wood is commonly burned indoors in stoves or fireplaces and outdoors in furnaces and campfires. It provides heat for cooking and residential heating, and can even power steam engines in some rural or industrial applications. When harvested sustainably, wood is considered a renewable energy source, although debates exist over its carbon neutrality. While burning wood only releases the carbon it absorbed during its growth, additional impacts—like deforestation, land-use change, and biomass decomposition—must be factored into its environmental assessment.

Peat as Solid Fuel

Peat is another naturally occurring solid fuel formed from partially decomposed plant material in waterlogged environments. Once dried, it can be burned for heat. Peat is traditionally used in countries like Ireland and Scotland, particularly in areas where wood or coal is scarce. Though sometimes considered a transitional fuel between biomass and coal, peat has a low calorific value, even when dried, and its extraction can damage sensitive ecosystems, making it environmentally controversial.

Fossil Solid Fuels: Coal and Coke

Coal is a sedimentary rock rich in carbon and energy, formed over millions of years from ancient plant material. It comes in various grades, including lignite, sub-bituminous, bituminous, and anthracite, each with different energy content and burning properties. Coal has historically fuelled industrial revolutions and remains a major global energy source, particularly for electricity generation. However, coal is also linked to significant environmental damage, including air pollution, greenhouse gas emissions, and health problems associated with mining and combustion.

A more refined form of coal is coke, produced by heating coal in the absence of air. Coke is a high-carbon, porous, and grey solid used primarily in metallurgical processes, especially in steelmaking. It burns hotter and cleaner than coal. There is also petroleum coke, or petcoke, which is derived from oil refining and used in similar industrial settings.

Smokeless Fuels

Smokeless fuels are designed to burn cleanly, producing minimal smoke and particulate emissions. They are often made by compressing anthracite coal into briquettes and are used in stoves and open fireplaces. Smokeless fuels are valued for their high combustion efficiency, long burn times, and reduced indoor and outdoor air pollution. Charcoal, produced through the restricted combustion of wood, is also classified as a smokeless fuel and is widely used for outdoor cooking such as barbecuing.

Specialized Solid Fuels: Rocket Propellants

In aerospace and defence applications, solid rocket propellants are carefully engineered compounds consisting of oxidizers like ammonium nitrate, mixed with high-energy powders, binders, and stabilizers. These fuels are dense, stable, and compact, making them easier to store and handle than liquid propellants. Their controlled and powerful combustion makes them suitable for missiles, spacecraft, and fireworks.

Natural Solid Fuels

These are fuels found in nature with minimal processing:

- **Wood** (logs, branches)
- **Twigs and brushwood**
- **Dried animal dung**
- **Crop residues** (e.g., rice husks, corn stalks, sugarcane bagasse)
- **Peat** (partially decomposed plant matter from wetlands)
- **Coal**
    - **Lignite** (brown coal – lowest rank)
    - **Sub-bituminous coal**
    - **Bituminous coal**
    - **Anthracite** (highest rank coal)
- **Charcoal** (from wood or biomass)
- **Forest residues** (fallen branches, leaves, bark)
- **Bamboo** (used in some regions)

Processed/Briquetted Solid Fuels

These are made by compressing or refining organic or waste materials for improved combustion and handling:

Fuel, Energy and Net Zero

- **Charcoal briquettes**
- **Biomass briquettes** (from sawdust, husks, shells, etc.)
- **Wood pellets**
- **Agricultural waste pellets**
- **Peat briquettes**
- **Refuse-derived fuel (RDF)** (processed from municipal solid waste)
- **Coal briquettes**
- **Turf briquettes** (compressed peat)

Industrial Solid Fuels

Used in industries, especially for power generation and metallurgy:

- **Coke** (derived from bituminous coal; used in steelmaking)
- **Petroleum coke (petcoke)** (by-product of oil refining)
- **Sawdust** (used in boilers and briquettes)
- **Wood chips** (used in biomass boilers)
- **Torrefied biomass** (thermally treated biomass with enhanced energy content)
- **Black liquor solids** (from paper pulp processing, dried and burned)
- **Sewage sludge cake** (dried waste material used in some industrial boilers)
- **Plastics-derived solid fuels** (thermally treated non-recyclable plastics)
- **Rubber residues** (such as shredded tires, sometimes co-fired with coal)

Experimental or Niche Solid Fuels

Used in research or specific applications:

- **Algae-derived solids** (dried algal biomass used for bioenergy research)
- **Sugarcane press mud** (used in some regions as solid fuel)
- **Spent coffee grounds** (dried and used as fuel in some eco-initiatives)
- **Cotton stalks, jute sticks** (used in rural energy schemes)
- **Charred nut shells and fruit pits** (e.g., coconut shells, olive pits)

This list highlights the diversity of solid fuels, from traditional biomass and fossil fuels to modern bioenergy materials and industrial by-products. Each type has its own characteristics in terms of energy content, combustion efficiency, emissions, and availability, which influence how and where it is used.

Energy Content (Calorific Value) of Solid Fuels

The calorific value, or heat of combustion, of solid fuels varies depending on their composition—mainly the proportions of carbon, hydrogen, moisture, and ash. It represents the amount of energy released when a certain mass of fuel is completely burned. This value is typically measured in megajoules per kilogram (MJ/kg).

Gaseous fuels like methane tend to have the highest calorific values, around 55.5 MJ/kg. In comparison, diesel fuel has a value of 44.8 MJ/kg, anthracite coal around 32.5 MJ/kg, and dry firewood about 21.7 MJ/kg. Peat, with its high moisture content and lower carbon concentration, has the lowest energy value, around 15 MJ/kg. These figures serve as general guidelines, as actual heat output depends heavily on how and where the fuel is burned.

Cost, Transport, and Availability

Solid fuels are generally less expensive and easier to transport than liquid or gaseous fuels. They are also more stable, making storage simpler. Coal, in particular, remains one of the most cost-effective fuels and is responsible for generating around 38% of global electricity due to its affordability and infrastructure.

Environmental and Health Concerns

Despite their advantages, solid fuels often cause greater environmental harm than liquid or gas fuels. Extraction methods can be destructive, and combustion tends to produce more carbon dioxide, nitrates, sulphates, volatile organic compounds (VOCs), and particulate matter. In regions where solid fuels are widely used, they are a major source of poor air quality and respiratory illnesses.

Even renewable solid fuels like wood and biomass can have environmental drawbacks if harvested unsustainably or burned inefficiently. The release of organic aerosols and secondary pollutants like ground-level ozone and secondary organic aerosols can contribute to long-term environmental and public health challenges. Therefore, cleaner-burning technologies, improved stoves, and sustainable harvesting practices are essential to reduce the impact of solid fuel use.

Fuel, Energy and Net Zero

Solid Fuel for Energy Production

Solid fuels continue to underpin energy use for billions globally, as highlighted by numerous studies demonstrating that, despite advancements in clean energy technologies, solid fuel remains indispensable in rural households, industrial sectors, and emerging economies. In many rural communities across Africa, Asia, Latin America, and the Middle East, households rely upon firewood, charcoal, agricultural residues, and animal dung not only because these fuels are locally and freely available but also due to longstanding cultural cooking practices that require the open-flame techniques and high heat provided by solid fuels [48, 49]. The World Health Organization estimates that nearly 2.4 billion people rely on biomass fuels for cooking and heating, emphasizing that infrastructural deficiencies—such as inadequate gas pipelines and unreliable electricity—force these communities to depend on readily accessible, low-cost fuels [48, 49]. Traditional cooking methods, which often demand prolonged boiling and direct heat, further cement the cultural reliance on solid fuels in these settings [48, 49].

In contrast, the industrial application of solid fuels exhibits an equally strong persistence. Heavily industrialized regions, particularly those endowed with significant coal reserves such as India, China, South Africa, Australia, and parts of Eastern Europe, continue to utilize coal-fired power plants to meet large-scale energy demands. Coal and coke are favoured in industries such as steel production and cement manufacturing because of their cost-effectiveness and ability to generate the high temperatures required for various industrial processes [50, 51]. These regions not only leverage domestic fuel availability to bolster energy security but also harness the mature, established infrastructure associated with coal-fired technologies, despite ongoing global transitions toward cleaner energy sources [50].

Emerging economies, notably in India and China, represent a hybrid energy landscape where rapid economic growth coexists with persistent reliance on solid fuels. While both nations are investing heavily in renewable energy, the integration of solid fuels remains critical to ensuring uninterrupted energy supply, particularly in the face of escalating energy demands. This continued reliance is driven by the scalability, cost efficiency, and established logistics of solid fuel-based energy systems [48, 49]. Furthermore, on both urban and rural scales, informal and local economies—including street food vendors and small restaurants—opt for solid fuels like charcoal and wood due to the dual benefits of economic affordability and operational simplicity, which circumvent the need for modern energy infrastructure [52].

Beyond these primary sectors, specialized applications further underscore the role of solid fuels. For instance, emergency and off-grid scenarios often require portable solid fuel stoves or tablets for heating and cooking, while sectors such as the military and aerospace continue to employ solid rocket propellants for their stability and storability advantages [52]. Thus, the confluence of economic, geographical, cultural, and infrastructural factors ensures that solid fuels remain integral to the global energy mix, even as the world incrementally shifts toward cleaner alternatives [48, 49, 52].

Solid Fuel Dependence

Dependence on solid fuels among households, particularly in rural and low-income regions of developing countries, is profoundly influenced by economic limitations, cultural practices, and local biomaterial availability. Solid fuels, including firewood, charcoal, and agricultural residues, continue to be the primary energy source for over 2.8 billion people globally, particularly in regions across Africa, Asia, and parts of Latin America [53]. This dependency primarily arises from several interrelated factors.

Firstly, access to modern energy sources such as electricity and gas is limited in many remote and off-grid communities. The World Health Organization notes that the lack of infrastructure leads to high reliance on solid fuels [53]. Economic constraints also play a crucial role, as solid fuels are often more accessible and less expensive than cleaner alternatives like liquefied petroleum gas (LPG) or electricity [54]. For example, studies show that in low-income settings, solid fuels are often freely available or significantly cheaper than their modern counterparts [53].

Cultural practices also perpetuate the use of solid fuels. Traditional cooking methods have been passed down through generations, creating a cycle where communities remain entrenched in the use of solid fuels [54]. The simplicity of cooking with solid fuels—requiring no complex infrastructure or expensive equipment—makes it a practical choice for many households [53, 55]. A study in Ethiopia illustrated these dynamics, detailing how economic and cultural factors influence the fuel choices that families make, often leading them to prefer what they can gather or afford easily [56].

Furthermore, industrial and commercial users rely on solid fuels for their efficiency and cost-effectiveness. Solid fuels such as coal and biomass are integral in sectors like cement and brick manufacturing and food processing due to their high-temperature output, which is crucial for operational efficiency [57]. In many regions, existing equipment and infrastructure already support the use of solid fuels, making it economically disadvantageous to switch to cleaner technologies, especially where recent economic or infrastructural developments have not reached [57].

In rural areas, agricultural activities also drive dependence on solid fuels. Biomass waste from crops is abundant, providing a cost-effective means for drying crops and heating spaces [53]. This practical advantage extends to various community facilities, including schools and clinics, where solid fuels serve as the primary energy source due to economic constraints [55]. Street vendors and small businesses likewise utilize solid fuels largely for their low operational costs and the unreliable availability of electricity [54].

The reliance on these fuels, however, presents significant health and environmental challenges. Indoor air pollution from solid fuel combustion is linked to respiratory diseases and contributes to over 4 million premature deaths each year [58]. Additionally, pervasive reliance on firewood and charcoal exacerbates deforestation and land degradation, creating a cycle of environmental depletion that further entrenches poverty in these communities [53].

Liquid Fuels: Gasoline, Diesel, Kerosene, Biofuels

Liquid fuels are energy-dense substances in liquid form that release energy through combustion or other chemical processes. These fuels are widely used to power internal combustion engines, turbines, generators, and heating systems, primarily by converting stored chemical energy into kinetic or mechanical energy. A defining feature of liquid fuels is that they take the shape of their container, and the flammable vapours (not the liquid itself) are what ignite during combustion.

Most of the liquid fuels in use today are derived from fossil fuels, particularly petroleum. However, an increasing number of alternative and renewable liquid fuels—such as ethanol, biodiesel, butanol, hydrogen, and ammonia—are being explored and adopted in efforts to reduce carbon emissions and promote sustainability. Liquid fuels play a critical role in transportation, as they power the majority of the world's vehicles, aircraft, ships, and agricultural equipment. They are also essential to industrial operations, military logistics, and power generation.

Liquid fuels are relatively easy to store, handle, and transport, especially compared to gaseous fuels. Their physical and chemical properties, such as flash point (the lowest temperature at which vapours ignite), fire point (temperature at which sustained combustion occurs), cloud point (the temperature at which wax crystals begin forming), and pour point (the lowest temperature at which the fuel can still flow), are crucial for safe handling and performance. These characteristics influence their usability in cold climates, storage requirements, and safety regulations.

Most commonly used liquid fuels are petroleum derivatives, refined from crude oil. These include:

- **Gasoline -** Gasoline (petrol) is a volatile mixture of hydrocarbons with a high energy content and is widely used in spark-ignition engines in cars and motorcycles. It is produced through the distillation and refining of crude oil. Gasoline's combustion relies on its vapor, not the liquid itself. Its octane rating measures resistance to premature combustion (knocking). Historically, additives like tetraethyl lead were used to increase octane ratings, but due to health concerns, modern refineries now rely on cleaner methods.
- **Diesel -** Diesel is a heavier, less volatile fuel used in compression-ignition engines. It is more energy-dense than gasoline and burns more efficiently in suitable engines. Diesel engines compress air to high temperatures, then inject the fuel to ignite it without a spark. Diesel fuels vary in sulphur content, with ultra-low sulphur diesel (ULSD) being the standard in many countries to reduce harmful emissions.
- **Kerosene -** Kerosene is used in jet engines, heating systems, and lamps. It burns cleaner than heavier fuels and is refined to meet specific freeze and smoke point standards. It has applications in both domestic and aviation sectors, with specific grades like Jet A, Jet A-1, and RP-1 (rocket-grade kerosene).
- **Liquefied Petroleum Gas (LPG) -** LPG is a compressed mixture of propane and butane, often used in cooking, heating, and increasingly as a motor fuel. It's denser than air and burns with a clean flame, though it doesn't match the energy density of gasoline or diesel.

- These fuels are widely used in transportation, heating, power generation, aviation, military, and industrial applications due to their high energy density, portability, and versatile handling properties.

Petroleum-Based Liquid Fuels (Fossil Fuels)

These are refined from **crude oil** and form the backbone of global transportation and industrial energy use.

- **Gasoline (Petrol)** – Used in spark-ignition internal combustion engines (cars, motorcycles).

- **Diesel** – Used in compression-ignition engines (trucks, buses, generators).

- **Kerosene** – Used in jet engines, lamps, and domestic heating.

- **Jet Fuel (Jet A, Jet A-1, JP-8)** – Aviation-grade fuel derived from kerosene.

- **Fuel Oil** – Includes distillate and residual oils; used in furnaces and marine engines.

- **Heavy Fuel Oil (HFO)** – Thick, high-viscosity fuel for ships and power plants.

- **Marine Diesel Oil (MDO)** – Blend of heavy fuel oil and lighter oils used in shipping.

- **Naphtha** – Used in petrochemical industries and some engines.

- **Liquefied Petroleum Gasoline (LPG Blends)** – Sometimes blended with gasoline for vehicles.

- **Light Distillate Fuel Oils** – Used for power generation and industrial burners.

Bio-Based Liquid Fuels (Biofuels)

Derived from biological sources such as crops, plant oils, or animal fats. Often used as replacements or additives for petroleum fuels.

- **Ethanol (Ethyl Alcohol)** – Commonly blended with gasoline (e.g., E10, E85); made from corn, sugarcane, or cellulose.

- **Methanol (Methyl Alcohol)** – Used in racing cars, fuel cells, and as a feedstock.

- **Butanol** – Alcohol fuel with higher energy content than ethanol; usable in gasoline engines.

- **Biodiesel (Fatty Acid Methyl Esters, or FAME)** – Made from vegetable oils or animal fats; used in diesel engines.

- **Hydrotreated Vegetable Oil (HVO)** – Advanced biodiesel with superior cold-weather performance.

- **Straight Vegetable Oil (SVO)** – Unprocessed oil (e.g., canola, sunflower) used in modified diesel engines.

- **Bioethanol** – Ethanol produced from organic materials via fermentation.

- **Biobutanol** – Advanced alcohol biofuel under development.

- **Renewable Diesel** – Chemically identical to fossil diesel, but made from biomass using hydrogenation.

Synthetic Liquid Fuels

These are produced through **chemical synthesis**, often to reduce reliance on fossil fuels or to create cleaner-burning alternatives.

- **Fischer-Tropsch Diesel (FT Diesel)** – Made from coal, natural gas, or biomass (Coal-to-Liquid or Gas-to-Liquid processes).

- **Synthetic Gasoline** – Produced via catalytic conversion of syngas or other intermediates.

- **E-fuels (Electrofuels)** – Made by combining captured CO_2 with green hydrogen using renewable electricity.

- **Methanol-to-Gasoline (MTG)** – Synthetic gasoline produced from methanol.

- **Dimethyl Ether (DME)** – Can be liquefied and used in diesel engines or LPG systems.

Specialized and Industrial Liquid Fuels

Used in specific industries such as aviation, rocketry, military, and chemical processing.

- **Aviation Gasoline (AvGas)** – High-octane fuel for piston-engine aircraft.

- **JP-4 / JP-5 / JP-8** – Military jet fuels with varying flashpoints and freezing points.

- **Rocket Propellants (RP-1)** – Highly refined kerosene used in rockets.

- **Nitromethane** – Used in drag racing and model engines.

- **Hydrazine and Derivatives** – Toxic, high-energy fuels for spacecraft and military propulsion.

- **Turpentine** – Distilled from tree resin; used historically as lamp fuel and solvent.

- **Alcohol-based racing fuels** – Blends of ethanol, methanol, and additives for performance engines.

Emerging and Experimental Liquid Fuels

Under research for future clean energy solutions.

- **Ammonia (as liquid fuel)** – Stored under pressure or at low temperature; emits no CO_2 when burned.

- **Liquid Hydrogen (LH2)** – Used in cryogenic rocket engines and proposed for aviation.

- **Algae-based bio-oils** – Extracted from algae and refined into biodiesel or other liquids.

- **Pine oil fuels** – Derived from forest residues; used in small engines and as a bio-additive.

- **Isobutanol and iso-paraffins** – Next-gen fuels derived from biomass with compatibility in existing engines.

Liquid fuels are highly versatile and come from a wide range of natural, biological, synthetic, and experimental sources. While petroleum fuels still dominate the market due to their energy density and infrastructure, biofuels, synthetic liquids, and cleaner alternatives are rapidly gaining ground as the world shifts toward sustainable energy systems. This evolving fuel landscape supports not only transportation but also industries, aviation, military technology, and space exploration.

Liquid Fuel for Energy Production

Liquid fuels continue to serve as a pivotal energy source across varied sectors and regions due to their intrinsic characteristics, including high energy density, portability, ease of storage, and established distribution networks. Their role is particularly pronounced in the transportation sector, which is the largest consumer globally and heavily relies on fuels such as gasoline, diesel, jet fuel, and marine oil in vehicles ranging from cars to airplanes and ships. This reliance stems from the energy-dense nature of liquid fuels, allowing vehicles to operate for extended distances with fewer refuelling stops [59].

Additionally, the infrastructure already in place for distribution, including refineries and fuel stations, enhances the convenience and efficiency of using these fuels significantly compared to alternative energy sources, particularly electricity, which requires extensive infrastructure for recharging [60].

Moreover, liquid fuels perform well in harsh conditions, including cold climates and remote terrains where electric grids are often insufficient or entirely missing [61]. While advancements in electric mobility are progressing, the aviation and shipping industries maintain a strong dependence on liquid fuels due to current technological challenges related to electric batteries, which struggle with energy density and recharging times, particularly for long-haul flights and shipping routes [62].

In industrial and commercial applications, liquid fuels such as diesel and fuel oil are key for operating heavy machinery, generators, and boilers where high-temperature energy is critical. They provide an on-demand energy source especially advantageous for remote operations, such as mining or construction, where conventional electricity may not be readily accessible [63]. This versatility extends to their role as backup energy systems, with diesel generators serving vital functions during outages. Furthermore, certain sectors require specialty liquid fuels for chemical feedstocks, demonstrating the broad application of liquid fuels beyond just transportation [64].

Households in both developing and developed nations also utilize liquid fuels heavily, particularly kerosene and liquefied petroleum gas (LPG). In developing countries, these fuels often replace less efficient and more polluting energy sources such as wood and coal for cooking and heating, given their convenience and relative cleanliness [65]. LPG has also gained traction in urban settings within developed nations where natural gas pipelines are absent, further showcasing the appeal of liquid fuels for daily energy needs [66].

The agricultural sector capitalizes on liquid fuels as well, with diesel being essential for powering tractors and irrigation systems in regions where electrical grids are sparse. This highlights not only the adaptability of liquid fuels in meeting energy needs across diverse settings but also the increasing interest in biofuels like biodiesel, which offer an environmentally sustainable option [67].

Remote and off-grid areas, particularly in disaster-prone regions, often depend exclusively on liquid fuels to provide energy solutions for hospitals, telecommunications, and emergency services, substantiating the necessity of these fuels in crisis situations [59]. In examining the future, despite growing interest in renewable energy sources, the high energy content, established global supply chains, and readiness of infrastructure favour the continued use of liquid fuels across various applications [68]. Shifting away from this dependence entails profound changes in technology, infrastructure, and policy frameworks to facilitate a sustainable transition.

Liquid Fuel Dependence

Liquid fuels such as gasoline, diesel, kerosene, jet fuel, biodiesel, ethanol, and liquefied petroleum gas (LPG) play an essential role in contemporary energy systems globally. Their significance spans across multiple sectors including transportation, industrial, military, commercial, agricultural, and residential users, making them critical to daily operations and economic stability.

The transportation sector is the largest consumer of liquid fuels, relying on fuels such as gasoline and ethanol blends for cars and motorcycles, diesel for trucks and buses, and kerosene-based jet fuels for aircraft. This dependency is primarily due to the high energy density of these fuels, which allows vehicles to travel long distances without frequent refuelling, thus ensuring efficiency and convenience in both personal and commercial travel [69, 70]. Moreover, the well-established global infrastructure for liquid fuels—including fuel stations, pipelines, and refineries—facilitates their easy access and storage. Importantly, internal combustion engines, which are prevalent across various types of vehicles, efficiently operate under diverse conditions provided by liquid fuels [69, 70]. While electric vehicles are gaining traction, sectors such as aviation and shipping remain heavily reliant on these fuels due to their operational requirements [70].

In industrial and commercial settings, liquid fuels like diesel and kerosene are fundamental for operations. They power boilers, industrial furnaces, and backup generators, providing a reliable energy source in both on-site transport and heavy machinery operations. These fuels are favoured for their ability to deliver high and consistent heat, crucial for many industrial processes, and are often more practical than gaseous alternatives or electric sources in remote locations with limited infrastructure [69, 71]. This is particularly relevant in sectors such as mining, construction, and agriculture, where uninterrupted operations hinge on dependable energy sources [69, 71].

The aviation sector's reliance on liquid fuels, particularly kerosene-based fuels like Jet A and JP-8, underscores the unique requirements of flight. These fuels provide high energy content crucial for flight and exhibit stability under extreme conditions [69, 72]. Current alternatives, such as electric aircraft, do not yet offer the range or capacity required for long-haul travels, demonstrating the continuing necessity of liquid fuels in this domain [69, 70].

In agriculture, diesel and biodiesel are vital for powering tractors and harvesters, making them indispensable for modern farming practices. Many rural areas lack reliable electric infrastructure, further cementing the role of diesel as a preferred energy source [71, 73]. The durability and efficiency of diesel engines enable farmers to efficiently utilize this energy source, significantly contributing to agricultural productivity [71, 73].

Liquid fuels also provide essential backup systems and support disaster relief and military operations in areas with unstable power supplies. Fuels like diesel are crucial for powering generators, providing immediate and dependable energy [69, 71]. Their transportability and storage advantages are beneficial in extreme environments, where conventional power sources may be impractical [69, 70].

Residential usage of liquid fuels such as kerosene and LPG is particularly prevalent in developing regions, where they are favoured for cooking and lighting due to unreliable electricity access. LPG, in particular, offers a cleaner alternative to traditional fuels such as wood or charcoal, improving indoor air quality [69, 71]. As urban areas in countries like India and Indonesia seek to transition to cleaner cooking options, LPG adoption is seen as a key strategy to enhance health and efficiency [71, 73].

Lastly, the maritime sector's dependence on heavy liquid fuels reflects its critical role in global trade. With over 80% of global trade by volume reliant on ships operating with these fuels, their efficiency and cost-effectiveness are indispensable for long voyages [69, 71]. The infrastructure supporting marine fuels ensures the feasibility of operations across international waters, maintaining the seamless flow of global commerce [69, 70].

Gaseous Fuels: Natural Gas, Hydrogen, Propane

Gaseous fuels refer to any type of fuel that exists in a gaseous state under normal atmospheric conditions and can be combusted to release energy. Unlike solid and liquid fuels, gaseous fuels are primarily composed of hydrocarbons such as methane, propane, and butane, or of gases like hydrogen, carbon monoxide, or a mixture of these. Due to their chemical composition and physical properties, gaseous fuels are easy to distribute through pipelines and offer high combustion efficiency, making them highly suitable for a broad spectrum of energy applications.

Although gaseous in nature, many gaseous fuels—such as liquefied petroleum gas (LPG) and autogas—can be stored and transported in liquid form under pressure or at low temperatures. Their gaseous form offers several benefits, including clean combustion, precise control of fuel flow, and reduced emissions of soot or ash. However, their use also comes with safety concerns. Because gaseous fuels are naturally odorless, odorants such as mercaptans are added to make leaks detectable and prevent potential explosions. Among all gaseous fuels, natural gas is currently the most widely used worldwide due to its availability, efficiency, and established infrastructure.

Gaseous fuels are commonly categorized based on their origin and method of production into two main types: manufactured gaseous fuels and naturally extracted gaseous fuels.

Manufactured gaseous fuels are produced through chemical or thermal conversion of solids, liquids, or other gases. These fuels were historically crucial to early urban energy systems and were typically produced in specialized facilities known as gasworks. Though many have been replaced by cleaner alternatives like natural gas, manufactured gaseous fuels still have relevance in certain industrial and renewable applications.

Some key types include:

- **Coal Gas** is formed through the pyrolysis of coal in an oxygen-limited environment. It was widely used for lighting, heating, and cooking before the rise of natural gas. Its composition includes methane, hydrogen, carbon monoxide, and various contaminants like tar and ammonia that require scrubbing before use.
- **Water Gas** is made by passing steam over hot coke, producing hydrogen and carbon monoxide. Once common, it is now largely obsolete.
- **Producer Gas** is generated by blowing a mixture of steam and air over hot coal or coke, resulting in a combustible gas composed of carbon monoxide, nitrogen, and hydrogen. It has a lower calorific value and is no longer widely used.
- **Syngas (Synthesis Gas)** is a modern and versatile gaseous fuel composed of hydrogen and carbon monoxide, often produced from natural gas, biomass, or coal. It is primarily used in the chemical industry to manufacture ammonia, methanol, and synthetic fuels.

- **Wood Gas** is created by gasifying wood, particularly during times when petroleum fuels are scarce. It contains combustible gases like hydrogen and methane and is sometimes used in off-grid applications.
- **Biogas** is produced from the anaerobic digestion of organic matter, such as agricultural waste, food waste, and sewage. Composed primarily of methane and carbon dioxide, it is renewable and increasingly used for electricity generation and heating.
- **Blast Furnace Gas** is a by-product of iron production, mainly consisting of nitrogen and carbon monoxide. It is often repurposed as fuel within steel plants.
- **Hydrogen**, while often a component of syngas, is increasingly being produced independently through steam reforming and electrolysis. As a clean-burning fuel, it is gaining prominence in fuel cells, green energy systems, and chemical manufacturing.

Manufactured gaseous fuels typically require additional processing to remove impurities like sulphur compounds, tar, or moisture, which makes their infrastructure and handling more complex than that of naturally occurring gaseous fuels.

Naturally Extracted Gaseous fuels are found in geological formations and can be extracted from the earth with minimal processing.

Natural Gas is the most common gaseous fuel, primarily composed of methane with small amounts of ethane, propane, butane, carbon dioxide, nitrogen, and trace elements. It is widely used in homes, industries, power generation, and as a raw material for chemicals.

Propane and Butane, often derived from natural gas processing or petroleum refining, are stored and distributed as LPG. These are used in heating, cooking, portable energy applications, and as transport fuels (autogas).

Regasified Liquefied Petroleum Gas refers to LPG that has been stored in a liquefied form and returned to a gaseous state for distribution or use.

Hydrogen-enriched Compressed Natural Gas (HCNG) is a blend of hydrogen and natural gas that offers cleaner combustion and higher efficiency, particularly in internal combustion engines and fuel cell vehicles.

Natural gas and similar fuels often contain water vapor and heavier hydrocarbons, which must be separated before use. This prevents condensation in pipelines and ensures consistent combustion quality, often requiring superheating and pressurization.

Gaseous fuels are used in domestic, industrial, commercial, transportation, and power generation sectors due to their clean combustion, controllability, and high thermal efficiency.

Naturally Occurring Gaseous Fuels

These are obtained directly from natural geological sources.

- **Natural Gas**

- Mainly composed of methane (CH_4)

- May also contain ethane, propane, butane, and small amounts of CO_2, N_2, and H_2S

- Used for cooking, heating, electricity generation, industrial processes, and vehicles (CNG)

- **Coalbed Methane (CBM)**

  - Methane extracted from coal seams

  - Used similarly to natural gas

- **Associated Petroleum Gas (APG)**

  - Natural gas found dissolved in or associated with crude oil

  - Can be used for power generation, reinjection, or flared

- **Methane Hydrates**

  - Methane trapped in ice crystals under seabeds and permafrost

  - Still experimental for fuel extraction

Manufactured or Processed Gaseous Fuels

These are created from chemical processing or thermal conversion of solid or liquid fuels.

- **Liquefied Petroleum Gas (LPG)**

  - A blend of propane (C_3H_8) and butane (C_4H_{10})

  - Stored under pressure as a liquid

  - Used for cooking, heating, vehicles, and backup power

- **Compressed Natural Gas (CNG)**

  - Compressed methane stored in high-pressure cylinders

  - Used as an alternative automotive fuel

- **Liquefied Natural Gas (LNG)**

  - Natural gas cooled to a liquid state for transport

  - Used in bulk power generation, heavy transport, and export/import trade

- **Producer Gas**

- o Made by gasifying coal, coke, or biomass in limited air
- o Contains CO, H_2, CH_4, and N_2
- o Used in metalworking furnaces and engines

- **Water Gas**
  - o Produced by passing steam over hot carbon (coal or coke)
  - o Composed of CO and H_2
  - o Often enriched with hydrocarbons for added energy (carburetted water gas)

- **Town Gas (Coal Gas)**
  - o An old type of gas made by distilling coal
  - o Contains H_2, CH_4, CO, and N_2
  - o Phased out in favour of natural gas in most countries

- **Synthetic Natural Gas (SNG)**
  - o Produced by methanation of syngas from coal, biomass, or industrial waste
  - o Used where natural gas supply is limited

- **Blast Furnace Gas**
  - o A by-product from iron smelting in blast furnaces
  - o Contains CO, CO_2, N_2, and small hydrocarbons
  - o Used internally for heating and power in steel plants

- **Coke Oven Gas**
  - o A by-product from the production of coke from coal
  - o Contains hydrogen, methane, carbon monoxide, and ammonia
  - o Used for heating or electricity in coke and steel plants

Renewable and Bio-Based Gaseous Fuels

These are derived from biological processes and are considered renewable.

- **Biogas**
  - o Produced from anaerobic digestion of organic matter (e.g., manure, food waste)
  - o Composed of methane and CO_2

- Used for cooking, heating, and electricity generation

- **Landfill Gas**

  - A type of biogas generated from decomposing organic waste in landfills

  - Contains methane, CO_2, and trace gases

  - Captured for electricity and heat

- **Sewage Gas**

  - Biogas produced from sewage sludge in wastewater treatment plants

  - Similar in composition and use to biogas

- **Syngas (Synthesis Gas)**

  - Made by gasifying biomass, coal, or waste

  - Mixture of H_2 and CO, sometimes with CH_4 and CO_2

  - Used for fuel, hydrogen production, or chemical synthesis

Advanced and Experimental Gaseous Fuels

These include hydrogen and ammonia, considered future fuels in clean energy systems.

- **Hydrogen (H_2)**

  - Can be produced from natural gas, water electrolysis, or biomass

  - Used in fuel cells, industrial processes, and space propulsion

  - Burns cleanly with zero CO_2 emissions, emitting only water vapor

- **Green Hydrogen**

  - Hydrogen produced via electrolysis using renewable electricity

  - Seen as a key player in decarbonizing hard-to-abate sectors

- **Ammonia (NH_3)**

  - Can be used as a fuel directly or as a hydrogen carrier

  - Under development for shipping, power plants, and fuel cells

Specialty and Rare Gaseous Fuels

- **Acetylene (C_2H_2)**

- o   Used in welding and cutting torches for its high flame temperature

- o   Highly reactive and stored dissolved in acetone

- **Dimethyl Ether (DME)**

  - o   Can be used as a gas or liquefied under low pressure

  - o   Clean-burning substitute for diesel or LPG

- **Propylene (C$_3$H$_6$)**

  - o   Sometimes used in heating and industrial processes

Gaseous fuels are diverse, ranging from naturally occurring methane to manufactured gases like syngas and bio-based sources like biogas. They are chosen for their clean combustion, flexibility, and efficiency across applications in:

- Residential (cooking, heating)

- Industrial (furnaces, boilers, turbines)

- Transportation (CNG, LPG, hydrogen vehicles)

- Power generation (combined heat and power plants)

- Future energy systems (green hydrogen, ammonia)

The composition and specification of gaseous fuels, especially natural gas, play a critical role in ensuring their safe, efficient, and compatible use across various residential, commercial, and industrial applications. Natural gas is composed predominantly of methane (CH$_4$), which accounts for approximately 93% of its volume. This simple hydrocarbon is the primary contributor to the gas's high energy content and clean-burning characteristics. In addition to methane, natural gas typically contains about 3% ethane (C$_2$H$_6$), along with a combined 4% of other constituents such as propane (C$_3$H$_8$), butane (C$_4$H$_{10}$), nitrogen (N$_2$), carbon dioxide (CO$_2$), and trace amounts of helium. The exact composition can vary depending on the gas source and the processing methods used before distribution.

To ensure uniform quality, safety, and appliance compatibility, gaseous fuels like natural gas must conform to specific technical parameters. For example, in the British National Transmission System, a series of regulatory specifications are in place to control the properties of the gas supplied through the national grid. One key metric is the gross calorific value, which refers to the total amount of heat released when a given volume of gas is burned completely. For natural gas, this value typically ranges from 37.0 to 44.5 megajoules per cubic metre (MJ/m^3).

Another crucial measure is the Wobbe Index, which indicates the interchangeability of fuel gases and helps ensure that burners and appliances operate correctly when switching between different gas sources. For British standards, the Wobbe Index of natural gas must fall within 47.2 to 51.4 MJ/m^3. Maintaining this range ensures that the gas delivers consistent heat output, regardless of slight variations in composition.

Chemical content is also strictly monitored to avoid corrosion, combustion inefficiencies, and health hazards. Hydrogen sulphide (H_2S), for example, must not exceed 5 milligrams per cubic metre, as it is both toxic and corrosive. Likewise, carbon dioxide levels are limited to 2% by volume, and the combined amount of inert gases, which includes nitrogen and other non-combustible components, must be kept below 7% to ensure the gas remains highly flammable and energy-dense.

In addition to these basic chemical limits, parameters such as the Soot Index and the Incomplete Combustion Factor (ICF) are used to evaluate how well the fuel performs in actual combustion systems. The Soot Index helps predict the gas's tendency to generate soot or carbon deposits during burning, which can foul appliances and reduce efficiency. The Incomplete Combustion Factor, on the other hand, provides a numerical indication of how likely the gas is to burn incompletely—producing undesirable by-products like carbon monoxide instead of carbon dioxide.

One of the major advantages of natural gas over many manufactured gaseous fuels is its higher calorific value, which is approximately double that of gases like producer gas or coal gas. This means that for the same volume, natural gas delivers significantly more energy, making it more cost-effective and suitable for large-scale applications such as power generation, heating, and industrial processing. The ability to meet these rigorous specifications is what makes natural gas one of the most dependable and widely used gaseous fuels in modern energy systems.

Gaseous Fuel for Energy Production

Gaseous fuels have gained prominence as a primary energy source across various sectors due to their inherent benefits such as clean combustion, high thermal efficiency, ease of transportation, and versatility in applications. These fuels, primarily natural gas, liquefied petroleum gas (LPG), biogas, and hydrogen, are utilized across residential, industrial, commercial, power generation, agricultural, and transportation sectors.

In residential settings, gaseous fuels serve as a critical energy source for cooking, water heating, and space heating. Natural gas, in particular, is prevalent in urban households within developed nations like the United States, Canada, the UK, and Australia, where it is delivered through extensive pipeline networks. For regions lacking this infrastructure, LPG cylinders become the norm, especially in rural areas of developing nations, including India and Bangladesh, where they facilitate essential cooking and heating functions [74]. The preference for gaseous fuels stems from their clean-burning properties, which lead to lower emissions of pollutants compared to traditional solid fuels like wood or coal. This characteristic not only enhances indoor air quality but also contributes to safety and convenience in household management [75].

Industrially, gaseous fuels are indispensable in processes such as steel manufacturing, cement production, and chemical synthesis, including ammonia and methanol. Their role encompasses high-temperature heating, steam generation, and acting as feedstocks in petrochemical production [76]. Gaseous fuels allow for meticulously controlled flame generation, which is essential in machinery requiring uniform heating. This efficiency, coupled with reduced maintenance needs arising from lower pollutant emissions, positions gaseous fuels as a preferred choice in high-demand industrial processes [77].

In the electricity generation sector, natural gas has become a cornerstone fuel source, powering both baseload and peaking power plants. The technological advancements in power plant design, particularly in open cycle and combined cycle gas turbines, have underscored the ability of natural gas facilities to respond swiftly to fluctuating demand, necessary in modern energy systems that integrate variable renewable sources like wind and solar [74]. Natural gas-fired plants emit significantly lower amounts of CO_2 and other pollutants compared to coal-fired counterparts, making them pivotal in strategies aimed at reducing carbon footprints while ensuring a reliable electricity supply [78].

The agricultural sector utilizes gaseous fuels such as biogas and natural gas for applications including irrigation, greenhouse heating, and machinery operation (Fedorov et al., 2022). Biogas, which can be derived from organic waste, exemplifies a sustainable energy solution that supports off-grid energy independence, especially on rural farms in countries like Germany and China [79]. The energy efficiency of gaseous fuels in agriculture underscores their critical role in promoting sustainable practices and reducing dependency on more costly fossil fuel alternatives.

Commercial entities, including restaurants and laundries, extensively use gaseous fuels for cooking, heating, and sterilization processes. Their controllability leads to enhanced operational efficiency and reduced emissions, contributing to improved air quality in urban settings [77]. This practicality aligns well with increasing regulatory demands for lower emissions in confined spaces [80].

In the transportation domain, gaseous fuels such as compressed natural gas (CNG) and hydrogen are gaining traction as cleaner alternatives to petrol and diesel. The logistics of using these fuels are supported by existing infrastructure and government incentives aimed at reducing local air pollution [81]. Countries like Brazil and India are at the forefront of adopting these fuels in public transportation systems, which underscores their potential in transitioning toward more sustainable urban mobility solutions [74].

Finally, gaseous fuels also find significance in laboratory settings where controlled combustion and precise heat are required for various scientific applications, highlighting the utility of these fuels across diverse sectors [79]. As the global energy landscape evolves, investments in hydrogen and biogas technologies stand to transform not only the energy supply chain but also contribute to circular economy models, where waste is reused as a resource for energy generation [82].

Fuel, Energy and Net Zero

Gaseous Fuel Dependence

The reliance on gaseous fuels as a primary energy source is widespread, spanning various sectors including households, industries, power generation, transportation, and agriculture. This dependency arises from the inherent efficiency, convenience, cleanliness, and cost-effectiveness offered by gases such as natural gas, liquefied petroleum gas (LPG), biogas, syngas, and hydrogen.

Urban and suburban households in developed countries such as the USA, UK, Australia, and Germany rely heavily on gaseous fuels, particularly for cooking and heating, while many rural households in developing nations utilize LPG cylinders. These fuels facilitate convenient cooking and instant heating, providing a controlled flame that is essential for daily cooking activities [83]. LPG and biogas significantly reduce indoor air pollution, a crucial factor especially for women and children in developing regions where traditional fuels like firewood can lead to health hazards [84]. Additionally, the cleaner combustion of these gaseous fuels results in minimal soot and smoke compared to solid fuels [85].

Industries, particularly in sectors like steel and chemical manufacturing, also depend on gaseous fuels to power various processes such as boilers and kilns. These fuels, including natural gas, provide the controllable high temperatures necessary for these operations while also contributing to lower emissions compared to coal and oil, thereby reducing environmental compliance costs [86]. The transition towards gaseous fuels is driven by the need for energy efficiency and operational reliability, as they generally lead to lower maintenance needs and extended equipment lifespans [87].

Electricity generation increasingly relies on natural gas through technologies like combined cycle gas turbine (CCGT) plants, particularly in regions such as the USA and Qatar. Gas turbines efficiently generate electricity while being flexible enough to support the intermittent nature of renewable sources like wind and solar [88]. Gaseous fuels emit significantly less CO_2 than traditional coal-fired plants, aiding in climate goals and providing a more sustainable energy mix across national grids [89].

In the transport sector, governments in countries such as India, Brazil, and Italy have incentivized the use compressed natural gas (CNG) and LPG for vehicular fuel to combat urban air pollution. Gaseous fuels present a cost-effective and cleaner alternative to petrol and diesel, contributing to reduced greenhouse gas emissions and enhancing urban air quality [89, 90]. Additionally, hydrogen fuel is progressively finding its place in long-haul transportation, promising lower emissions and contributing to the shift towards cleaner transport solutions [91].

In agriculture, smallholder farmers utilize biogas generated from animal manure and crop residues for various needs including cooking and lighting. This practice not only provides energy independence but also aligns with sustainable waste management practices, enhancing energy security in regions lacking access to traditional electricity sources [92]. The dual use of agricultural waste for energy production illustrates the potential for transitioning to a circular economy model, benefiting both the environment and local communities [84].

Restaurants and service industries rely on gaseous fuels for consistent heating and food preparation, which allows for greater operational efficiency and cost management. The clean-burning characteristics of gaseous fuels lead to lower pollution output, making these fuels attractive options for businesses striving for sustainability [93].

Gaseous fuels such as methane are utilized in laboratories for controlled heating and sterilization. The precision offered by these fuels is essential in sensitive experiments, where consistency and cleanliness are critical to success [94]. The ability to manage combustion processes effectively establishes gaseous fuels as indispensable in research and development environments [85].

Part 2

Where Fuel Comes From

Chapter 4

Fossil Fuels

Having explored the nature of fuel and the variety of energy sources available to humanity, we now turn to one of the most influential and controversial categories of fuels—fossil fuels. As the dominant energy source for over a century, fossil fuels have shaped global economies, geopolitical strategies, industrial processes, and lifestyles. However, they are also the primary drivers of greenhouse gas emissions and climate change. In this chapter, we delve into the origins, characteristics, extraction methods, and applications of coal, oil, and natural gas. We also begin to unravel the complex legacy these fuels leave behind and examine the growing tensions between their historic utility and the urgent need for a sustainable energy transition. This chapter lays the foundation for understanding why fossil fuels remain central to energy debates—and why moving beyond them is so critical to achieving net zero.

How Coal, Oil, and Natural Gas Are Formed

Coal, oil, and natural gas are all fossil fuels—energy-rich substances formed from the remains of ancient plants, animals, and microorganisms that lived millions of years ago. Although they are all hydrocarbons, they formed in different environments under different geological conditions, which led to their distinct physical and chemical properties.

Fossil fuels, comprising coal, oil, and natural gas, are naturally occurring materials that have formed from the remains of ancient organisms subjected to immense heat and pressure over millions of years. The organic matter from these organisms is buried under sediment where geological processes convert it into solid (coal), liquid (oil), or gaseous (natural gas) fuels [95, 96]. This transformation process, known as diagenesis, results in the formation of fossil fuels, which are critical to contemporary energy production [96].

Among the various types of fossil fuels, coal is predominantly used for electricity generation. It is a solid fuel formed primarily from peat that has transformed under heat and pressure [96]. In contrast, oil serves as a vital resource for transportation, heating, and industrial processes, acting as the backbone of modern energy systems [95, 97]. Natural gas, the gaseous counterpart, is widely utilized in heating and electricity generation and is noted for its cleaner-burning properties compared to coal and oil [96, 97].

The combustion of fossil fuels is a significant source of greenhouse gas emissions, particularly carbon dioxide (CO_2), which contributes to climate change [98]. This relationship arises because burning fossil fuels releases stored carbon from millions of years ago into the atmosphere, causing an increase in atmospheric CO_2 levels [98]. Studies indicate that combustion is not only a source of CO_2 but also other harmful pollutants such as nitrogen oxides and particulate matter, which further exacerbate the greenhouse effect and contribute to global warming [99, 100].

Research has shown that fossil fuel combustion accounts for a substantial portion of elemental carbon emissions in urban areas, with estimates suggesting it contributes upwards of 80% of black carbon, a critical component in climate change dynamics [101]. Additionally, regions heavily reliant on fossil fuels for energy generation demonstrate marked increases in air pollutants, impacting both public health and environmental quality [102, 103].

Coal Formation

Coal formation is a process that spans millions of years, involving the transformation of ancient plant material into a dense, carbon-rich rock used widely as a fuel source. This transformation occurs through a series of stages driven by biological decay, burial under sediments, and geological pressure and heat. Coal is considered a fossil fuel and is mostly derived from terrestrial vegetation that once thrived in lush, swampy environments during prehistoric times.

The story of coal begins roughly 300 to 400 million years ago, especially during the Carboniferous period, when the Earth was covered with vast swamp forests rich in ferns, mosses, trees, and other plant life. When these plants died, their remains began to accumulate in thick layers in wet, marshy regions. Because these environments were low in oxygen, the decay of plant matter was slow and incomplete, which allowed large quantities of organic material to build up over time.

This accumulation of partially decayed vegetation led to the formation of peat, a soft, fibrous, water-saturated material that marks the first stage in coal formation. Peat is rich in carbon and retains the structure of the original plant material. In many parts of the world today, peat is still harvested and used as a fuel, although it is considered the least energy-dense and most moisture-laden form in the coal formation process.

As time passed, layers of sediment—such as mud, silt, and sand—gradually buried the peat, increasing the pressure and temperature on the underlying organic matter. This process, called compaction, began to squeeze out water and further reduce the volume of the material. Chemical changes also started to occur, breaking down complex organic molecules and rearranging them into simpler forms that are more carbon-rich.

Over millions of years, the combination of increased pressure, heat, and time transformed peat into coal through a geological process known as coalification. As coalification progressed, the carbon content increased while moisture and volatile components decreased. This gradual enrichment of carbon led to the formation of several distinct grades of coal, each with varying properties and energy content.

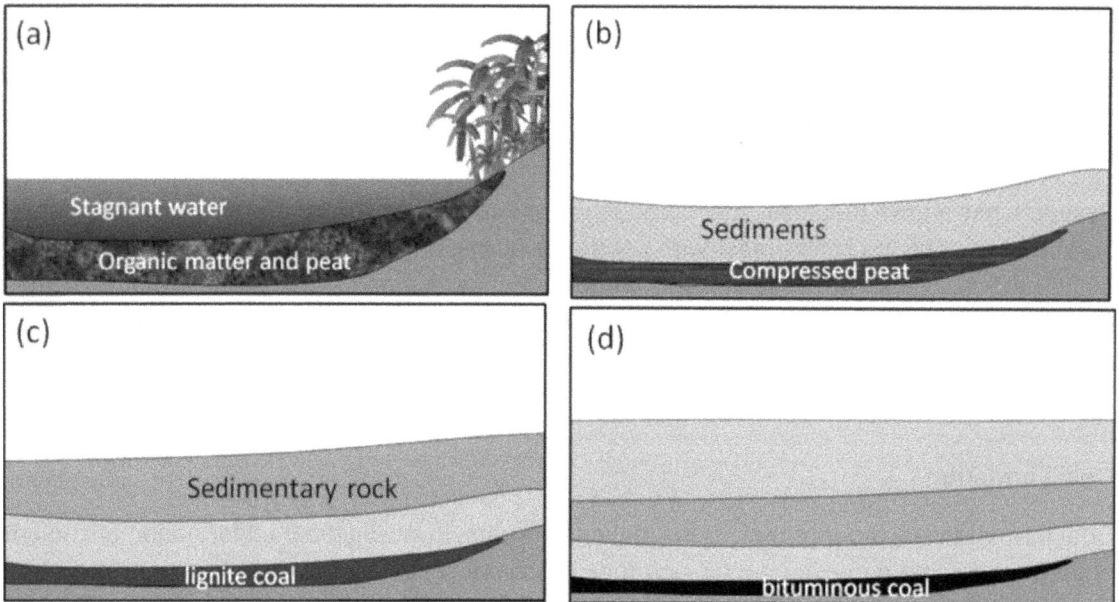

Figure 5: Formation of coal: (a) accumulation of organic matter within a swampy area; (b) the organic matter is covered and compressed by deposition of a new layer of clastic sediments; (c) with greater burial, lignite coal forms; and (d) at even greater depths, bituminous and eventually anthracite coal form. Steven Earle, CC BY-SA 4.0, via Wikimedia Commons.

Fuel, Energy and Net Zero

The first product of coalification is lignite, also known as brown coal. Lignite has a high moisture content and low carbon concentration, making it the least energy-dense form of coal. It is typically used in power generation close to where it is mined, due to its low energy value and difficulty in transport.

With continued burial and heating, lignite can transform into sub-bituminous coal, which has slightly more energy content and is often used in industrial boilers and power plants. Further coalification results in bituminous coal, which is one of the most commonly used coals today. Bituminous coal has a higher carbon content, burns hotter, and is widely used for electricity generation, steel production, and in various industrial processes.

The final and highest grade of coal is anthracite. It forms under even greater temperature and pressure and contains the highest percentage of carbon, giving it a hard, glossy appearance and the ability to burn hotter and cleaner than other coal types. Anthracite is the most energy-efficient coal and is often used for residential heating and in metallurgical applications.

Coal formation is a complex natural process that begins with the accumulation of plant debris in swampy environments and ends with the creation of various coal types through the effects of burial, pressure, and heat over geological time. Each stage of coal—from peat to anthracite—represents a deeper level of transformation and energy concentration, reflecting the length and intensity of the coalification process.

Coal is classified into different ranks based on two primary factors: its gross calorific value and its carbon content, specifically the percentage of fixed carbon available for combustion. These characteristics determine the amount of energy a type of coal can release when burned, and how efficiently it performs as a fuel. The higher the rank, the greater the carbon content and energy yield.

At the top of this ranking system is anthracitic coal, commonly referred to as anthracite. This form of coal is recognized by its hard, glossy texture and high energy output. It contains the highest percentage of fixed carbon—making it the most efficient and clean-burning form of coal. Anthracite also has the lowest moisture and volatile content, meaning it burns more completely and produces less smoke, ash, and pollutants than lower-grade coals.

Just below anthracite in quality is bituminous coal, which also has a relatively high carbon content and energy yield. Bituminous coal is widely used in electricity generation and steel production due to its favourable combustion characteristics and availability. It is dark, compact, and contains more moisture and volatile compounds than anthracite, but it still burns efficiently and generates substantial heat.

Further down the ranking is subbituminous coal, which has less carbon and energy content than bituminous coal. It is typically dull black or dark brown and contains more moisture. As a result, it delivers a lower calorific value when burned. However, subbituminous coal is still used extensively in power plants, particularly in the United States, due to its lower sulphur content and cleaner emissions compared to bituminous coal.

At the lowest end of the coal quality scale is lignite, often called brown coal. Lignite has the lowest fixed carbon content and the highest moisture content, which makes it the least energy-dense and least efficient type of coal. It is typically soft, crumbly, and brown in colour. While lignite burns easily, its low energy output and high emissions make it less desirable, though it is still used in regions with abundant local supplies.

The transformation of coal from one rank to another is part of a natural process driven by heat, pressure, and time. It begins with the formation of peat, a soft, partially decomposed plant material found in waterlogged environments. Through lithification, peat is compacted and solidified into lignite. As the geological pressure and temperature increase over millions of years, lignite undergoes further transformation into subbituminous coal, then into bituminous coal.

At very high heat and pressure, bituminous coal is further compressed and structurally altered into anthracite. Unlike the lower-grade coals, which are classified as sedimentary rocks due to their formation from compacted sediments, anthracite is classified as a metamorphic rock. This is because its physical and chemical structure has been significantly altered by intense geological forces. Anthracite is denser and no longer displays the layered, sheet-like structure seen in sedimentary coal.

In extreme conditions, if anthracite continues to be subjected to even greater heat and pressure, it can undergo another transformation—into graphite, a crystalline form of pure carbon. Graphite is not used as a fuel but is valuable in industrial applications, including lubricants, batteries, and as a material in high-temperature crucibles and electrodes.

Thus, coal's quality and classification reflect its journey through geological time. Each step in the coalification process results in a denser, more carbon-rich, and more energy-efficient fuel. Understanding these ranks helps determine how coal is used in different industries and how it impacts energy production and the environment.

Oil Formation

Oil formation is a process that took place over millions of years, involving the transformation of ancient microscopic marine organisms, such as algae and zooplankton, into crude oil. These organisms lived in the Earth's prehistoric oceans and formed the foundational organic material necessary for the oil we extract today. Unlike coal, which formed from land-based plant matter, oil originates mainly from life that existed in marine environments.

The first step in this transformation began when these tiny organisms died and settled on the seafloor, particularly in areas where oxygen levels were low. In these anoxic environments, the decomposition of organic matter was extremely slow, allowing the dead biological material to accumulate and mix with fine particles of mud and silt. Over time, this created thick, organic-rich layers of sediment. These areas often became sediment basins, which are crucial to oil formation because they provide both the material and the conditions needed to preserve it.

As these sediments accumulated, they were gradually buried under additional layers of sand, silt, and clay. This process, known as sedimentation, increased the pressure and temperature acting on the lower layers. The weight of the overlying material compacted the sediments, squeezing out water and pushing the organic matter deeper into the Earth's crust. Over millions of years, this burial and compaction created the ideal geological environment for chemical transformation.

With increasing pressure and heat, the organic matter was chemically altered into a solid, waxy substance known as kerogen. Kerogen is an intermediate product in the formation of oil and is found in source rocks, especially in organic-rich shale formations. The presence of kerogen marks a critical stage in oil development, as it represents stored chemical energy that can be released under the right thermal conditions.

As the temperature in the source rock increased—typically reaching between 60 and 120 degrees Celsius—the kerogen began to break down into simpler hydrocarbon molecules, resulting in the formation of liquid petroleum or crude oil. This stage is known as catagenesis, a key part of the oil formation process. If temperatures rise too high, beyond 150–160°C, the hydrocarbons can break down further into natural gas.

Once oil formed, it didn't stay in the source rock. Because oil is less dense than surrounding water and rock, it began to migrate upward through porous rock layers, such as sandstone or limestone. These reservoir rocks allowed oil to move freely, until it encountered impermeable rock layers, known as cap rocks, which blocked its further movement. The oil then accumulated in traps, forming oil reservoirs. These traps are what oil companies seek when drilling for petroleum.

The formation of oil is a slow, multi-stage process that begins with the death and burial of marine organisms and ends with the accumulation of liquid hydrocarbons in underground reservoirs. It involves biological deposition, geological pressure and heat, chemical transformation into kerogen, and finally the generation and trapping of oil. This entire process spans tens to hundreds of millions of years, making crude oil a non-renewable resource on human timescales.

Natural Gas Formation

Natural gas formation is a geological process closely related to oil formation but typically occurs under higher temperatures and often over longer geological timescales. Like oil, natural gas is considered a fossil fuel and originates from the remains of ancient marine organisms, though in some cases, it also forms from terrestrial plant material. The result is a cleaner-burning hydrocarbon fuel, primarily composed of methane (CH_4).

The process begins with the accumulation of organic material—mainly microscopic marine life such as algae and zooplankton—on the seafloor. When these organisms died, they settled into layers of sediment rich in mud and silt. In oxygen-poor environments, this organic material did not fully decompose. Instead, it became preserved and buried under increasing layers of sediment, where over millions of years it was subjected to intense pressure and heat.

As the layers deepened, the temperature within the Earth's crust increased. At around 60–120°C, the organic matter transformed into crude oil. However, when the temperatures rose further, typically above 120°C, the kerogen and heavier hydrocarbons broke down into simpler, lighter hydrocarbons—most notably methane, which is the main component of natural gas. This thermal maturation stage is essential for gas formation and is more advanced than the process that yields oil. The resulting gas is often referred to as thermogenic natural gas.

Natural gas can also form from plant-based material, especially in coal seams. This type of gas is known as coal bed methane. It forms during the coalification process, where plant debris is transformed into coal. As the coal forms, methane is generated as a by-product and becomes trapped within the pores and fractures of the coal seam. Coal bed methane is increasingly tapped as an energy source, especially in countries with significant coal reserves.

Another source of natural gas is associated gas, which is found in the same reservoirs as oil. When oil and gas form together, gas typically accumulates above the oil due to its lower density. In contrast, non-associated gas exists in reservoirs where no oil is present, consisting almost entirely of methane. These natural gas reservoirs can be tapped independently of oil fields and are often the target of large-scale gas production.

As with oil, once natural gas forms, it begins to migrate upward through layers of porous rock. The gas moves until it reaches an impermeable layer, which acts as a cap rock, trapping the gas beneath it. These natural gas traps can be structural (like anticlines), stratigraphic (caused by changes in rock types), or a combination of both. Due to its lightness, natural gas is usually found above oil in shared reservoirs or in its own isolated pockets closer to the surface.

Natural gas is formed through the breakdown of organic material under high heat and pressure deep within the Earth's crust. Whether formed in marine sedimentary basins, coal seams, or as part of oil formations, it migrates through rocks and accumulates in traps, where it can be extracted for use. Its cleaner combustion, abundant availability, and versatility make it one of the most important energy resources in the world today.

Global Fossil Fuel Reserves

Coal: The total global proven recoverable coal reserves are estimated at approximately 1.16 trillion short tons. The dominance in coal reserves is evident, as the United States, Russia, Australia, China, and India collectively hold about 75% of these reserves, with the United States accounting for about 22% alone [104]. This significant concentration raises questions about energy security, particularly in countries with substantial coal reserves.

Oil: Proven global oil reserves are estimated at around 1.7 trillion barrels, primarily concentrated in a few countries. Saudi Arabia, Russia, and the United States lead in this metric, with Iran also being a major contributor [105, 106]. The intertwined relationship between these countries often influences global oil prices and energy stability, emphasizing the geopolitical implications of oil reserves [107].

Natural Gas: As of early 2021, the world holds approximately 7,299 trillion cubic feet (Tcf) of proven natural gas reserves. Significant reserves are found in the United States, Russia, and countries within the Middle East [105]. The production from shale gas sources has substantially contributed to the increase in recoverable natural gas reserves in recent years [108].

Global Consumption Rates

Coal: In 2023, global coal-fired power generation saw an increase of nearly 2%, totalling approximately 10,400 terawatt-hours (TWh). This resurgence is indicative of coal's continued relevance in the global energy mix despite environmental concerns [104].

Oil: Oil consumption in 2023 reached about 100 million barrels per day, reflecting a 2% increase from the previous year. The growing demand for oil underscores its importance in global energy consumption and the extent to which economies rely on fossil fuel infrastructures [105, 106].

Natural Gas: Consumption of natural gas remained relatively stable in 2023, maintaining a 22% share of global electricity production. This consistency points towards natural gas's role in transitioning from more polluting fossil fuels like coal and oil, and signifies its strategic importance in the global energy sector [105].

Reserves-to-Production Ratios (R/P)

The reserves-to-production (R/P) ratio provides insights into how long reserves can last at current production rates:

- **Coal:** Approximately 133 years.
- **Oil:** Around 51 years.
- **Natural Gas:** Roughly 53 years [104].

These ratios indicate significant disparities in the longevity of fossil fuel resources, suggesting that while coal may provide a more extended supply, the more immediate consumption rates of oil and gas emphasize the need for a transition towards renewable energy sources to mitigate eventual shortages [106].

Use and Reliance

Fossil fuels have profoundly shaped economic development and modern energy systems since their introduction during the Industrial Revolution. The transition from manual labour to fossil fuel reliance marked a pivotal moment in energy history, allowing for unprecedented advancements in various industrial sectors, transportation, and living standards [109, 110]. Currently, fossil fuels — which encompass coal, oil, and natural gas — constitute approximately 80% of total global primary energy consumption [109]. This dominance is largely attributable to their high energy density, transportability, and the substantial infrastructure developed around them [110, 111].

Consumption patterns of fossil fuels illustrate significant growth since the mid-20th century, with total consumption increasing nearly fourfold since 1950 and doubling since 1980 [109]. Initially dominated by coal, global energy consumption has shifted, with oil and natural gas gaining prominence [112]. In many regions, particularly in developed economies, coal usage has begun to decline; however, it remains substantial in rapidly developing countries like India and China [113, 114]. Conversely, oil continues to be the principal energy source for transportation globally, while natural gas is increasingly favoured over coal in electricity generation due to lower carbon emissions [113, 115].

Geospatial disparities in fossil fuel consumption reveal critical nuances. High per capita consumption is observed in countries such as the United States and Australia, influenced by energy-intensive lifestyles [116]. In contrast, nations with growing energy demands, like China and India, exhibit moderate to low per capita consumption due to varying levels of industrialization and infrastructure development [104]. As a result, global patterns highlight not only the unequal distribution of fossil fuel resources but also the complexities involved in energy transitions, especially in low and middle-income countries where fossil fuels still underpin economic growth [117].

The extraction and combustion of fossil fuels contribute significantly to environmental degradation, particularly through greenhouse gas emissions. The Intergovernmental Panel on Climate Change (IPCC) attributes a large percentage of anthropogenic emissions to fossil fuel usage, underscoring their role in climate change [38, 117]. Notably, fossil fuels accounted for about 69% of global greenhouse gas emissions as of 2010 [38]. This has catalysed calls for a transition to low-carbon energy sources, as evidenced by international agreements such as the Paris Agreement, which seek to curb fossil fuel dependency [118].

The discussion surrounding fossil fuel reserves is also critical. Current reserves for coal, oil, and gas are significantly concentrated in a few countries, notably the U.S., China, and several nations in the Middle East and Russia [112, 113]. This concentration not only affects global energy security but also policy decisions regarding energy sustainability [119].

Looking ahead, the global community faces the daunting challenge of transitioning from fossil fuels while balancing economic realities and climate goals [111]. The push for renewables, electric vehicles, and carbon capture technologies reflects a broader acknowledgment that to combat climate change effectively, most fossil fuel reserves must remain untapped [117, 118]. Therefore, strategic planning and international cooperation are essential for sustainable energy transitions that prioritize both environmental integrity and economic development [120, 121].

Extraction Techniques: Mining, Drilling, Fracking

Fuel extraction techniques are the methods used to obtain fossil fuels—such as coal, oil, and natural gas—from the Earth. The three most common techniques are mining, drilling, and fracking. Each method is suited to different types of fuels and geological conditions.

Mining (Primarily for Coal)

Mining is the primary method used to extract coal, a solid fossil fuel, from deposits located either near the Earth's surface or deep underground. It involves the removal of geological material to access coal seams, which are layers of coal embedded in rock. The process is essential for energy production and industrial use but is also associated with significant environmental and safety concerns. There are two major types of mining techniques used in coal extraction: surface mining and underground mining.

Coal extraction methods vary significantly depending on whether mining is conducted underground or at the surface, as well as factors such as coal seam thickness, overburden characteristics, and geological formations. The choice of mining technique is driven by technical and economic feasibility, which considers aspects like seam depth, material strength, climate, topography, and land access. For instance, the most efficient method for surface mining is typically the use of electric shovels or draglines, while longwall mining is the preferred and most economical method for deep underground coal seams. Longwall mining involves a mechanised system with two rotating drums equipped with carbide-tipped bits that shear coal from a long wall of seam, enabling continuous and high-volume extraction.

Coal extraction begins with a detailed analysis of multiple environmental, technical, and economic factors. These include the continuity and thickness of the coal seam, the nature of surrounding geological layers, surface drainage, groundwater conditions, and access to labour and infrastructure. Additionally, the anticipated coal market—particularly buyer demands for quantity and quality—and the need for initial capital investment influence whether a project proceeds. Many coals, whether extracted from surface or underground mines, require processing in a coal preparation plant (or wash plant) to remove impurities and improve combustion quality.

Surface mining—also known as open-pit or strip mining—is employed when coal seams are relatively close to the surface of the Earth. In this method, large volumes of overburden (the soil, vegetation, and rock covering the coal) are removed to expose and extract the coal underneath. Surface mining is generally considered more economical than underground mining because it allows for the recovery of a larger portion of the coal seam with fewer workers and less complex infrastructure. However, it has considerable environmental drawbacks. The large-scale removal of earth materials causes extensive land degradation and destruction of natural habitats, which can take decades to recover. Furthermore, the visual and ecological scars left by surface mining can permanently alter landscapes.

Surface mining, particularly in the context of coal extraction, encompasses a variety of methods that enable the extraction of resource-rich deposits located close to the Earth's surface. The predominant techniques include open-pit mining, strip mining, contour mining, and mountaintop removal. Each of these methods has distinctive operational practices and environmental implications.

Open-pit mining is recognized for its efficacy in accessing larger coal deposits, particularly within the context of bituminous coal production. This technique is prevalent in regions such as the United States and Australia, where surface mining strategies account for a significant proportion of coal production—up to two-thirds in the U.S. and approximately 80% in Australia [122]. The initial phases involve the removal of overburden, which includes soil and rock layers above the coal seam, typically executed using heavy machinery like draglines and power shovels [123]. This substantial removal process, while cost-effective, is associated with notable environmental consequences, including habitat destruction and ecosystem alteration [124]. The compression of terrain into a flat, plateau-like surface, particularly in methods like mountaintop removal, exemplifies the profound impact of these practices [124].

Figure 6: Open Pit Mining. ПАО «Гайский ГОК»/Rinat Gareev, CC BY-SA 4.0, via Wikimedia Commons.

Strip mining, as a specific subset of surface mining, involves the systematic removal of overburden in elongated strips to expose coal seams [122]. The process is characterized by its sequential approach, where the overburden from the first strip is typically stored outside the mine, while subsequent strips utilize the excavation from previous ones for waste disposal—a practice known as in-pit dumping. This method is ideally suited to flatter terrains, leading to operational developments that can extend over several decades [123]. The coal extracted through strip mining is not only transported for immediate use but is often washed and processed to meet quality standards for various applications.

In contrast, contour mining is strategically employed in hilly or mountainous regions, where the mining follows the terrain's natural contours [123]. Historical practices of dumping spoil downslope led to significant environmental risks, such as erosion and landslides—a legacy that modern mining methodologies aim to mitigate through backfilling and leaving unmined buffers [123]. While effective for certain steep terrains, contour mining becomes economically unfeasible beyond a specific stripping ratio, prompting the use of auger mining for further extraction of coal [125].

Mountaintop removal is a highly controversial technique where entire mountain ridges are removed to facilitate coal extraction, primarily in steep terrain [122]. This method amalgamates elements of both area and contour mining; however, its environmental ramifications are substantial, raising concerns over hydrology, ecosystem disruption, and landscape sustainability [124]. The management of the overburden, which is often deposited into surrounding valleys, is critical to preventing ecological damage and maintaining water quality [124].

Figure 7: Landform complexes produced at the Samples Mountaintop Removal project on Cabin Creek, West Virginia Cabin Creek, USA. Library of Congress, Public Domain, via LOC's Public Domain Archive.

Underground mining, on the other hand, is used when coal seams are located deeper beneath the surface and cannot be accessed economically through surface mining. This method involves the construction of tunnels and shafts to reach and extract the coal from underground deposits. While it causes less visible damage to the land surface, it poses higher risks to worker safety due to the potential for cave-ins, gas explosions, and other hazards. It is also more expensive and labour-intensive, requiring specialized equipment and safety protocols to prevent accidents and ensure efficient extraction.

Underground coal mining remains a significant method for extracting coal, especially in regions where coal seams lie too deep for economical surface mining methods. This approach currently accounts for approximately 50% of the world's coal production, highlighting its importance within the industry [126]. The complexities involved in underground mining require extensive engineering and technological solutions to ensure efficiency, ventilation, and safety. Factors influencing the choice of underground mining methods include geological conditions, seam thickness, and economic viability [127].

One widely employed technique in underground mining is the room and pillar method, where miners carve out a grid of "rooms" within the coal seam, leaving behind "pillars" of coal for roof support. This system preserves the structural integrity of the mine while facilitating initial coal extraction [127]. The pillars typically constitute around 30% of the total coal seam, serving as safety measures against collapses [127]. When deemed necessary, a retreat mining phase follows, wherein miners extract coal from these pillars, thus enhancing recovery rates—but at increased risk for cave-ins [127].

Figure 8: Longwall shearer and armored face conveyor operating at the Twentymile underground coal mine. Peabody Energy, Inc., CC BY-SA 4.0, via Wikimedia Commons.

Among underground mining methods, longwall mining stands out as one of the most productive, responsible for nearly half of underground coal production. It employs a specialized machine called the longwall shearer, which efficiently cuts coal from extensive faces while hydraulic supports maintain the stability of the mine roof during operations [126]. This method optimizes coal recovery, achieving rates between 80% to 90% depending on surrounding geological conditions [128]. Continuous mining is another prevalent technique, using a continuous miner to extract coal in a cyclic process that integrates equipment repositioning with roof stabilizing activities [127].

Alternative techniques, such as blast mining, have largely fallen out of favour due to safety and efficiency concerns. Studies indicate that traditional blast mining now contributes to less than 5% of U.S. underground production [129]. The shift towards more modern and automated systems has seen the introduction of technologies designed to enhance safety and operational efficiency, including real-time monitoring sensors and robotic controls [130]. Such advancements are pivotal for addressing the daring nature of retreat mining, which is regarded as one of the more perilous methods of underground mining due to unpredictable roof collapses [127].

Underground coal mining is an essential method for coal extraction, particularly for deep seams that are uneconomical to access via surface methods. Each mining method offers unique advantages and challenges, closely interwoven with geological and operational dynamics. The evolution of technology within this domain continues to promote higher recovery rates while rigorously addressing safety concerns, thereby optimizing production outcomes in the coal mining sector [127, 131].

Both surface and underground mining contribute significantly to environmental degradation. One of the major concerns is land degradation, where mined areas are stripped of vegetation and topsoil, making it difficult for ecosystems to recover. Mining also leads to habitat destruction, displacing wildlife and disrupting biodiversity. Another serious issue is acid mine drainage, a process where sulphide minerals exposed by mining react with water and air to form sulfuric acid. This acidic runoff can contaminate nearby rivers and groundwater, posing risks to aquatic life and human health. Lastly, methane emissions from coal seams are a significant environmental hazard. Methane is a potent greenhouse gas, and its release during mining contributes to climate change and atmospheric pollution.

While mining is a crucial method for coal extraction, providing fuel for electricity and industrial processes, it must be managed carefully due to its significant environmental and safety impacts. Modern mining operations increasingly incorporate environmental restoration efforts and improved safety practices to reduce these harms, though challenges remain.

Drilling (Primarily for Oil and Natural Gas)

Drilling for oil and natural gas involves creating wells to access subsurface reservoirs. Two main types are vertical drilling, which drills straight down, and directional/horizontal drilling, which allows access to deposits further from the drill site. The process includes site selection, drilling, casing and cementing the well, and extracting the oil or gas. Environmental impacts include risks of spills, contamination, and ecosystem disruption.

Drilling is a method used to access subsurface oil and natural gas deposits that are trapped in porous rock formations beneath layers of earth and sediment. Unlike coal, which is extracted as a solid, oil and gas are fluids and require specialized techniques to tap into their reservoirs. Drilling is essential to modern energy production and occurs both onshore and offshore, depending on the location of the fuel deposits. The goal of drilling is to create a pathway (a well) through which oil or gas can be brought to the surface for collection, processing, and distribution.

There are two main types of drilling: vertical drilling and directional (or horizontal) drilling. Vertical drilling is the traditional approach and involves drilling a straight borehole directly down to the oil or gas reservoir. This method is simple and effective for deposits located directly beneath the drilling site. However, it has limitations in terms of reach and efficiency, particularly when resources are located at an angle or spread out horizontally across a geological formation.

To overcome these limitations, directional and horizontal drilling techniques were developed. Directional drilling allows the well to be angled and curved to access deposits that are not directly below the drilling rig. Horizontal drilling, a form of directional drilling, turns the wellbore horizontally after reaching a certain depth, allowing operators to follow the layer of oil- or gas-rich rock over a long distance. This technique enables multiple wells to be drilled from a single surface location, reducing land disruption and increasing yield from a single reservoir.

The drilling process begins with site selection and geological surveys. Geophysicists and engineers use seismic imaging and geological data to locate promising deposits. Once a site is selected, a rotary drilling rig is set up to bore into the ground. This process creates a wellbore, which must be stabilized to prevent collapse or contamination. To achieve this, steel casing pipes are inserted into the borehole and sealed in place with cement. These casings prevent fluids from seeping into surrounding rock layers and protect groundwater supplies.

Once the well is secure, oil or gas can be extracted. In some cases, natural underground pressure is sufficient to push the fluids to the surface. If not, pumps are installed to assist in bringing the resources up through the well. The extracted materials are then sent to processing facilities for refinement and transport.

Despite its importance, drilling poses significant environmental risks. One of the most serious threats is the risk of oil spills, which can occur during drilling or transportation and lead to devastating ecological damage, particularly in marine environments. Groundwater contamination is another concern, especially if well casings fail or are improperly installed, allowing oil, gas, or drilling fluids to enter drinking water aquifers. Additionally, drilling operations can cause air pollution, notably from flaring—the burning of excess natural gas—which releases carbon dioxide, methane, and other pollutants. Finally, both onshore and offshore drilling can disrupt land and marine ecosystems, leading to habitat destruction, noise pollution, and long-term environmental degradation.

Drilling is a complex but critical technique for extracting oil and natural gas, providing the energy that powers industries, transportation, and households globally. While innovations in drilling technology have improved efficiency and reduced some environmental impacts, significant challenges remain in ensuring that these operations are conducted safely, responsibly, and with minimal harm to ecosystems and communities.

Directional drilling, also known as slant drilling, is an advanced technique that enables the extraction of subsurface resources by allowing wells to be drilled along a non-vertical path. This methodology allows operators in the energy sector to access natural resources that are not directly beneath the drilling site. The evolution of directional drilling has enhanced resource recovery, reduced ecological footprints, and facilitated drilling in challenging environments, such as urban and offshore locations [132]. Consequently, this practice plays a critical role in modern drilling operations across various industries.

Figure 9: Mining Drill at NCL's Dudhichua Coal Mine. Rosehubwiki, Kuber Patel, CC BY-SA 4.0, via Wikimedia Commons.

Applications of directional drilling can be categorized into four primary types: oilfield directional drilling, utility installation, horizontal directional drilling (HDD), and surface-in-seam (SIS) drilling.

1. **Oilfield Directional Drilling**: This is the predominant application of directional drilling within the petroleum industry. It allows multiple wellbores to be drilled from a single surface location, often referred to as a "pad." This method significantly optimizes land use while decreasing infrastructure costs and environmental disturbances, especially in ecologically sensitive areas and urban locations. Oilfield directional drilling permits sidetracking of wells to circumvent unforeseen obstacles or to access additional hydrocarbon reserves beneath urban centres or marine environments [132, 133].

2. **Utility Installation Directional Drilling**: In civil engineering and infrastructure projects, this method is vital for laying underground utilities such as water pipes and telecommunications cables without the need for open trenching. This technology is particularly beneficial in densely populated urban areas, helping to reduce the disruptions associated with traditional digging methods [134]. The ability to drill beneath highways, rivers, and existing buildings adds significant value, as innovations such as steerable drill heads enable accurate navigation along predetermined paths [135].

3. **Horizontal Directional Drilling (HDD)**: HDD is a specialized form of utility installation that entails a precise, shallow drilling technique to navigate long distances horizontally. This method generally comprises three stages: the initial drilling of a pilot hole, enlargement of this hole, and subsequent installation of utilities. The advantages of HDD include its minimal ecological footprint, making it particularly suitable for environmentally sensitive projects [136, 137]. It effectively crosses significant obstacles while minimizing surface disruption, contributing to its widespread adoption in utility installations [138].

4. **Surface-in-Seam (SIS) Drilling**: This technique is predominantly employed in coal seam gas extraction, where horizontal drilling within a coal seam optimizes gas recovery efficiency. By intersecting previously drilled vertical wells, SIS drilling improves gas flow and recovery rates while minimizing surface disturbances [134]. This precision drilling necessitates advanced technology and skilled operation to effectively target the small windows of opportunity created by existing well infrastructure.

The advent of directional drilling has transformed the methodologies for resource extraction and utility installation beneath the Earth's surface. Its core principle of guiding boreholes along controlled non-vertical trajectories has expanded its applications across the energy and construction sectors. Whether by maximizing oil and gas recovery, facilitating urban infrastructure projects, or optimizing coal seam methane extraction, directional drilling continues to represent a significant advancement in drilling technology.

Fracking (Hydraulic Fracturing)

Fracking, also known as hydraulic fracturing, is a technique that has significantly changed the energy landscape by enabling the extraction of oil and natural gas from dense shale formations previously deemed uneconomical. This process involves drilling a well vertically and then horizontally to maximize exposure to the hydrocarbon-bearing rock and injecting a high-pressure mixture of water, sand, and chemicals to create fissures, which allows hydrocarbons to flow more freely to the surface.

The fracking procedure begins with drilling to considerable depths, with horizontal segments potentially extending several kilometres, thereby enhancing extraction efficiency. This technique has led to a substantial increase in the domestic oil and gas supply, particularly in the United States, which has reduced dependency on foreign energy sources and contributed to lower energy prices [139]. The shift towards natural gas has also facilitated a transition from coal to cleaner energy sources, potentially resulting in a decrease in overall greenhouse gas emissions associated with electricity generation [140, 141].

However, the advantages of fracking come with serious environmental and social concerns. One significant issue is the extensive water usage; each well can consume millions of litres of water, which may strain local resources, particularly in arid regions [142]. Additionally, the risk of groundwater contamination has been underscored in various studies, especially regarding methane leaks from faulty well casings or migratory pathways through inadequately sealed sections, threatening aquifer integrity [143, 144]. Research indicates that methane emissions can negate the climate benefits of natural gas compared to coal when assessing their greenhouse potential [141, 145].

Fracking has also been associated with increased seismic activity, as the injection of high-pressure fluids can alter subsurface stress dynamics, leading to minor earthquakes in areas previously thought stable [146]. This raises questions about the long-term sustainability and safety of hydraulic fracturing operations, especially in densely populated areas [147]. The public perception of fracking is often complicated by local resistance due to potential health impacts and infrastructure concerns, such as noise pollution and increased traffic [148].

While hydraulic fracturing has unlocked vast reserves of hydrocarbons and changed energy production dynamics, the environmental ramifications require careful regulatory measures and community engagement strategies. Policymakers must balance the economic benefits of fracking with the necessity of protecting the environment and community health [149]. The complexities of this issue highlight the interplay between energy needs, environmental stewardship, and public health, emphasizing the need for effective governance in this field.

Global Applications

Coal mining is predominantly undertaken in countries richly endowed with coal reserves. China stands out as the largest producer and consumer of coal, accounting for about 50.4% of global production in 2020, and it relies heavily on coal for its electricity needs [150]. India follows China, similarly depending on coal generated from both underground and open-cast mining to meet its energy demands, with substantial reserves facilitating this. Australia plays a pivotal role as a major exporter, utilizing advanced mining techniques to extract high-grade coal primarily for export to Asia. In the United States, mining occurs chiefly in regions like Wyoming and West Virginia, where extensive coal deposits are present. South Africa also relies on coal as a primary energy source, powering a large portion of its national grid and engaging in export activities.

The contribution of coal mining to energy consumption is significant, as it primarily fuels thermal power stations, providing electricity for residential, commercial, and industrial uses [151]. Furthermore, metallurgical coal plays a vital role in steel production, a cornerstone industry for manufacturing. Despite a global decline in coal's share of energy production in favor of cleaner alternatives, it remains crucial in heavily industrialized countries and those still developing, impacting both their energy strategies and environmental practices [152].

Drilling is the foremost technique used for oil and natural gas extraction across several key regions. The Middle East, including nations like Saudi Arabia and Iraq, houses some of the world's largest conventional oil reserves, making it a focal point for drilling activities. In the United States, Texas and North Dakota are notable drilling hotspots, employing both onshore and offshore drilling strategies. Russia maintains its status as a leading oil producer through extensive operations, particularly in Siberia. Canada engages in both conventional drilling and the exploitation of oil sands. Brazil has gained recognition for its offshore deepwater drilling operations, characterized by sophisticated technology.

The role of drilling in the global energy sector cannot be overstated, with crude oil and natural gas serving as the backbone for various fuels, including gasoline, diesel, and heating oil. Natural gas, in particular, supports electricity generation and is critical as an industrial feedstock. These fossil fuels underlie the transportation sector, which includes aviation and automotive industries, marking their profound impact on the global economy [153, 154].

Fracking has emerged as a transformative technology in oil and gas extraction, particularly for tapping unconventional resources such as shale gas and tight oil. The United States leads in fracking technology, notably in the Permian Basin and Marcellus Shale. Canada is also active in fracking, particularly in Alberta, while Argentina explores the vast Vaca Muerta shale basin, which is among the largest globally. China's efforts to develop its shale gas resources face technical challenges, yet the country is making notable strides.

The impact of fracking on the energy supply chain has been remarkable, significantly increasing domestic oil and gas production in North America and stabilizing global prices. Additionally, the transition facilitated by fracking from coal to natural gas for electricity generation has reduced carbon dioxide emissions per unit of energy produced, despite ongoing concerns regarding methane leaks from these processes [152].

The combined efforts of mining, drilling, and fracking account for a substantial portion of global energy consumption. These techniques are pivotal in powering electricity grids, fuelling transportation systems, and supporting industrial processes. They facilitate economic growth and job creation in resource-rich nations, underpinning infrastructure development and overall societal advancement.

Nevertheless, the legacy and future of fossil fuel extraction methods must confront significant environmental challenges, including climate change and ecological degradation. Transitioning to renewable energy systems while managing the environmental impacts associated with mining, drilling, and fracking will be crucial as the global community strives for sustainable energy solutions moving forward.

Although the shift toward renewable energy is gaining momentum, mining, drilling, and fracking continue to be integral to the global energy economy. Their relevance to energy access, industrial development, and modern living standards remains profound, but their associated environmental costs necessitate an urgent call for innovative and sustainable extraction practices.

Major Producing Countries

Fossil fuel production, encompassing coal, oil, and natural gas, continues to play a significant role in the global energy system, despite an increasing focus on renewable energy sources. Fossil fuels remain predominant in the world's energy supply due to their high energy density, compatibility with existing infrastructure, and their critical role in sustaining industrial economies.

1. Crude Oil Production

The United States is currently the leading crude oil producer, with an estimated production of approximately 12.6 million barrels per day in 2024, largely due to advancements in shale oil extraction techniques in regions such as Texas and North Dakota [155]. Following the U.S. is Saudi Arabia, which is expected to produce around 10.6 million barrels per day, primarily from large, low-cost conventional fields, showcasing its significant position within OPEC [155]. Russia maintains a substantial output estimated at 9.3 million barrels per day, despite facing economic sanctions, highlighting its resilience and continued relevance in global markets [155, 156]. Canada produces around 5.3 million barrels per day mainly from oil sands, and Iraq, Iran, China, and Brazil each contribute significant volumes, further emphasizing the importance of these nations in global oil dynamics [155].

The global significance of oil extends beyond production metrics; it is integral to transportation, petrochemicals, and industrial energy, serving as the most traded commodity in the world. The strategic and geopolitical implications of oil production underscore its importance, as nations leverage their fossil fuel resources in international relations and economic strategies [156, 157].

2. Natural Gas Production

In terms of natural gas, the United States again leads with a production rate of approximately 100 billion cubic feet per day, primarily due to advancements in hydraulic fracturing techniques that have significantly increased output from shale deposits [155, 158]. Russia follows, producing around 65 billion cubic feet per day, largely due to extensive reserves in Siberia and a vast pipeline network exporting to Europe and Asia [155, 158]. Iran holds a robust position as well, producing between 25 and 30 billion cubic feet per day, mainly from the South Pars/North Dome field, which is one of the world's largest gas reserves [155].

Countries like China and Qatar are also major players, with Qatar recognized for its leadership in liquefied natural gas (LNG), underscoring the growing importance of LNG trade in diversifying global energy distribution [158]. Furthermore, natural gas plays critical roles in electricity generation and heating, often regarded as a transitional fuel due to its comparatively lower carbon emissions relative to coal [159].

3. Coal Production

When examining coal production, China is the dominant player, with annual output ranging from 3.8 to 4.1 billion tonnes, accounting for over half of the world's coal supply, predominantly for electricity generation and industrial applications, such as steelmaking [156, 160]. India follows, using coal for over 70% of its electricity needs, indicating its reliance on this fossil fuel as consumption continues to rise [156]. Other significant contributors include Indonesia, the United States, and Australia, reflecting coal's ongoing relevance in global energy discussions, despite its environmental implications [156].

Fuel, Energy and Net Zero

The energy transition is further complicated by the continued reliance on coal in emerging economies, where it remains essential for electricity generation and industrial demand [156, 160]. The environmental concerns surrounding coal, alongside international efforts for a phase-out, illustrate the tension between energy needs and sustainable practices [160].

Global Trends and Shifts

Fossil fuel production remains under pressure from energy transition forces, with many countries striving to diversify towards renewable sources. However, fossil fuels still hold a dominant position in regions such as Asia and the Middle East, where economic reliance on these resources is profound [155, 160]. Geopolitical dynamics also play a significant role; countries heavily engaged in fossil fuel exports, including Russia, Saudi Arabia, and Qatar, maintain considerable influence over global energy markets [156, 157]. Lastly, the financial stability of several nations is closely tied to their fossil fuel exports, further demonstrating the sector's pivotal role in global economics despite the transition towards renewables [160].

For most of human history, societies relied on basic energy sources—manual labour, animal power, and biomass (wood, crop residues). This limited the scale and speed of economic and technological development. However, the Industrial Revolution, which began in the late 18th century, marked a transformative moment in energy history. It introduced fossil fuels—coal, oil, and natural gas—as powerful, energy-dense resources that revolutionized transport, industry, agriculture, and living standards. These fuels have since become central to the global energy system, enabling massive infrastructure, industrial expansion, and economic growth [161].

Today, fossil fuels remain the dominant global energy source, accounting for roughly 80% of global primary energy consumption [161]. Their convenience, transportability, and high energy return have made them foundational to modern energy systems. Coal has historically powered electricity generation and steelmaking; oil fuels the global transport sector; and natural gas is critical for electricity, heating, and industrial uses.

However, the combustion of fossil fuels releases carbon dioxide (CO_2) and other greenhouse gases, making them the largest contributors to climate change. Additionally, their extraction and use produce air pollutants like particulate matter and nitrogen oxides, linked to millions of premature deaths annually from respiratory and cardiovascular diseases. These environmental and health consequences have fuelled calls for a transition to low-carbon energy sources, such as renewables and nuclear power [161].

Fossil fuel consumption has surged since the mid-20th century, increasing eightfold since 1950 and nearly doubling since 1980. Initially, coal dominated global consumption, but over time, oil and then gas have become increasingly important. Coal consumption is now declining in many advanced economies, but remains high in rapidly developing countries. In contrast, oil and gas consumption continues to grow, particularly in transport, manufacturing, and electricity generation [161].

- Coal remains prominent in China, India, and Southeast Asia.

- Oil is globally consumed, with high usage in the U.S., Middle East, and transport-heavy economies.

- Natural gas is replacing coal in many regions due to its lower emissions per unit of energy, especially in North America and Europe.

Global fossil fuel consumption is unevenly distributed across countries, with total usage often reflecting national population size, economic development, and industrial activity. The United States, China, and India are among the largest consumers of fossil fuels in absolute terms due to their vast populations and energy-intensive economies. However, when consumption is measured on a per capita basis, a different pattern emerges [161].

Countries such as the United States, Australia, and Germany exhibit high per capita fossil fuel consumption. This is largely the result of energy-intensive lifestyles, widespread use of private vehicles, and significant industrial output. In contrast, countries like China, the United Kingdom, and South Africa show moderate levels of per capita consumption, reflecting their status as either rapidly growing economies or nations in the midst of an energy transition away from coal and toward more diversified energy sources.

Lower per capita consumption is most common in countries such as India, as well as across much of Africa and Southeast Asia. These regions typically have limited infrastructure, lower rates of industrialization, and reduced access to energy resources, which collectively constrain their overall energy use on a per-person basis [161].

Globally, approximately four-fifths of primary energy continues to come from fossil fuels, although this share is gradually declining as renewable energy sources expand. While fossil fuels remain the foundation of the global energy system, their dominance is slightly less pronounced in the electricity sector, where a more diverse mix of sources is emerging [161].

Coal remains the largest single source of electricity worldwide, with countries such as China, India, and South Africa continuing to rely heavily on coal-fired power generation. Natural gas is the second-largest source of electricity and is widely used in regions including the United States, Europe, and the Middle East, valued for its flexibility and relatively lower emissions compared to coal. Oil, on the other hand, plays only a minor role in electricity production but remains the primary fuel for the global transport sector [161].

The electricity sector is also where renewable energy is making the most significant progress. Countries in Europe, as well as Australia and parts of the United States, are rapidly increasing their reliance on wind, solar, and other low-carbon energy sources, gradually reshaping the global electricity mix toward more sustainable systems [161].

Coal production is led by China, India, the United States, Indonesia, and Australia. Among these, China is not only the top producer but also the largest consumer by a wide margin, followed by India and the United States. Coal continues to play a vital role in electricity generation and steel production, particularly in developing and industrializing nations. However, in many OECD countries, coal is being phased out due to its significant environmental and health impacts, as governments implement policies to reduce carbon emissions and air pollution [161].

Fuel, Energy and Net Zero

Oil production is dominated by the United States, Saudi Arabia, Russia, Canada, and Iraq. The United States also ranks as the largest oil consumer, followed closely by China and India. Oil remains the most widely used energy source globally and is especially critical to the transportation sector, fuelling cars, trucks, ships, and airplanes. Its importance to global trade and industry ensures that oil continues to shape geopolitical relationships and economic policies around the world [161].

Natural gas production is highest in the United States, Russia, Iran, Qatar, and Canada. Consumption is similarly concentrated in the United States, China, Russia, Iran, and Japan. Natural gas is often viewed as a "transition fuel" because it emits less carbon dioxide per unit of energy than coal when burned. As a result, many countries are turning to gas as a bridge between coal-heavy systems and renewable energy, using it to reduce emissions while maintaining energy reliability and economic stability [161].

Coal reserves are most abundant in the United States, Russia, China, Australia, and India. These countries possess the geological formations and economic infrastructure necessary to continue coal extraction for many decades. However, global usage patterns are beginning to shift as environmental concerns and clean energy transitions lead many nations to reduce their dependence on coal [161].

Oil reserves are largest in Venezuela, Saudi Arabia, Canada, Iran, and Iraq. The Middle East remains the central hub for global oil production and trade, due to both the volume and accessibility of its reserves. Nonetheless, substantial oil resources also exist in North and South America, particularly in Canada and Venezuela, where unconventional sources like oil sands contribute significantly to supply [161].

Natural gas reserves are heavily concentrated in Russia, Iran, Qatar, Turkmenistan, and the United States. The growing use of liquefied natural gas (LNG) technology has allowed for the expansion of gas trade beyond the constraints of pipeline networks. This has made natural gas a more globally accessible energy resource, increasing its role in the international energy market and facilitating its adoption as a transitional fuel in various countries [161].

The world is at a crossroads. While fossil fuels continue to support economic growth and energy security, their environmental impacts are becoming increasingly untenable. Global agreements such as the Paris Agreement aim to limit global warming by reducing fossil fuel use. Countries are committing to net-zero targets, phasing out coal, and investing in renewables, electric vehicles, hydrogen, and carbon capture technologies [161].

However, the transition must balance climate goals with economic and social realities—especially in low- and middle-income countries where fossil fuels still underpin development.

Fossil fuels have driven human advancement more than any other energy source. From powering factories and transportation to enabling digital economies, they have shaped modern civilization. Yet, the same fuels are the largest contributors to climate change and environmental degradation. A balanced, just, and strategic energy transition—grounded in clean technology, global cooperation, and sustainable development—is essential to addressing the fossil fuel legacy while securing our energy future [161].

Chapter 5

Renewable and Alternative Fuels

With a firm understanding of how fuels are combusted and how efficiently they can be used, the conversation now shifts from traditional fuels to the new frontiers of energy. Chapter 5 explores the evolving landscape of cleaner, more sustainable energy options that are reshaping the global response to climate change and resource depletion.

This chapter introduces a range of fuel alternatives—including biofuels, green hydrogen, synthetic fuels, and electricity from renewable sources such as wind, solar, and hydro. It delves into the science, production methods, and applications of these energy carriers while also considering their environmental benefits, limitations, and economic implications.

As the world confronts the posited urgent need to reduce greenhouse gas emissions, these alternatives represent not just technical solutions but a fundamental reimagining of energy systems. Chapter 5 provides a critical transition point in the book, moving from understanding how we use current fuels to exploring how we can fuel the future.

Biofuels: Ethanol, Biodiesel, Algae-based Fuel

Biofuels are renewable energy sources derived from biomass—organic material from plants and animals. They are seen as a more sustainable alternative to fossil fuels, particularly in the transport sector, where electrification is still limited in many areas. Among the most well-known biofuels are ethanol, biodiesel, and algae-based fuel. Each has distinct production methods, sources, benefits, and challenges.

Biofuels, including ethanol, biodiesel, and algal-based fuels, present viable renewable alternatives to traditional fossil fuels, significantly contributing to energy sustainability. Ethanol, predominantly derived from corn, serves as an additive to gasoline, thereby enhancing octane levels and reducing greenhouse gas emissions when blended in vehicles [162]. Biodiesel, on the other hand, can be sourced from various feedstocks, including vegetable oils and animal fats, facilitating its use as a direct substitute for diesel fuel [163]. Both biofuels exemplify the transition towards renewable energy sources, reducing dependence on finite fossil fuel resources.

Algae, often categorized as a third-generation feedstock, are gaining recognition for their potential in biofuel production. Various algal species exhibit biomass yields that are substantially superior — 20 to 30 times higher — compared to first and second-generation feedstocks, such as vegetable crops [164]. Notably, algae possess the capability to produce biodiesel and bioethanol, with oil content averaging approximately 30% [165]. The capability of algae to grow in diverse environmental conditions, including wastewater, further enhances their attractiveness as a sustainable feedstock that does not directly compete with food crops [166].

Research indicates that algal biofuels could effectively replace fossil transportation fuels in truck and aircraft applications, particularly in sectors where electrification is not currently feasible [167]. The environmental advantages of algal biofuels include their carbon neutrality; algal cultivation can utilize CO_2 emissions from industrial sources, which contributes to greenhouse gas mitigation efforts [168]. Moreover, life-cycle assessments of various biofuels suggest that while first-generation biofuels may compete with food for land and resources, algae can be produced without leading to biodiversity loss or resource depletion [169].

To ensure the economic viability of algal biofuel production, advancements in genetic engineering and biotechnology are essential. Enhanced lipid production through these technologies can significantly increase the efficiency of biofuel yields, thereby making algal biofuels more competitive against fossil fuels [170]. Studies have highlighted the necessity to streamline production processes and optimize algal species selection, which can help in producing high-quality biofuels at lower costs [171].

The ongoing exploration and refinement of biofuels, particularly those derived from algae, signify a promising path towards sustainable energy solutions that could alleviate dependence on fossil fuels while addressing environmental concerns. The multifaceted applications of biofuels underscore the urgent need for innovation in production technologies and careful management of renewable resources [172].

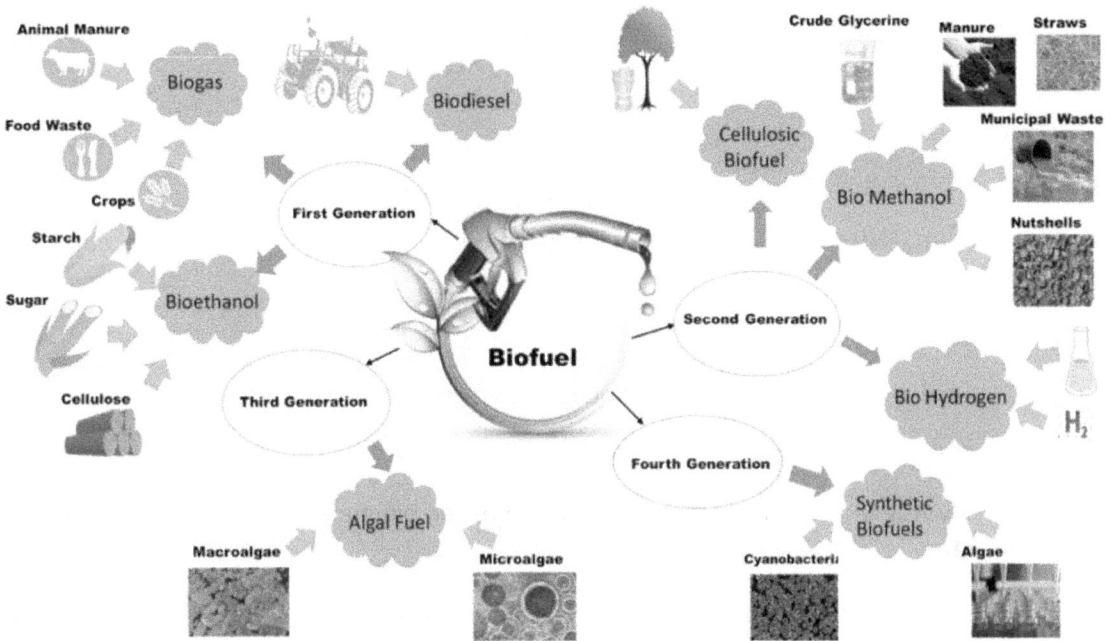

Figure 10: Types and generation of biofuels. Muhammad Rizwan Javed, Muhammad Junaid Bilal, Muhammad Umer Farooq Ashraf, Aamir Waqar, Muhammad Aamer Mehmood, Maida Saeed and Naima Nashat, CC BY-SA 4.0, via Wikimedia Commons.

Figure 10 provides a visual overview of biofuel classification, production sources, and generation types, showcasing how various raw materials are converted into different forms of biofuels through technological advancement. It organizes the information into four generations of biofuels, each based on the source material and processing technology.

At the centre of the image is "Biofuel", represented by a fuel nozzle and a green plant motif, symbolizing its renewable and environmentally friendly nature. Arrows radiate from the centre, linking biofuels to their generation types and the feedstocks and products associated with each.

First-Generation Biofuels

Feedstocks: Food crops (corn, sugarcane, starch crops), animal manure, and vegetable oils.

Biofuel Types:

- **Bioethanol:** Produced from sugar, starch, and crops.

- **Biodiesel:** Made from vegetable oils and animal fats.

- **Biogas:** Generated from food waste and manure through anaerobic digestion.

83

Characteristics: These are the earliest biofuels, derived from food-grade feedstocks using relatively simple fermentation or transesterification processes. They are commercially available but often criticized for contributing to the food vs. fuel debate.

Second-Generation Biofuels

Feedstocks: Non-food biomass such as cellulose, municipal waste, straws, manure, nutshells, and crude glycerine.

Biofuel Types:

- **Cellulosic Biofuel:** Made from lignocellulosic biomass (wood, crop residues).

- **Bio Methanol:** Derived from organic waste and cellulose.

- **Bio Hydrogen:** Produced through gasification or biochemical processing of organic waste.

Characteristics: These biofuels use non-edible biomass, addressing the limitations of the first generation by avoiding food crops and utilizing agricultural waste. They require more complex technology and are still scaling commercially.

Third-Generation Biofuels

Feedstocks: Algae—macroalgae, microalgae, and cyanobacteria.

Biofuel Type:

- **Algal Fuel:** Includes algal biodiesel and other algal-derived oils.

Characteristics: Algae offer a high oil yield per acre, can be grown on non-arable land with saline or wastewater, and do not compete with food crops. This generation is still under development but has high future potential due to its scalability and low environmental impact.

Fourth-Generation Biofuels

Feedstocks: Genetically engineered algae and microorganisms, including cyanobacteria.

Biofuel Types:

- **Synthetic Biofuels:** Biofuels created using advanced biotechnological processes, such as synthetic biology and carbon capture.

Characteristics: These are the most advanced biofuels, aiming to be carbon-neutral or even carbon-negative by using genetically modified organisms and integrating carbon capture and storage. They are still in experimental or pilot stages.

Figure 10 effectively maps the evolution of biofuels from food-based to advanced synthetic fuels. It shows how technological progress allows for more sustainable feedstocks and improved environmental performance, moving from basic fermentation to complex biochemical and synthetic processes. The diagram also emphasizes the diverse input sources available for biofuel production, highlighting the transition from agricultural products to waste and engineered organisms.

Ethanol

Ethanol is an alcohol-based fuel typically made by fermenting sugars or starches from crops like corn, sugarcane, or wheat. It is commonly used as a fuel additive to reduce emissions and improve combustion in petrol (gasoline) engines.

In countries like the United States, ethanol is mostly produced from corn, while Brazil uses sugarcane, which offers a better energy return and lower greenhouse gas emissions. Ethanol can be used in various blends with gasoline:

- E10 (10% ethanol, 90% gasoline) is widely used around the world.

- E85 (up to 85% ethanol) is used in flex-fuel vehicles designed to tolerate higher ethanol concentrations.

Advantages of ethanol include reduced emissions of carbon monoxide and particulates, lower reliance on crude oil, and support for agricultural economies. It burns cleaner than gasoline and is biodegradable.

Ethanol (C_2H_5OH) is increasingly recognized as a renewable fuel source produced from various plant materials, collectively referred to as "biomass." As an alcohol, ethanol serves a dual purpose in the fuel industry; it acts as both a blending agent with gasoline, contributing to higher octane ratings, and as a means to mitigate harmful emissions such as carbon monoxide and other pollutants generated during combustion. The most common fuel blend containing ethanol is E10, which comprises 10% ethanol and 90% gasoline, widely approved for use in the majority of conventional vehicles in the United States. Additionally, blends such as E15 (15% ethanol) and E85 (51%-83% ethanol) cater to flexible fuel vehicles, showcasing a growing acceptance of ethanol in various fuel formulations [173, 174].

Ethanol predominantly arises from fermenting sugars and starches, with corn starch being the principal feedstock in the United States, accounting for approximately 95% of its ethanol production as of 2016 [175]. This process generally focuses on the fermentation of carbohydrate-rich substances to yield ethanol through the action of microorganisms such as yeast and bacteria [176]. The fermentation method consists of breaking down plant-based carbohydrates into simpler sugars, which can subsequently be converted into ethanol [176]. This biotechnological approach not only supports the production of renewable energy but also offers potential pathways for advancements in cellulosic ethanol production, leveraging non-edible plant materials that comprise much of agricultural biomass [177, 178].

Moreover, ethanol's production mechanisms include not only fermentation but also synthetic methods such as the hydration of ethylene. This process entails reacting ethylene with steam under high temperature and pressure. Regardless of the production method, ethanol is typically extracted as a dilute aqueous solution, necessitating methods like fractional distillation to achieve higher purity levels. This purification is crucial, especially considering that anhydrous ethanol is often denatured with substances like methanol to prevent its consumption as a beverage [179]. The utility of ethanol extends beyond automotive applications; its characteristics as a solvent make it valuable in various industrial applications, including the chemical synthesis and manufacture of alcoholic beverages [180].

Despite the advantages of ethanol, there are some U.S. regulations and commercial challenges surrounding its use, particularly concerning the balance between food production and fuel needs. Nevertheless, ongoing research and policies aim to expand the role of non-food biomass, such as agricultural residues and waste materials, in ethanol production to maximize sustainability and minimize competition with food resources [181]. Ethanol not only presents a cleaner alternative to fossil fuels, significantly reducing pollutants like carbon dioxide and hydrocarbons, but also promotes energy independence and sustainability in the transportation sector [182].

Thus, ethanol emerges as a critical player within the scope of renewable fuels, balancing the need for cleaner energy while contributing to environmental sustainability through reduced emissions and the efficient use of biomass resources.

Challenges include competition with food production, since many ethanol crops are also food crops. Large-scale corn ethanol production also consumes significant water and fertilizers, and can contribute to deforestation and land-use change in some regions. Additionally, ethanol has a lower energy density than gasoline, meaning more fuel is needed to travel the same distance.

Figure 11: Biofuel pumps (B20, left, and E85, right) at a gasoline station near Pentagon City, Arlington, Virginia. Mariordo Mario Roberto Duran Ortiz, CC BY-SA 3.0, via Wikimedia Commons.

Biodiesel

Biodiesel is a renewable, biodegradable fuel made from vegetable oils (such as soybean, canola, palm), used cooking oil, or animal fats. It is produced through a chemical process called transesterification, where oils or fats are reacted with alcohol (usually methanol) in the presence of a catalyst to produce fatty acid methyl esters (FAME)—the chemical name for biodiesel—and glycerine as a by-product.

Biodiesel can be used in pure form (B100) or blended with petroleum diesel in varying proportions:

- B5 (5% biodiesel) and B20 (20% biodiesel) are the most common blends for commercial use.

- It is compatible with many existing diesel engines, especially at lower blend levels.

Advantages of biodiesel include reduced emissions of greenhouse gases, particulate matter, and sulphur compounds compared to petroleum diesel. It supports waste recycling when made from used cooking oil or non-food feedstocks and enhances energy security.

Challenges stem from the sustainability of feedstocks. Some sources, like palm oil, are linked to deforestation, habitat loss, and biodiversity threats. Biodiesel can also gel in cold temperatures, requiring modified storage or additives in cold climates. There are also concerns about competition with food crops and land use, similar to ethanol.

Biodiesel is increasingly recognized as a renewable and biodegradable alternative to conventional petroleum diesel. This liquid fuel is produced primarily via transesterification, a chemical process that involves the reaction of vegetable oils, animal fats, or recycled cooking grease with alcohol—typically methanol. The output of this process includes fatty acid methyl esters (FAME), which constitute biodiesel, and glycerol, a by-product that finds applications in the pharmaceutical and cosmetic industries [183, 184]. Biodiesel can be utilized directly in compression-ignition engines or blended with traditional diesel in various ratios such as B5 (5% biodiesel), B20 (20% biodiesel), or B100 (pure biodiesel), and it shows compatibility with existing diesel infrastructure [185, 186].

The history of biodiesel dates back to the early 20th century; Rudolf Diesel famously demonstrated his engine running on peanut oil at the 1900 Paris Exposition. Interest in vegetable oil fuels surged during periods of fuel scarcity, leading to the establishment of standardized production methods in the 1930s. Today, biodiesel is valued for its high lubricity, elevated flash point (generally above 130°C), and low sulphur content, making it an appealing substitute from both performance and environmental perspectives [185, 186]. Biodiesel's combustion characteristics often lead to lower emissions of harmful pollutants such as carbon monoxide and particulate matter compared to petroleum diesel, particularly when derived from waste oils or fats [187, 188].

Although biodiesel presents many benefits, its cold-weather performance can be a concern, particularly the phenomenon of gelling at higher temperatures than petroleum diesel, especially in the case of B100. To ameliorate these issues, additives like cold flow improvers are frequently used in colder climates, while lower blends such as B5 or B20 are employed [189]. Despite having slightly lower energy content relative to conventional diesel by approximately 9%, biodiesel's favourable combustion efficiency and excellent lubricating properties can make it a viable alternative in real-world applications [190, 191].

Environmental benefits are significant for biodiesel, contributing to reductions in carbon emissions and potentially achieving up to an 86% decrease in greenhouse gases when produced from waste oils [192]. However, sustainability matters remain controversial; the cultivation of feedstocks like palm oil has been linked to deforestation and land-use change, making it imperative to consider responsible sourcing practices [193, 194]. The debate over "food vs. fuel" raises additional concerns regarding the implications of diverting food crops for fuel production, affecting food prices and availability [195].

Biodiesel's applications extend beyond transportation; it is used in heating systems, power generators, and even in oil spill clean-up operations due to its biodegradability and solvent properties [196]. Its versatility is exemplified by successful implementations in public transport fleets and even in railways, as seen in initiatives by cities like Halifax and UK franchises [197].

Figure 12: Biodiesel Fuel Gas Pump for Cars. Robert Couse- Baker, CC BY 2.0, via Wikimedia Commons.

The biodiesel market has expanded globally, particularly within the EU and the United States, where policies like the Renewable Fuel Standard have encouraged production. Developments in second-generation biofuels from non-food sources such as algae are explored for their potential to mitigate competition with food resources [195, 198]. Despite optimistic prospects, challenges remain regarding biodiesel's technical integration into engines, particularly concerning fuel system compatibility, moisture absorption, and gelling [199, 200].

Biodiesel's contributions extend beyond emission reductions to influence energy security, fostering greater reliance on domestically available, renewable resources. Various governments have implemented mandates and incentives to advance biodiesel usage, thus aiding job creation within rural areas and supporting shifts toward sustainable energy portfolios [197, 201]. However, successful deployment requires careful consideration of environmental impacts, production practices, and the careful management of feedstock sourcing to maximize benefits while minimizing drawbacks.

Algae-Based Fuel

Algae-based fuel is an emerging biofuel made from microalgae—single-celled organisms that can produce large amounts of lipids (fats), which are extracted and converted into biodiesel or other fuel types. Unlike traditional biofuel crops, algae do not compete with food production and can be grown in non-arable areas using saline or wastewater.

Algae can produce 30 to 100 times more oil per acre than conventional crops like soybeans. Moreover, they absorb carbon dioxide during photosynthesis, offering potential carbon sequestration benefits. Algae can also be genetically modified to enhance oil yield or produce other valuable co-products, such as proteins or pigments, supporting biorefinery models.

Advantages include high productivity, use of marginal land and water sources, and reduced environmental footprint compared to land-based crops. Algae fuel can also be refined into a range of fuels, including jet fuel, diesel, and even ethanol, offering flexibility.

Algal biofuels, derived from algae, represent a promising alternative to conventional fossil fuels due to the high lipid content in many algal species. These biofuels can be cultivated on non-arable land and utilize seawater, making them particularly appealing in contrast to traditional biofuel sources such as corn and sugarcane, which require fertile land and freshwater resources [202, 203]. Algal biofuels exhibit potential in displacing conventional fuels, as the oil yields from algae can exceed those of terrestrial crops by a factor of ten to twenty [204, 205]. When made from macroalgae, this biofuel is often referred to as seaweed fuel, highlighting its versatility and applicability in various biofuel categories including biodiesel and biogasoline [203].

The historical exploration of algae as a viable source of biofuels can be traced back to 1942 when researchers like Harder and Von Witsch championed the notion of cultivating microalgae for lipid recovery [167]. Post World War II, interest waned until a resurgence during the oil crises of the 1970s led to significant federal initiatives in the United States, notably the Department of Energy's Aquatic Species Program, which investigated algae for fuel production [206]. This program documented that certain strains could achieve a lipid content of up to 70% of their dry weight, a significant factor for biofuel viability [204]. Despite proving that large-scale production was feasible, economic challenges, such as high production costs relative to fossil fuels, stalled progress, leading to the program's discontinuation in 1996 [207].

While the Aquatic Species Program ceased, global interest in algal biofuels has fluctuated, with renewed vigour in the 2000s prompted by escalating oil prices [208]. Recent studies highlight that algae can be transformed into multiple fuel types through various processes: the extraction of lipids can produce biodiesel, while carbohydrates can be fermented into ethanol [209]. Moreover, algal biomass can serve as feedstock for biogas production via anaerobic digestion, further enhancing its appeal in a comprehensive bioenergy strategy [210].

Research indicating the biochemical pathways and genetic enhancements necessary for optimizing lipid production is ongoing. For instance, techniques such as metabolic engineering are being deployed to foster rapid growth rates and higher lipid yields [211]. Additionally, innovations in algal cultivation, including integrating photovoltaic systems, aim to improve the cost-efficiency of algal biofuel production by maximizing lipid content and biomass yield [212].

Despite the substantial prospects for algal biofuels, current challenges remain, chiefly their high production costs and the technical complexities associated with processing algal biomass [213]. As stated in recent evaluations, funding in algal biofuel research has historically been inconsistent, and projections suggest that large-scale commercial viability may be two decades away unless significant advancements in technology and funding occur [214]. The commitment to overcoming these barriers will be critical in the quest for sustainable energy solutions forged through algae.

Challenges include high production costs, technical complexity, and scaling issues. Growing algae requires controlled conditions (light, temperature, nutrients) and efficient harvesting and extraction methods, which are not yet cost-competitive with fossil fuels or other biofuels. However, research is ongoing to improve yields, reduce costs, and integrate algae production with carbon capture and wastewater treatment.

Biofuels as a Replacement for Fossil Fuels

Biofuels, including ethanol, biodiesel, and algae-based fuels, are pivotal in the transition from fossil fuels, particularly within the transportation and heating sectors. Their integration into existing infrastructures presents a multifaceted challenge influenced by technological, economic, environmental, and infrastructure readiness factors. Each biofuel type offers distinct advantages along with inherent limitations that can hinder their potential for widespread adoption.

Ethanol, primarily produced from corn and sugarcane, is a significant component of fuel blends, with countries like the U.S. and Brazil utilizing blends such as E10 and E85. Ethanol's integration with internal combustion engines, particularly in flexible-fuel vehicles, supports its role in reducing petroleum dependence and lowering harmful emissions including carbon monoxide and particulates [215]. However, it has a lower energy content compared to gasoline, necessitating larger quantities to achieve the same distance, thus affecting the overall efficiency of transportation systems [216]. Moreover, the competition for arable land between food and fuel production raises food security concerns, compounded by the high inputs required for ethanol production regarding water, land, and fertilizers [217]. Thus, while ethanol can enhance fuel mixtures, its capacity to completely replace fossil fuels is limited, especially as electric vehicles become more prevalent [218].

Biodiesel, derived from vegetable oils and animal fats, is recognized for its manifold environmental advantages, including reduced greenhouse gas emissions and improved engine lubricity [219]. It is compatible with most diesel engines without necessitating significant modifications, making it a practical short- to medium-term substitute for conventional diesel fuels [220]. Despite these positives, the scalability of biodiesel production is hindered by feedstock availability and constraints related to land and resources [221]. Additionally, cold-weather performance and fuel stability remain concerns, alongside environmental implications tied to large-scale crop cultivation, such as deforestation associated with palm oil production [222]. Therefore, biodiesel can supplement but is unlikely to entirely supplant traditional diesel fuels globally.

Biofuel energy production, 2022

Total biofuel production is measured in terawatt-hours (TWh) per year. Biofuel production includes both bioethanol and biodiesel.

Our World In Data

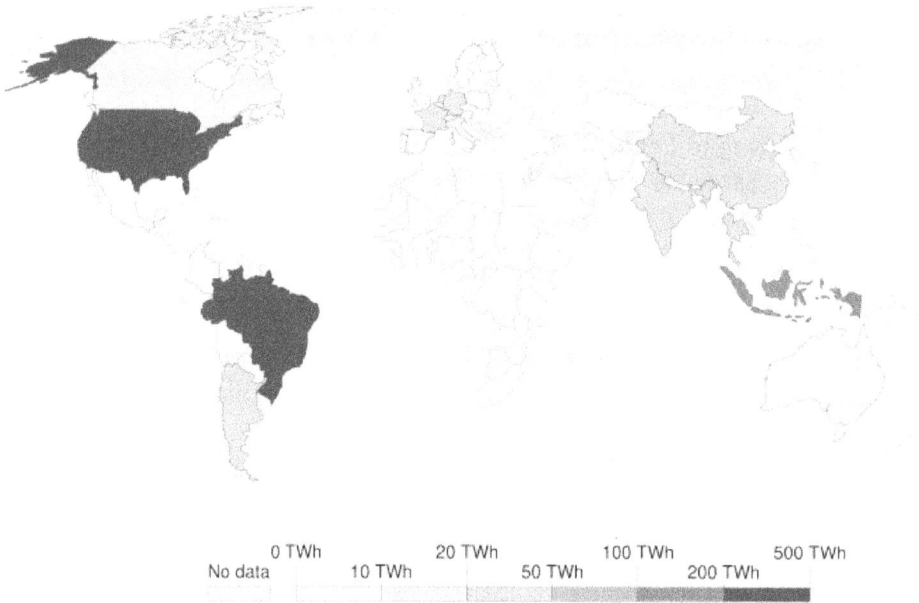

| | 0 TWh | 20 TWh | 100 TWh | 500 TWh |
| No data | 10 TWh | 50 TWh | 200 TWh | |

Source: Energy Institute Statistical Review of World Energy (2023)   OurWorldInData.org/renewable-energy • CC BY

Figure 13: Total biofuel production is measured in terawatt-hours (TWh) per year. Biofuel production includes both bioethanol and biodiesel. Our World In Data, CC BY 4.0, via Wikimedia Commons.

Algae-based fuels represent a promising frontier in biofuel technology, leveraging microalgae's capacity to thrive on non-arable land and their superior oil yields compared to conventional crops [223]. The versatility of algae allows for the production of various fuel types, including biodiesel and ethanol, while also sidestepping food versus fuel debates due to their minimal land footprint and potential for use with saline water or wastewater [224]. However, the commercial viability of algae-based fuels remains hampered by high production costs, the complexity of harvesting processes, and the energy requirements for processing [224]. Addressing these technological and economic challenges is crucial for realizing the potential of algae fuels in sectors that are difficult to electrify, such as aviation and maritime transport [225].

Biofuels hold significant potential in facilitating a transition away from fossil fuels, particularly in transportation and heating contexts where electrification is not feasible. They are especially relevant for applications in road freight, aviation, and rural heating, contributing to a diversified low-carbon future [224]. Yet, the full replacement of fossil fuels by biofuels remains unlikely in the near term due to pressing challenges such as land use, resource constraints, and competing clean energy technologies. Consequently, biofuels should be perceived as a component of a broader energy strategy, integrating with renewables and efficiency improvements to curtail global reliance on fossil fuels [226].

The costs associated with biofuels—such as ethanol, biodiesel, and algae-based fuels—can be understood through several key dimensions, including production expenses, infrastructure and distribution needs, environmental and social impacts, and market competitiveness. Comparing these with fossil fuels involves not only evaluating direct economic costs but also considering externalities such as pollution, health effects, and long-term environmental damage.

Biofuel production costs vary considerably depending on the type and source of the fuel. Ethanol, especially from corn or sugarcane, generally has lower production costs than other biofuels. In countries like the United States, ethanol remains relatively affordable due to established supply chains and government subsidies, though it still tends to be more expensive per unit of energy than gasoline. Biodiesel is often made more economically from waste oils or animal fats, while biodiesel derived from virgin vegetable oils carries higher production costs. Despite being a cleaner fuel, biodiesel still costs more per gallon than traditional diesel, primarily due to feedstock and processing demands. Algae-based fuels, while promising due to their high yields and minimal land use, are currently much more expensive—up to ten times the cost of conventional diesel—because of the technological complexity and energy-intensive processes involved in cultivating, harvesting, and extracting oils from algae. Fossil fuels, by contrast, benefit from decades of infrastructure development, large-scale production, and economies of scale. Crude oil, for instance, had an average production cost of around $30–$50 per barrel in 2023, while natural gas and coal remain cheap to extract in resource-rich regions. However, these figures often exclude environmental degradation, carbon emissions, and long-term ecological costs.

Infrastructure and distribution expenses also differ between biofuels and fossil fuels. While biofuels can sometimes use existing logistics systems, challenges remain. Ethanol, due to its corrosive nature and tendency to absorb water, cannot be transported through conventional pipelines and must instead rely on more expensive truck or rail transport. Biodiesel blends up to B20 can be delivered using standard diesel infrastructure, but higher-concentration blends like B100 require separate storage and handling systems. Algae fuels, still in the developmental stage, would need entirely new infrastructure to support widespread distribution. Fossil fuels hold a significant advantage here, with an extensive and efficient global network of pipelines, refineries, ports, and storage facilities already in place.

The environmental and social costs—often referred to as externalities—present another layer of complexity. Biofuels, especially those derived from waste or non-edible crops, offer reduced greenhouse gas emissions compared to fossil fuels. However, these benefits can be undermined by negative land use changes, such as deforestation for palm oil plantations, or by intensive farming practices that lead to water scarcity, soil erosion, and fertilizer runoff. The food versus fuel debate also raises concerns, as the use of arable land for fuel crops can affect food prices and availability. Fossil fuels, meanwhile, remain the largest contributors to global carbon emissions, and their extraction and combustion lead to significant air pollution, ecosystem disruption, and health problems. Deepwater drilling, tar sands extraction, and fracking carry substantial ecological and social risks. These impacts are rarely reflected in the market price of fossil fuels, making them appear less costly than they truly are when all factors are considered.

In terms of market competitiveness, biofuels often rely on policy support to remain viable. In the United States, mandates such as the Renewable Fuel Standard require blending of biofuels into conventional fuels, which helps sustain demand. The European Union promotes biodiesel through blending quotas and tax incentives. Algae-based fuels currently benefit mostly from research grants and pilot projects, as they are not yet economically feasible at scale. Fossil fuels also receive widespread subsidies and fiscal incentives, which further complicate fair price comparisons. According to the International Monetary Fund, global fossil fuel subsidies, including the social and environmental costs that are not priced into the market, exceeded $7 trillion in 2022. These hidden subsidies give fossil fuels a major price advantage over renewables and biofuels.

When comparing the two, biofuels generally have higher direct production and distribution costs, lower energy content per unit, and more complex logistical needs—especially for ethanol and algae fuels. However, they offer substantial environmental benefits when sustainably sourced, and they reduce reliance on geopolitically volatile oil supplies. Fossil fuels maintain lower upfront costs due to long-standing infrastructure and government support but impose far higher costs on public health, climate stability, and ecosystem health over time.

While biofuels are not yet a full substitute for fossil fuels on a global scale, they offer valuable economic and environmental benefits in specific sectors and regions. As technological innovation reduces costs and as governments increasingly recognize the true costs of fossil fuels, biofuels could become a more central part of a diversified and sustainable energy system. Their role will be most effective when integrated with electrification, hydrogen, synthetic fuels, and aggressive efficiency measures in the broader strategy to transition away from carbon-intensive energy sources.

Hydrogen: Production and Uses

Hydrogen is the most abundant element in the universe and is increasingly viewed as a key component of the clean energy transition. However, pure hydrogen gas (H_2) does not naturally exist in large quantities on Earth and must be produced from other substances, primarily water and hydrocarbons. Once produced, hydrogen can be used in a wide range of applications, from energy storage and transport to industrial processes and power generation.

Hydrogen is increasingly being recognized as a crucial component of future energy systems, especially if it can be produced without releasing carbon dioxide. Its versatility allows it to serve both as an energy carrier and a feedstock in industrial processes, which makes it particularly valuable in a decarbonized economy. Hydrogen is already being used in various sectors, including transport, refining, and chemical manufacturing, but its role is expected to expand significantly in the coming decades [227].

In transportation, hydrogen is beginning to play a more prominent role, despite the technical challenges associated with its storage and use. Because hydrogen gas requires high-pressure containment or liquefaction, storing it onboard vehicles remains a complex issue. Nevertheless, it is being adopted in fuel cell electric vehicles (FCEVs), particularly for buses, heavy-duty trucks, and trains where battery-electric systems may not be suitable due to weight and range limitations. The refining industry is also making increasing use of hydrogen, especially when upgrading heavy crude oils and tar sands into liquid fuels. These processes require significant volumes of hydrogen to remove impurities and improve fuel quality [227].

Hydrogen's role is not limited to direct fuel applications. It can be chemically combined with carbon dioxide to produce methanol or dimethyl ether (DME), both of which are viable liquid transport fuels. These synthetic fuels offer a way to store and transport energy more conveniently than pure hydrogen, especially across long distances. Moreover, hydrogen has promising applications in heavy industry, such as serving as a cleaner substitute for coke in steelmaking and other metallurgical processes. This is particularly significant, as these industries are among the most challenging to decarbonize using electricity alone [227].

Currently, most hydrogen is produced from fossil fuels through methods like steam methane reforming (SMR) or coal gasification, processes that emit large quantities of carbon dioxide. This "grey hydrogen" is abundant and relatively cheap but undermines climate goals. A cleaner form, "blue hydrogen," uses the same processes but incorporates carbon capture and storage to mitigate emissions. However, the long-term goal is to produce "green hydrogen" via electrolysis powered by renewable electricity, or through high-temperature thermochemical processes using heat from nuclear reactors. Electrolysis splits water into hydrogen and oxygen without carbon emissions, provided the electricity is clean. Using nuclear energy during off-peak hours or employing high-temperature gas-cooled reactors in the future could make hydrogen production more continuous and efficient [227].

Hydrogen is already a significant industrial product, with global demand reaching around 74 million tonnes in 2018. About half of this was used for ammonia production in fertilizers through the Haber-Bosch process, and another large share went into oil refining. Additional demand came from mixed gases used in methanol production and direct reduced iron (DRI) in steelmaking. This broad industrial base provides a foundation for scaling up hydrogen usage, although transitioning to low-carbon hydrogen requires substantial technological and infrastructural shifts [227].

Like electricity, hydrogen is an energy carrier rather than a primary energy source. Its potential as a substitute for oil in transport is particularly noteworthy, although it lacks the energy density and ease of handling that liquid hydrocarbons provide. Nevertheless, hydrogen compares favourably to batteries in certain contexts, particularly for heavy-duty applications and long-range transportation. Because electricity and hydrogen are interconvertible, hydrogen can be used to store intermittent renewable power through a "power-to-gas" strategy. This involves producing hydrogen by electrolysis during periods of excess wind or solar generation and injecting it into natural gas pipelines or converting it to other fuels like ammonia [227].

Ammonia is gaining attention as a hydrogen carrier because it is more energy-dense by volume and easier to transport and store. While hydrogen's energy density by mass is very high—between 120 and 142 megajoules per kilogram—it has a low volumetric energy density, making storage a challenge. Liquid ammonia, in contrast, offers a more compact and manageable form for global trade and long-term energy storage. This is particularly relevant for decarbonizing shipping, aviation, and other sectors requiring dense energy carriers [227]..

The shift toward a hydrogen economy is already underway, supported by global chemical and refining industries. However, the scale of transformation required is immense. Producing enough hydrogen to meet future energy needs could demand more energy than the world currently generates for electricity. Transport is expected to drive the largest share of hydrogen demand in regions like the EU and South Korea by 2050, as these economies transition from diesel to hydrogen-powered vehicles. Other major applications include residential heating, replacing natural gas, and rail transport [227].

If hydrogen is to serve the shipping and aviation sectors effectively, a global infrastructure for refuelling will be essential. This will require international coordination and investment far beyond regional plans currently in development. A report by LucidCatalyst in 2020 emphasized that hydrogen and synthetic fuels are the only viable solutions for hard-to-decarbonize sectors, provided they can be produced affordably and at scale from non-fossil sources [227].

An abundant supply of low-cost, clean hydrogen would not only support transportation and industry but also enhance global food security by improving access to nitrogen fertilizers. Realizing this potential depends on major investments in technology, infrastructure, and policy—particularly efforts to bring hydrogen production costs below $1 per kilogram. As these goals are met, hydrogen is likely to become a cornerstone of the global transition to a low-carbon energy future [227].

Demand for Hydrogen

National and regional plans for hydrogen production and use have expanded significantly in recent years, reflecting global momentum toward decarbonizing energy systems. In 2020, global hydrogen use reached approximately 115 million tonnes, with the bulk used in oil refining (39 Mt) and ammonia production for fertilizers (32 Mt), followed by methanol production (14 Mt) and syngas applications such as chemical manufacturing and iron reduction [227]. Looking ahead, the International Energy Agency (IEA) projects that hydrogen demand could double to over 200 million tonnes by 2030 and rise further to 530 million tonnes by 2050. A growing share of this hydrogen—estimated at 70% by 2030 and 90% by 2050—is expected to be "low-carbon," derived from electrolysis using renewable or nuclear energy, or from fossil sources with carbon capture and storage (CCS) [227].

Hydrogen's expanding role is especially important in hard-to-electrify sectors such as steel production, shipping, and aviation. Reports such as the Energy Transitions Commission's "Making the Hydrogen Economy Possible" forecast that clean hydrogen could supply up to 800 million tonnes annually by 2050, accounting for 17% of total final energy demand [227]. This would require up to 30,000 TWh of electricity annually just for hydrogen production, in addition to 90,000 TWh for direct electrification across global energy systems [227]. Clean hydrogen, particularly "green hydrogen" from electrolysis powered by renewable electricity, is expected to dominate future supply, with estimated production costs falling to around $2/kg by 2030 if supported by 50 GW of global electrolyser capacity [227].

Hydrogen's carbon footprint depends heavily on its production method. "Grey hydrogen" from methane emits CO_2, while "brown hydrogen" from coal gasification emits even more. "Blue hydrogen" involves capturing those emissions with CCS, offering a cleaner alternative. "Green hydrogen" is produced with renewable electricity via electrolysis, and "yellow" or "pink hydrogen" uses nuclear electricity. "Turquoise hydrogen" is produced through pyrolysis of methane, yielding solid carbon instead of CO_2. Only green, yellow, and future thermochemical nuclear-based hydrogen are reliably zero-carbon. The French Parliamentary Office for Scientific and Technological Assessment emphasized that only nuclear and hydropower can produce low-carbon hydrogen cost-effectively due to their reliability and lower cost compared to intermittent renewables [227].

Europe is at the forefront of hydrogen development. The European Commission's 2020 Hydrogen Strategy calls for installing at least 6 GW of renewable-powered electrolysers by 2024 and 40 GW by 2030 to produce 10 million tonnes of green hydrogen [227]. These targets are part of a broader effort to decarbonize industry and transport, especially aviation and steelmaking. The EU plans to invest between €180–470 billion in renewable hydrogen and €3–18 billion in low-carbon hydrogen from fossil sources with CCS by 2050. Additional infrastructure plans include the European Hydrogen Backbone—a proposed 22,900 km hydrogen pipeline network to integrate regional hydrogen markets [227]. Germany's national hydrogen strategy, backed by €9 billion, includes 5 GW of electrolyser capacity by 2030 and aims to import hydrogen to meet future demand. France, Spain, and the Netherlands have similar strategies involving significant electrolyser deployments and integration with transport and industry [227].

In Eastern and Central Europe, several nations—led by France—have urged the EU to recognize hydrogen from nuclear sources as "green." This recognition would expand eligibility for funding and strategic support. The European Commission is considering draft rules to classify nuclear-based hydrogen under renewable targets, which could support broader hydrogen adoption and energy security [227].

The UK's Hydrogen Strategy targets 5 GW of low-carbon hydrogen production by 2030, primarily through blue hydrogen, to supply 20% of its 2050 target. It includes support for nuclear-powered electrolysis and high-temperature steam electrolysis, with ambitions to produce up to 225 TWh of hydrogen annually by mid-century. Other European countries like Slovakia, Italy, and the

Netherlands are also advancing their hydrogen agendas, with a mix of renewable and nuclear strategies for hydrogen production and gas grid blending initiatives [227].

Outside Europe, China is the world's largest hydrogen producer at around 22 million tonnes annually, primarily from coal. Although it has not set aggressive green hydrogen targets, China is scaling up its hydrogen infrastructure, including fuel cell vehicles and refuelling stations. By 2050, it envisions hydrogen comprising 10% of its energy supply. Japan has taken a leadership role in hydrogen strategy, emphasizing ammonia as an energy carrier and aiming for widespread fuel cell use in buildings and transport. Its Ministry of Economy, Trade, and Industry (METI) is spearheading efforts to import zero-carbon hydrogen and ammonia, and aims for hydrogen and ammonia to provide 10% of electricity by 2050 [227].

South Korea's hydrogen roadmap envisions hydrogen comprising over 20% of the national energy mix by 2050, with a focus on fuel cell electric vehicles (FCEVs), especially for buses and trucks. The country emphasizes blue hydrogen but is also exploring clean hydrogen imports and infrastructure development [227].

Russia is pursuing hydrogen through both blue and turquoise production. Rosatom and Gazprom are working on large-scale projects involving methane pyrolysis and nuclear-powered electrolysis, including plans to produce hydrogen on Sakhalin Island for export to Asia. Rosatom began small-scale hydrogen production at the Kola nuclear power plant in 2022 and is planning further expansion [227].

The United States has taken a multi-sectoral approach with the Department of Energy's H2@Scale initiative, aiming to integrate hydrogen across electricity, transport, industry, and storage. The U.S. produces 11 Mt of hydrogen annually, mostly from natural gas. Hydrogen is seen as essential for decarbonizing transport, refining, and ammonia production. The DOE's Hydrogen Program Plan and various demonstration projects support hydrogen production from nuclear, renewables, and fossil fuels with CCS. A key focus is enabling flexible operation of nuclear plants to support hydrogen production using high-temperature electrolysis [227].

Australia's National Hydrogen Roadmap includes gasification of lignite with CCS in Victoria's Latrobe Valley and plans for hydrogen export to Japan. While current production is minimal, commercial-scale projects aim to reduce costs to about US$2/kg by the 2030s. Electrolysis using renewable electricity may be used as an interim approach, with Japan monitoring progress closely [227].

Hydrogen strategies around the world reflect a shared understanding of its importance in achieving net-zero targets. While the sources, technologies, and scale differ, the common goal is to expand hydrogen use in sectors that are difficult to electrify and to transition to cleaner forms of hydrogen production. The challenge lies in reducing costs, building infrastructure, and coordinating policy to support a global hydrogen economy [227].

Hydrogen Uses

Hydrogen has become a critical element in discussions about transitioning to a low-carbon future, and its applications span a wide range of industrial, energy, transport, and chemical processes [227].

One major use is as a supplement in natural gas grids. Hydrogen can be blended into existing natural gas pipelines at concentrations up to about 15–20% without significant modification. However, higher concentrations require changes to infrastructure and end-use appliances due to hydrogen's lower energy content and its tendency to cause embrittlement in steel. New trunk pipelines, like Nord Stream in Europe, may allow up to 70% hydrogen transport. Projects across Germany, the Netherlands, France, the UK, and the USA are piloting hydrogen blending and developing large-scale power-to-gas systems that convert renewable electricity into hydrogen for injection into gas grids.

In the petroleum refining industry, hydrogen is essential for desulfurization and catalytic cracking of heavy crude oils, especially from tar sands. It converts complex hydrocarbons into lighter, energy-dense fuels like gasoline and diesel. This process requires high pressures and temperatures and consumes a significant share of global hydrogen production, particularly to upgrade lower-grade crude oil into usable transport fuels [227].

Hydrogen is also a key component in ammonia production via the Haber-Bosch process, which combines hydrogen with nitrogen from the air. Ammonia synthesis consumes more than half of the world's pure hydrogen production, largely for use in nitrogen-based fertilizers essential for global food production. The Haber-Bosch process is energy-intensive and traditionally relies on hydrogen from natural gas. However, future production may increasingly use low-carbon hydrogen to reduce environmental impact. Ammonia is also being explored as a fuel and hydrogen carrier for storage and transport [227].

Another application of hydrogen is in producing methanol and its derivative dimethyl ether (DME), which are potential alternative fuels. Methanol has lower energy density than gasoline or diesel but can be used in fuel cells, internal combustion engines, or as a feedstock for synthetic hydrocarbons. DME, made by dehydrating methanol, is already used as a propane substitute. Methanol and DME can be synthesized using CO_2 and hydrogen, ideally from carbon-neutral sources such as nuclear or renewable electricity. These fuels are easier to store than hydrogen gas and can serve as transitional fuels in transport and heating sectors [227].

Hydrogen's most transformative role may be in fuel cell electric vehicles (FCEVs). These vehicles use hydrogen fuel cells to generate electricity, producing only water as a by-product. While early hydrogen-powered cars used internal combustion engines, modern FCEVs use proton exchange membrane (PEM) fuel cells due to their high efficiency and power density. Heavy-duty transport sectors—like buses, trucks, and trains—are prime candidates for FCEVs because they require fast refuelling and long range. Manufacturers like Toyota, Honda, Hyundai, and emerging companies such as Nikola and Ballard are leading efforts in this space. However, the lack of refuelling infrastructure and the high cost of fuel cells and hydrogen storage remain significant challenges [227].

Figure 14: Hyundai ix35 Fuel Cell electric vehicle at Hydrogen Gas Station at Shell, Schmiestrasse, Wuppertal. Dr. Artur Braun, CC BY-SA 4.0, via Wikimedia Commons.

Fuel cells are also being explored for stationary energy applications such as backup power and off-grid energy systems. Solid oxide and molten carbonate fuel cells operate at high temperatures and are used in industrial and residential settings for combined heat and power (CHP) systems. Japan has deployed thousands of residential fuel cell CHP units with government support. Hydrogen fuel cells provide a clean and efficient power source in remote areas or where electricity grid reliability is an issue [227].

Figure 15: Hyundai NEXO Hydrogen Fuel Cell Car beim Autosalon Genf 2018. Dr. Artur Braun (Arturbraun), CC BY-SA 4.0, via Wikimedia Commons.

In the maritime sector, hydrogen and ammonia are being tested as fuels for ships. Ammonia has a higher volumetric energy density and easier storage properties compared to hydrogen, making it suitable for long-distance shipping. Countries like Japan and companies like ABB and Siemens are exploring ammonia and hydrogen-powered vessels, while the International Maritime Organization is pushing for a significant reduction in shipping-related emissions by 2050 [227].

Hydrogen is increasingly being used as a reductant in steelmaking. Traditional steel production relies heavily on coke from coal, emitting large amounts of CO_2. Using hydrogen to directly reduce iron ore (producing sponge iron) can significantly cut emissions. Projects like HYBRIT in Sweden and Infrabuild in Australia are pioneering hydrogen-based steelmaking, aiming to produce "green steel" by replacing coke with hydrogen. This shift could transform one of the world's most carbon-intensive industries [227].

Finally, hydrogen plays a role in various chemical manufacturing processes beyond ammonia and methanol. It is used as a reducing agent, in hydrocracking, and in the synthesis of other chemical compounds. Future developments may see hydrogen derived from nuclear or renewable sources supporting a broader transition to sustainable chemistry [227].

Overall, hydrogen's versatility makes it a key pillar in decarbonization strategies. It serves not only as a clean fuel but also as a vital industrial feedstock and energy storage medium. While infrastructure, production costs, and technology readiness pose challenges, the global momentum toward a hydrogen economy suggests it will play a growing role across sectors in the coming decades [227].

Hydrogen Production Methods

Hydrogen production is central to the development of a low-carbon energy economy, and its classification by colour codes reflects the methods used and their environmental impacts. Currently, the most common form of hydrogen is *grey hydrogen*, produced by steam methane reforming (SMR) of natural gas. This process involves reacting methane with steam at high temperatures, producing hydrogen and carbon dioxide. Grey hydrogen is responsible for significant CO_2 emissions and constitutes about 95% of global hydrogen production.

In contrast, *blue hydrogen* also uses SMR or autothermal reforming, but it includes carbon capture and storage (CCS) to mitigate emissions. Although it results in fewer emissions than grey hydrogen, its effectiveness depends on the capture rate and storage reliability. Blue hydrogen is often seen as a transitional technology toward cleaner alternatives.

Green hydrogen represents the ideal future state in terms of sustainability. It is produced by electrolysing water using electricity from renewable sources such as solar, wind, or hydropower. This method emits no greenhouse gases, but it remains costly due to the capital investment in electrolysers and the price of renewable electricity. Nevertheless, advancements in electrolyser technologies and falling renewable energy prices are expected to make green hydrogen more competitive over time.

Figure 16: Storage tank for liquid hydrogen fuel located just to the Northeast of Kennedy Space Center's former shuttle launch pad 39-A. The image shows the tank and fuel lines on this permanent facility. TomFawls, CC BY-SA 3.0, via Wikimedia Commons.

Other hydrogen types include *turquoise hydrogen*, produced by methane pyrolysis, which yields hydrogen and solid carbon without CO_2 emissions. *Pink hydrogen* is generated via electrolysis powered by nuclear energy, offering a zero-carbon solution with reliable baseload power. *Yellow hydrogen* refers to hydrogen made via electrolysis using grid electricity, which may or may not be renewable, thus its carbon footprint varies.

Despite the environmental benefits of green hydrogen, nearly all hydrogen today is produced from fossil fuels. Global production is around 75 million tonnes of pure hydrogen annually, with another 45 million tonnes used in synthesis gas for industries like steel and methanol production. This contributes to approximately 830 million tonnes of CO_2 emissions annually, making it a significant source of greenhouse gases.

Steam methane reforming remains the dominant method. It involves reacting methane with steam at temperatures between 700°C and 1000°C to produce hydrogen and carbon monoxide. A subsequent water-gas shift reaction converts carbon monoxide into additional hydrogen and CO_2. Coal gasification is another common method, particularly in countries with abundant coal reserves. Both these processes generate grey hydrogen unless equipped with CCS to form blue hydrogen.

In recent years, pyrolysis, particularly methane pyrolysis for turquoise hydrogen, has gained attention. This method thermally decomposes methane into hydrogen and solid carbon using molten metal reactors or plasma systems. Since no CO_2 is generated, it offers an appealing pathway, although it is still under development and not yet commercially widespread.

Electrolysis, although still a small contributor to global hydrogen supply (around 2%), is projected to play a major role in future hydrogen production. It involves splitting water into hydrogen and oxygen using electricity. The energy requirement is significant, around 50–55 kWh per kilogram of hydrogen. There are three main types of electrolysers: alkaline, polymer electrolyte membrane (PEM), and solid oxide electrolysis cells (SOECs). Alkaline electrolysers are the most mature, PEM electrolysers are more compact and efficient but expensive due to the use of precious metal catalysts, and SOECs operate at high temperatures, allowing for better efficiency but lower commercial readiness.

SOECs are particularly promising when integrated with high-temperature heat sources like nuclear reactors. They can also co-electrolyse steam and CO_2 to produce synthesis gas for fuels. A 2.6 MW SOEC system is under development in Rotterdam, and Denmark is launching a plant with 500 MW/year of SOEC production capacity. This technology may significantly enhance the viability of hydrogen as a clean energy carrier, especially in combination with nuclear energy.

Despite its potential, hydrogen production via electrolysis is still expensive. Cost estimates vary, with grey hydrogen typically around €1.50/kg, blue hydrogen at €2.50/kg, and green hydrogen ranging between €3 to €6/kg, depending on technology and location. However, costs for green hydrogen are falling. Some projections suggest that prices may drop below $1/kg by 2030 if capital costs decline and renewable electricity becomes cheaper.

Globally, hydrogen production from electrolysis is gaining momentum. By mid-2021, global electrolyser capacity reached 300 MW, with plans to increase it to over 54 GW by 2030. This would enable production of up to 8 million tonnes of green hydrogen annually. Europe leads in planned capacity, with major projects like BP's 250 MW electrolysis plant in Rotterdam and Ørsted's pilot plant at Heide. Large-scale initiatives in Saudi Arabia, Oman, and Australia aim to harness renewable energy for hydrogen and ammonia exports, targeting millions of tonnes per year.

The integration of surplus renewable electricity into hydrogen production is also a growing trend. Projects in Germany, the Netherlands, the UK, and France are developing PEM and alkaline electrolyser plants powered by wind and solar. These initiatives aim to balance electricity grids, reduce curtailment, and enable seasonal energy storage.

In parallel, nuclear power is being considered for off-peak electrolysis. Projects like EDF Energy's trial at Heysham in the UK and installations at US nuclear plants are exploring high-temperature steam electrolysis to improve efficiency. If successful, such integration could provide a low-carbon and stable hydrogen supply.

Electrolyser costs are expected to decrease significantly by 2030. Alkaline units, currently around €600/kW, are projected to drop to €400/kW, while PEM units could fall from €900 to €500/kW.

Solid oxide units may see even steeper declines. These cost reductions, coupled with increasing renewable energy deployment, are crucial for scaling up green hydrogen production.

The method of hydrogen production has significant implications for its environmental impact and cost. While grey hydrogen remains dominant, blue hydrogen serves as a transitional solution, and green hydrogen represents the long-term goal. Innovations in electrolysis, integration with renewable and nuclear energy, and large-scale infrastructure projects are rapidly shaping the future hydrogen economy. As costs decline and production scales up, hydrogen is expected to play a critical role in decarbonizing industry, transport, and energy systems globally.

Storing and Transporting Hydrogen

Storing and transporting hydrogen is a complex but critical part of building a viable hydrogen economy. Because hydrogen gas has a very low volumetric energy density—one kilogram occupies about 11.1 cubic metres at standard temperature and pressure—it is difficult to store and transport in its gaseous state. Additionally, hydrogen molecules are extremely small, making them prone to leakage, and the gas is highly flammable. To address these challenges, hydrogen is often compressed or liquefied, or alternatively, converted into compounds such as ammonia or liquid organic hydrogen carriers[227].

Compressed hydrogen is the most common form used for portable storage, especially in vehicles. Compression to 35 MPa (megapascals) increases the energy density to 23 kilograms per cubic metre, while requiring about 4.4 kWh of energy per kilogram of hydrogen. This level of compression is typically sufficient for transport and use in hydrogen-powered cars and buses. Large-scale stationary storage, however, requires hydrogen to be stored cryogenically as a liquid, which is much more energy-intensive. To liquefy hydrogen, it must be cooled to –253°C, consuming approximately 13 kWh per kilogram—roughly 30% of its energy content—and adding significant cost, often more than $2 per kilogram [227].

For long-distance distribution, existing gas pipeline networks offer an opportunity. In Europe, Hydrogen Europe estimates that 50,000 kilometres of natural gas pipelines could be repurposed for hydrogen at a cost of €25 billion [227]. This concept underpins the European Hydrogen Backbone (EHB) initiative, aiming to connect 40,000 kilometres across 21 countries by 2040. Germany and the Netherlands have already outlined plans to retrofit thousands of kilometres of pipeline, combining this infrastructure with salt cavern storage for balancing supply and demand [227]. In the UK, Project Union proposes converting existing natural gas infrastructure to meet around 25% of Britain's gas demand with hydrogen [227].

Because of the limitations in directly storing and transporting hydrogen, an alternative is to convert hydrogen into *ammonia* (NH_3), which acts as a highly effective carrier. Ammonia can be liquefied at –33°C and stored or transported under pressure (800–1000 kPa), similar to liquefied petroleum gas (LPG) [227]. It has a relatively high energy density (about 5.18 kWh per kilogram), and handling and transport infrastructure is well established due to its widespread use in agriculture. Converting hydrogen into ammonia via a refined Haber-Bosch process requires 14 kWh per kilogram of ammonia, while reversing the process to extract hydrogen—typically via thermal catalytic decomposition—requires around 8 kWh per kilogram of hydrogen. Although the conversion adds approximately $1 per kilogram to hydrogen's cost, it allows for more manageable large-scale storage and global shipping [227].

Ammonia can be used directly in certain fuel cells or internal combustion engines, especially in the marine sector, or it can be "cracked" back into hydrogen for use in applications like PEM fuel cells that require high-purity hydrogen. According to the International Renewable Energy Agency (IRENA), ammonia is expected to account for 74% of clean hydrogen demand in shipping by 2050, reflecting its importance as a hydrogen carrier. This would translate to 183 million tonnes of ammonia used annually in the shipping sector alone [227].

Another method of hydrogen storage involves *metal hydrides*, where hydrogen is chemically bonded to metals or alloys. One promising approach uses sodium borohydride ($NaBH_4$), which, when catalysed, releases hydrogen and forms sodium borate ($NaBO_2$), which must then be reprocessed. Metal hydrides offer high energy density and safety advantages, as hydrogen is stored in solid form, but they are still under development and not yet commercially widespread [227].

A further option is to use *liquid organic hydrogen carriers (LOHCs)*. These are chemical compounds—such as toluene—that react with hydrogen to form methylcyclohexane (MCH), which can be transported and stored at ambient temperatures and pressures. Upon reaching the point of use, MCH is catalytically dehydrogenated to release hydrogen and regenerate toluene for reuse. Japanese companies, including Chiyoda and Mitsubishi, are exploring LOHCs for long-distance hydrogen transport, particularly by sea. LOHC systems offer a balance between the safety of chemical storage and the convenience of liquid fuels, although the energy required for hydrogenation and dehydrogenation must be factored into cost and efficiency considerations [227].

While hydrogen has a range of storage and transport options, each method involves trade-offs between energy density, infrastructure compatibility, safety, and cost. Compressed and liquefied hydrogen are most common for near-term use, particularly in vehicles and stationary storage. Ammonia and LOHCs provide promising solutions for large-scale, long-distance distribution. Ultimately, the choice of hydrogen storage and transport method will depend on the specific application, the scale of deployment, and the availability of supporting infrastructure[227].

Hydrogen Feasibility

The feasibility of hydrogen as a fuel is an intricate subject that intertwines both environmental sustainability and production methods. Hydrogen is lauded for its clean-burning properties, primarily emitting only water vapor when utilized as a fuel. However, its environmental viability is critically dependent on its production method. Currently, the global production of hydrogen stands at approximately 94 million tonnes annually, with the predominant method being steam methane reforming (SMR) of natural gas, which produces what is termed grey hydrogen. This conventional method is associated with significant carbon dioxide emissions, thus compromising the overall benefits of hydrogen as a green fuel [228-230].

To address the carbon emissions linked with grey hydrogen, blue hydrogen has emerged as a transitional solution. Blue hydrogen utilizes the same production process as grey hydrogen but integrates carbon capture and storage (CCS) technologies to mitigate emissions. Despite these advancements, the climate benefits of blue hydrogen are becoming increasingly scrutinized. Recent life cycle assessments indicate that blue hydrogen may have a greenhouse gas footprint that exceeds that of burning natural gas, mainly due to the emissions associated with upstream methane leakage. This is particularly pronounced in regions like the United States [229, 231]. Furthermore, research suggests that the efficiency of CCS can vary widely based on the technology employed, thus adding another layer of complexity to its overall environmental advantage [229, 230].

Projects such as H21 in Europe suggest that advancements in capture technology can yield significant reductions in emissions compared to natural gas. However, concerns remain about the sustainability of relying on fossil fuels [229, 230]. Ultimately, while blue hydrogen offers some emission reductions, it is viewed as a temporary measure that still relies heavily on fossil fuel infrastructure and incomplete carbon capture technologies, making it insufficient for a long-term solution [228, 229].

In contrast, the prospects of green hydrogen, produced through the electrolysis of water using renewable energy sources, present a more sustainable pathway. The process generates hydrogen and oxygen without emitting carbon dioxide. However, despite its environmental superiority, the economic viability of green hydrogen remains a considerable obstacle. As of 2020, production costs for green hydrogen were significantly higher, estimated between \$2.50 to \$6.80 per kilogram, compared to between \$1 and \$1.80 for grey and blue hydrogen [228-230]. This disparity is influenced by the technological maturity of production methods and the capital investment required for renewable energy infrastructures.

Pink hydrogen—which utilizes nuclear energy for electrolysis—is also gaining traction and is beginning to see commercial viability, exemplified by recent developments in Sweden. However, the adoption of hydrogen in the energy landscape must go beyond reducing direct emissions. It involves overcoming significant barriers such as high capital costs, variability in renewable energy availability, and the development of necessary infrastructure [232, 233].

Thus, while hydrogen represents a technically feasible and environmentally clean energy carrier, realizing its potential on a large scale hinges upon transitioning from fossil fuel-based production methods. Grey hydrogen cannot be sustained environmentally, blue hydrogen presents limited benefits, and while green and pink hydrogen exemplify the future of decarbonization, widespread deployment will necessitate significant cost reductions and infrastructure development, alongside robust policy support.

Synthetic Fuels and E-Fuels

Synthetic fuels (synfuels) and electrofuels (e-fuels) are important alternatives to conventional fossil fuels in the transition towards sustainable energy. These fuels are primarily produced from renewable resources, significantly reducing reliance on petroleum. Synthetic fuels can be produced through processes such as Fischer–Tropsch synthesis, which converts carbon monoxide (CO), carbon dioxide (CO_2), and hydrogen (H_2) into hydrocarbon fuels via chemical reactions. This method allows for the creation of fuels that can compete with traditional energy sources such as gasoline, diesel, and jet fuel [234, 235].

Electrofuels, a specific category of synthetic fuels, are distinguished by their production process that relies on electricity, ideally derived from renewable sources. The production of e-fuels typically involves generating hydrogen through the electrolysis of water, which is then combined with captured CO_2 to synthesize hydrocarbons. This process is commonly referred to as power-to-liquid (PtL) or power-to-gas (PtG) [235, 236].

The primary objective of these technologies is to produce carbon-neutral or low-carbon energy sources that can help mitigate greenhouse gas emissions, particularly in sectors that are difficult to fully electrify, such as aviation, shipping, and heavy transport. The integration of e-fuels into transportation is viewed as crucial due to their ability to match the energy density of fossil fuels while minimizing ecological impacts [237-239].

However, several challenges must be addressed to make synfuels and e-fuels viable on a larger scale. These challenges include the high production costs compared to conventional fuels, the energy demands of hydrogen production through electrolysis, and the inefficiencies associated with carbon capture. Furthermore, the scalability of production and the required infrastructure for widespread adoption are significant hurdles [240, 241]. The economic feasibility of these fuels is closely tied to supportive policies and investments in renewable energy technologies and the public's acceptance of alternative fuelling solutions [242].

Recent trends suggest that successful collaboration between research institutions and industry is vital for advancing synthetic and electrofuels. Innovations to enhance catalysis and improve the efficiency of electrolysis could reduce production costs, thereby making these fuels more competitive with traditional fossil fuels [243, 244]. Additionally, ongoing advancements might allow for the use of biomass not just as a feedstock for biofuels but also in the production of synthetic fuels, which could further enhance their environmental benefits [236, 245].

Production Process

The production of synthetic fuels and e-fuels involves a multi-step process that centres around converting renewable energy into a storable and transportable chemical form. The core inputs in this process are hydrogen and carbon dioxide (CO_2), which are then synthesized into hydrocarbon-based fuels through well-established chemical processes. Here's a detailed explanation of each stage:

Hydrogen Production: The foundation of e-fuel synthesis begins with the production of hydrogen, primarily achieved through electrolysis of water. In this process, electricity is used to split water molecules (H_2O) into hydrogen (H_2) and oxygen (O_2). When this electricity is sourced from renewable energy such as wind, solar, hydropower, or nuclear power, the resulting hydrogen is classified as green hydrogen (or pink hydrogen in the case of nuclear-derived electricity). Electrolysis is a clean process in terms of emissions, provided the energy source is carbon-free. The most common types of electrolysis technologies include alkaline electrolysis, proton exchange membrane (PEM) electrolysis, and solid oxide electrolysis. Each has different operational characteristics and efficiencies, but all aim to provide a scalable and clean source of hydrogen for fuel synthesis.

Carbon Capture: Once hydrogen is produced, the next essential component is carbon dioxide, which acts as a carbon source for synthesizing hydrocarbon fuels. CO_2 can be obtained through direct air capture (DAC), which extracts it from the atmosphere using chemical sorbents, or from industrial processes, such as emissions from cement plants or steel mills. Capturing CO_2 from these sources is critical to ensuring that the final fuel is carbon-neutral or carbon-negative, particularly when the CO_2 is originally atmospheric. The carbon captured is not simply stored, as in carbon capture and storage (CCS), but is instead utilized to create energy-dense synthetic fuels.

Fuel Synthesis: With hydrogen and carbon dioxide secured, the next step involves chemical synthesis to produce fuels. Several well-established industrial processes can be employed:

- Fischer–Tropsch synthesis is a catalytic chemical process that converts hydrogen and carbon monoxide (CO) into liquid hydrocarbons, which can be refined into synthetic diesel, jet fuel, or gasoline. This process operates at high temperatures (150–300°C) and pressures and uses metal catalysts (typically iron or cobalt). It is especially suitable for creating long-chain hydrocarbons, ideal for aviation and maritime fuel.

- Methanation involves reacting hydrogen with CO_2 to produce synthetic methane (CH_4), a substitute for natural gas. This reaction is exothermic and typically occurs at moderate temperatures (250–400°C) in the presence of a nickel catalyst. Synthetic methane can be injected into existing natural gas infrastructure or used for heating and electricity generation.

- Methanol synthesis is another pathway where hydrogen reacts with CO_2 (or CO) to produce methanol (CH_3OH), a versatile liquid fuel. Methanol can be used directly as a fuel, blended into gasoline, or further processed into dimethyl ether (DME) or gasoline-equivalent hydrocarbons. It also serves as a feedstock for various chemicals and plastics.

In all of these processes, the use of renewable electricity and captured CO_2 ensures that the synthetic fuels are produced with minimal or net-zero greenhouse gas emissions, provided the entire lifecycle is properly managed. The result is a set of fuels that can be used in existing combustion engines, aircraft, and shipping fleets—offering a transitional solution toward a lower-carbon economy, particularly in sectors that are difficult to electrify.

Indirect Conversion Synthetic Fuels Manufacturing Processes

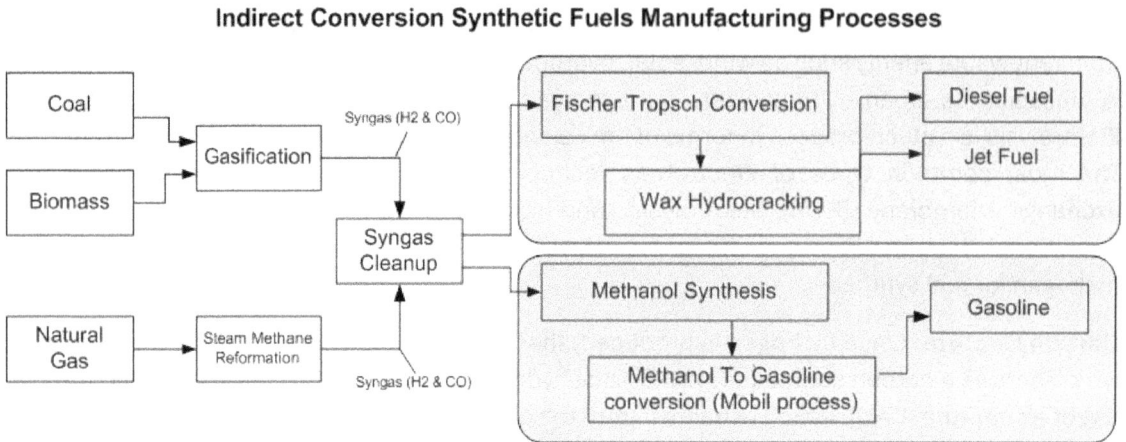

Figure 17: Simplified process flow diagram of basic indirect synthetic fuels manufacturing processes. Sfj4076, CC BY-SA 3.0, via Wikimedia Commons.

Types of Synthetic and E-Fuels: Uses, Availability, and Prominence

E-Diesel / Synthetic Diesel: E-diesel, or synthetic diesel, is a liquid fuel produced from renewable hydrogen and captured carbon dioxide via the Fischer–Tropsch synthesis process. Chemically, it is very similar to conventional diesel and can be used in existing diesel engines without modification. Its main advantage lies in being sulfur-free and offering a high cetane number, resulting in cleaner and more efficient combustion. E-diesel is particularly promising for decarbonizing heavy-duty transport sectors such as freight trucks, buses, and marine vessels, where battery electrification may not be practical due to energy density limitations. However, large-scale production is currently limited by high costs, complex infrastructure needs, and limited pilot projects. It is more prominent in research and demonstration programs in Germany and the EU, supported by climate policy goals.

E-Gasoline / Synthetic Gasoline: E-gasoline is another type of synthetic hydrocarbon fuel intended to replace conventional petrol. It is often derived from methanol through processes like methanol-to-gasoline (MTG) or directly through modified Fischer–Tropsch pathways. E-gasoline is fully compatible with modern internal combustion engines and existing fuel infrastructure. Its major appeal is the potential to extend the life of the current automotive fleet while reducing lifecycle emissions, especially in regions with insufficient electric vehicle (EV) penetration. While interest is growing, particularly in Europe and Japan, e-gasoline remains less commercially available than other e-fuels due to more complex synthesis requirements and the rise of EVs in the passenger vehicle market.

E-Kerosene / Synthetic Jet Fuel: E-kerosene, also known as synthetic aviation fuel or sustainable aviation fuel (SAF), is a vital candidate for decarbonizing the aviation industry. It is produced primarily through Fischer–Tropsch synthesis using renewable hydrogen and captured

CO_2. This fuel must meet stringent aviation standards (e.g., ASTM D7566) and can be blended with fossil-based jet fuel or used at higher concentrations in advanced aircraft. E-kerosene has gained significant prominence due to the aviation sector's limited options for decarbonization and international carbon reduction mandates like CORSIA. Several countries and airline partnerships are investing in SAF development, and multiple test flights have demonstrated its feasibility. However, availability is currently limited, and production volumes remain far below demand.

Synthetic Methane (CH_4): Synthetic methane, also known as e-methane, is created via methanation—a reaction between renewable hydrogen and captured CO_2. This fuel is chemically identical to natural gas and can be injected directly into existing gas grids or used in gas-powered vehicles, heating systems, and power generation. Synthetic methane is a useful transitional fuel in countries with extensive gas infrastructure and can help reduce reliance on fossil-based natural gas. It is more readily available than some other e-fuels, particularly in Germany, where power-to-gas projects have begun feeding e-methane into public gas systems. Its main limitation is relatively low production efficiency and energy losses in the conversion chain.

Methanol and Dimethyl Ether (DME): Methanol (CH_3OH) is a liquid synthetic fuel derived from hydrogen and carbon dioxide, and can also be produced from biomass or CO_2-rich industrial emissions. It is versatile and can be used as a fuel in internal combustion engines, blended into gasoline, or converted into more complex fuels like DME. Methanol is already widely used as a chemical feedstock and is gaining interest as a marine fuel due to its liquid nature, ease of storage, and cleaner combustion profile. Dimethyl ether (DME), derived from methanol dehydration, is a gaseous fuel with physical properties similar to liquefied petroleum gas (LPG), making it suitable for household cooking, heating, and modified diesel engines. Both fuels are being actively explored as alternative fuels in China, Sweden, and Iceland, with methanol having the broader commercial footprint due to its established industrial applications.

Ammonia (NH_3): Ammonia is a nitrogen-hydrogen compound that functions as both a hydrogen carrier and a potential fuel in its own right. It can be produced using renewable hydrogen and nitrogen extracted from the air, typically via the Haber–Bosch process. As a hydrogen carrier, ammonia offers higher energy density per volume than hydrogen gas and can be stored and transported more easily—especially in liquefied form. Ammonia is being considered as a maritime fuel and for use in high-temperature power generation systems, with research underway into using it in combustion engines and fuel cells. Japan and South Korea are leading the charge in ammonia fuel development, with plans for co-firing ammonia in thermal power stations and dedicated ammonia-fuelled vessels. However, challenges remain around combustion emissions (e.g., NOx) and safe handling due to its toxicity.

Each synthetic or e-fuel presents unique advantages and challenges. While availability remains limited due to high costs and infrastructure needs, these fuels offer significant potential to decarbonize hard-to-electrify sectors such as aviation, shipping, heavy transport, and industry. Government support, technological advances, and global climate targets are accelerating their development, with e-kerosene, e-diesel, and methanol currently among the most prominent and scalable options.

Advantages and Challenges of Synthetic and E-Fuels

One of the most compelling advantages of synthetic and e-fuels is their potential for carbon neutrality. When these fuels are produced using renewable electricity and carbon dioxide captured from the atmosphere—via processes like direct air capture or from industrial sources—the overall lifecycle emissions can be near-zero or net-zero. This means that the CO_2 released when the fuel is burned is effectively balanced by the CO_2 removed during its production. This closed carbon loop offers a significant advantage over traditional fossil fuels, which release long-stored carbon into the atmosphere, contributing to global warming. If scaled properly, synthetic fuels could become a key component of strategies aimed at decarbonizing sectors where emissions reductions are most difficult.

Another major benefit is their compatibility with existing infrastructure. Unlike hydrogen gas or battery electric solutions, synthetic and e-fuels can be used directly in current internal combustion engines, aviation turbines, marine engines, and even residential heating systems. This backward compatibility eliminates the need for a complete overhaul of global transport and energy systems. Fuel distribution networks, refuelling stations, storage tanks, and pipeline infrastructure can be repurposed or slightly modified, allowing for a smoother transition to a low-carbon economy. This aspect makes synthetic fuels especially appealing in industries such as aviation, shipping, and heavy transport, where electrification is not yet a viable alternative.

E-fuels also offer a practical solution to one of the biggest challenges facing renewable energy: energy storage. Wind and solar power are intermittent by nature, generating electricity only when the wind blows or the sun shines. Synthetic fuels allow this excess or off-peak electricity to be stored in chemical form as liquid or gaseous fuels. This stored energy can then be used on demand, helping to stabilize electrical grids, support seasonal storage needs, and improve energy security. By transforming renewable energy into a dense, transportable form, e-fuels act as a buffer between variable generation and continuous demand.

Finally, long-distance transport and storage of energy is significantly more feasible with synthetic fuels than with hydrogen gas. While hydrogen has a high gravimetric energy density, its low volumetric density and cryogenic or high-pressure storage requirements make it difficult and expensive to transport over long distances. In contrast, liquid e-fuels—such as e-diesel, e-kerosene, methanol, or ammonia—can be handled using existing shipping, trucking, and pipeline systems with relatively minor modifications. They are safer to store and less prone to leakage than hydrogen gas, making them an attractive option for global energy trade and distribution.

The advantages of synthetic and e-fuels—carbon neutrality, infrastructure compatibility, renewable energy storage, and ease of transport—make them a critical part of the future low-carbon energy mix. These benefits are especially relevant for applications where direct electrification is currently impractical or cost-prohibitive.

The development and integration of electrofuels face several significant challenges that inhibit their widespread adoption as sustainable alternatives to fossil fuels. These challenges include low efficiency, high costs, complexities in CO_2 sourcing, and limitations in scalability.

Firstly, a major challenge associated with e-fuels is their energy efficiency. The entire power-to-fuel-to-motion process involves multiple conversion stages—namely, electrolysis, fuel synthesis, and combustion—which collectively yield energy efficiencies significantly lower than direct electrification methods. Current estimates suggest that the overall energy efficiency can be as low as 10–20% in some cases due to substantial conversion losses at each step of the process [246, 247]. This inefficiency not only makes e-fuels less attractive for immediate application but also poses a significant obstacle to achieving carbon neutrality in the transportation sector.

Secondly, the financial aspect of e-fuel production presents a considerable barrier to commercial viability. At present, costs associated with e-fuels are estimated to be two to ten times higher than those of traditional fossil fuels, depending on various factors including technology and location [248, 249]. This increase in costs arises from several elements, including the expensive nature of electrolysis equipment, the lack of economies of scale, and the expenses tied to capturing CO_2 [247, 248]. Such high operating costs hinder competitiveness in a market where fossil fuels are historically cheaper and more established, limiting broader adoption of e-fuels.

Compounding these economic hurdles is the challenge of sourcing CO_2 for e-fuel production. The methods of capturing carbon dioxide, especially from the atmosphere, are energy-intensive and often expensive. While sourcing CO_2 from industrial processes may seem feasible, the sustainability and economic viability of these sources over the long term remains a point of contention [250, 251]. This creates uncertainty regarding the reliability of e-fuel supply, complicating the economic landscape even further.

Finally, the scalability of e-fuel production is a pressing concern. Currently, production is mostly limited to relatively small-scale operations, and significant investments in infrastructure, technology, and innovation are required to achieve commercial-scale production. This necessitates not only financial commitment but also advancements to effectively scale up production processes without sacrificing efficiency or expensively scaling costs disproportionately [237, 247]. The integration of supportive regulatory frameworks and a conducive policy environment will be essential to overcoming scalability challenges, ensuring that e-fuels can effectively meet demand in various transportation sectors.

In recent years, several innovative projects and initiatives have emerged in the realm of sustainable fuels, particularly focusing on alternatives to traditional fossil fuels to aid in decarbonization efforts across various sectors, including aviation and automotive industries. Notable among these projects is the collaboration between Porsche and Siemens Energy, which has initiated the Haru Oni project in Chile. This project aims to produce e-methanol and e-gasoline from renewable wind power, targeting applications in motorsports and potentially aviation, thus exemplifying practical steps toward sustainable fuel production [252, 253].

Airbus is another significant player in this space, exploring the feasibility of e-kerosene as a synthetic aviation fuel. This effort aligns with the global emphasis on shifting to low-carbon alternatives in the aviation sector to meet burgeoning climate targets. E-kerosene, produced using renewable energy sources, showcases the potential to drastically reduce the carbon footprint of air travel [237, 247]. Furthermore, Audi has previously undertaken pilot projects in Germany aimed at creating e-diesel and e-gasoline, which are crucial components in transitioning toward more sustainable transportation solutions [254]. These developments are critical for demonstrating the technical viability of electrofuels and their implications in real-world applications.

At the policy level, the European Union is actively considering mandates for synthetic aviation fuels, in line with its Fit for 55 and ReFuelEU initiatives. These legislative efforts aim to promote the adoption of sustainable aviation fuels (SAF) across the EU, ensuring that the aviation sector contributes to the bloc's climate objectives. Such policies could establish a framework that incentivizes the production and use of e-fuels in aviation, fostering a transition to greener alternatives [255].

The convergence of innovative projects like Haru Oni, the e-kerosene exploration by Airbus, Audi's e-fuel developments, and EU regulatory initiatives represents a significant movement toward sustainable fuel solutions. These efforts intertwine technological advancements with critical policy frameworks necessary for fostering a transition to sustainable energy in the transportation sector. As the global community pursues deep reductions in greenhouse gas emissions, these projects and associated policy measures potentially represent pivotal steps in this essential journey toward sustainability.

The discourse surrounding synthetic and e-fuels is increasingly important as the world pivots towards stringent carbon reduction goals, particularly in sectors where electrification may not offer a viable solution. These fuels are often not viewed as replacements for direct electrification in passenger vehicles, but rather as complements in industries where shifting to electric systems presents significant obstacles. This narrative is particularly salient in aviation, shipping, heavy road transport, backup power, remote applications, and the chemical and plastics industries.

In the aviation sector, e-fuels have garnered attention as potential game changers for reducing carbon emissions, especially for long-haul flights where battery technology presents limitations. The study by Speizer et al. [256] suggests that e-fuels could be deployed most effectively under economic conditions that favour higher carbon pricing, thereby making biofuels coupled with Carbon Capture and Storage (CCS) an attractive option for the aviation industry. Furthermore, recent assessments indicate that e-fuels can significantly reduce lifecycle aviation emissions, illustrating their vital role in aviation decarbonization, with reductions of up to 94% compared to 2019 levels being possible with the right fuel pathways and operational changes [257].

Shipping, another critical hard-to-decarbonize sector, stands to benefit from synthetic fuels. E-fuels, including e-ammonia and e-methanol, offer promising alternatives to conventional marine fuels, which are currently heavily reliant on fossil sources [258]. Research indicates that these fuels can support the shipping industry's decarbonization in line with International Maritime Organization (IMO) targets, despite existing technological and market barriers [259]. The high energy density of liquid fuels makes them indispensable for maritime applications, and their production technologies are evolving to better leverage renewable energy sources [260].

Heavy road transport also presents challenges for electrification, especially due to the energy demands associated with large freight vehicles. The analysis by Muratori et al. supports the notion that biofuels, particularly when coupled with CCS, emerge as one of the most promising alternatives for mitigating emissions in the freight sector [261]. This conclusion aligns with the broader recognition that synthetic fuels could help bridge the gap until batteries or hydrogen fuel cells become more economically viable for heavy-duty applications.

Moreover, synthetic fuels could play a crucial role in backup power systems and remote applications where grid access is limited or unreliable. Applications in chemical and plastics production further underscore the versatility of synthetic fuels, as these can serve as essential feedstock for producing hydrocarbons sustainably [262]. As such, the ability of synthetic fuels to provide a carbon-neutral alternative while maintaining energy density and compatibility with existing infrastructure positions them as a pivotal element in the global energy transition.

While electrification continues to advance rapidly within passenger cars, sectors such as aviation, shipping, and heavy transport present unique challenges that synthetic and e-fuels are well suited to address. Their role extends beyond mere transportation fuels; they are integral to achieving a holistic and effective approach to decarbonizing some of the most difficult sectors.

The transition to sustainable energy sources is significant in addressing climate change, particularly in sectors that are challenging to electrify. Synthetic and e-fuels are increasingly recognized as viable alternatives for decarbonizing these hard-to-abate sectors: aviation, shipping, heavy road transport, backup power, and chemical industries. Unlike electrification, which is often promoted for personal vehicles, synthetic fuels can address emissions where direct electrification is not feasible.

In aviation, emissions are rising faster than appropriate climate targets allow. According to a study, adopting synthetic fuels from biomass, green hydrogen, and atmospheric CO_2 could reduce lifecycle emissions from aviation by up to 94% by 2050, even amid increasing demand [257]. Similarly, in shipping, decarbonization involves alternative fuels such as hydrogen and ammonia, viewed as critical strategies due to the high energy demands and liquid fuel dependency of maritime vessels [263]. The infrastructure for these fuels, including liquefied natural gas (LNG) and hydrogen, is being explored as part of a transition strategy to meet the energy density requirements of large-scale transport [264].

Heavy road transport presents distinct challenges; however, synthetic fuels provide a pathway that eases the transition without requiring major infrastructure overhauls. These fuels can replace conventional fuels in existing engines, thus minimizing the costs associated with converting to electric vehicles, particularly in regions lacking extensive charging infrastructure [265]. Additionally, the chemical industry can utilize CO_2 captured from industrial processes as a feedstock for synthetic fuels, integrating waste management with energy transition efforts [266].

For backup power and remote applications, hydrogen and synthetic fuels offer reliable energy storage solutions that contribute to grid stability, especially where renewable energy sources need to balance demand fluctuations. Hydrogen is characterized by its high energy density and versatility, which enables it to serve as a clean energy carrier and support backup systems [267, 268]. This is particularly pertinent for regions reliant on intermittent renewable sources, providing a sustainable alternative to traditional fossil fuel backups [269].

Lastly, the integration of synthetic fuels into the chemical and plastics industries presents opportunities for decarbonization. With processes like Fischer-Tropsch synthesis for electrofuels, industries can harness renewable electricity to produce fuels and chemicals, thereby reducing reliance on fossil fuels [270]. A comprehensive strategy that embraces these various pathways could effectively contribute to global decarbonization goals, particularly in sectors where direct electrification is not currently feasible.

Synthetic fuels and e-fuels hold real promise for decarbonizing industries and transport systems that are resistant to electrification. However, they are currently limited by cost, energy inefficiency, and lack of infrastructure. Their future hinges on large-scale deployment of renewable energy, improvements in electrolysis and fuel synthesis technologies, and strong policy support through carbon pricing, mandates, and subsidies. If these conditions are met, synthetic fuels could be a key pillar in achieving net-zero emissions by mid-century.

Chapter 6

Fuel Processing and Refining

Having explored the nature of different fuel types and their origins, the next step in understanding the fuel journey is to examine how raw fuels are transformed into usable energy sources. Chapter 6, *Fuel Processing and Refining*, takes readers inside the critical industrial processes that convert crude oil, natural gas, biomass, and other raw materials into the refined products that power modern life.

This chapter delves into the key stages of fuel processing—from distillation and cracking in oil refineries to upgrading techniques used in biofuel production. It also introduces the chemical and physical changes involved in refining, the role of catalysts, and how refining processes are tailored to meet environmental standards, efficiency targets, and market demands.

By uncovering how fuels are processed and purified, Chapter 6 connects the scientific principles of chemistry with the practical needs of the energy economy. It provides the technical foundation needed to appreciate how complex and energy-intensive fuel transformation can be, and why refining remains a pivotal step in the global energy system—even as cleaner alternatives gain ground.

How Crude Oil Becomes Usable Fuel

Crude oil becomes usable fuel through a process called refining, which separates and transforms its components into various products like gasoline, diesel, and jet fuel. This involves removing impurities, separating different hydrocarbons based on their boiling points, and using various chemical processes to create specific products.

Crude oil, in its natural state, is a thick, dark, and often viscous liquid composed of a complex mixture of hydrocarbons and other organic compounds. To become usable fuel for transportation, heating, electricity generation, and industrial purposes, crude oil must go through a series of processing steps that transform it into refined petroleum products like gasoline, diesel, jet fuel, and heating oil. This transformation occurs primarily in oil refineries.

The refining process of crude oil involves several stages aimed at purifying and transforming crude oil into usable products. This response outlines key processes involved in the refining of crude oil: the removal of impurities, distillation, conversion processes, blending and treatment, and the end products generated.

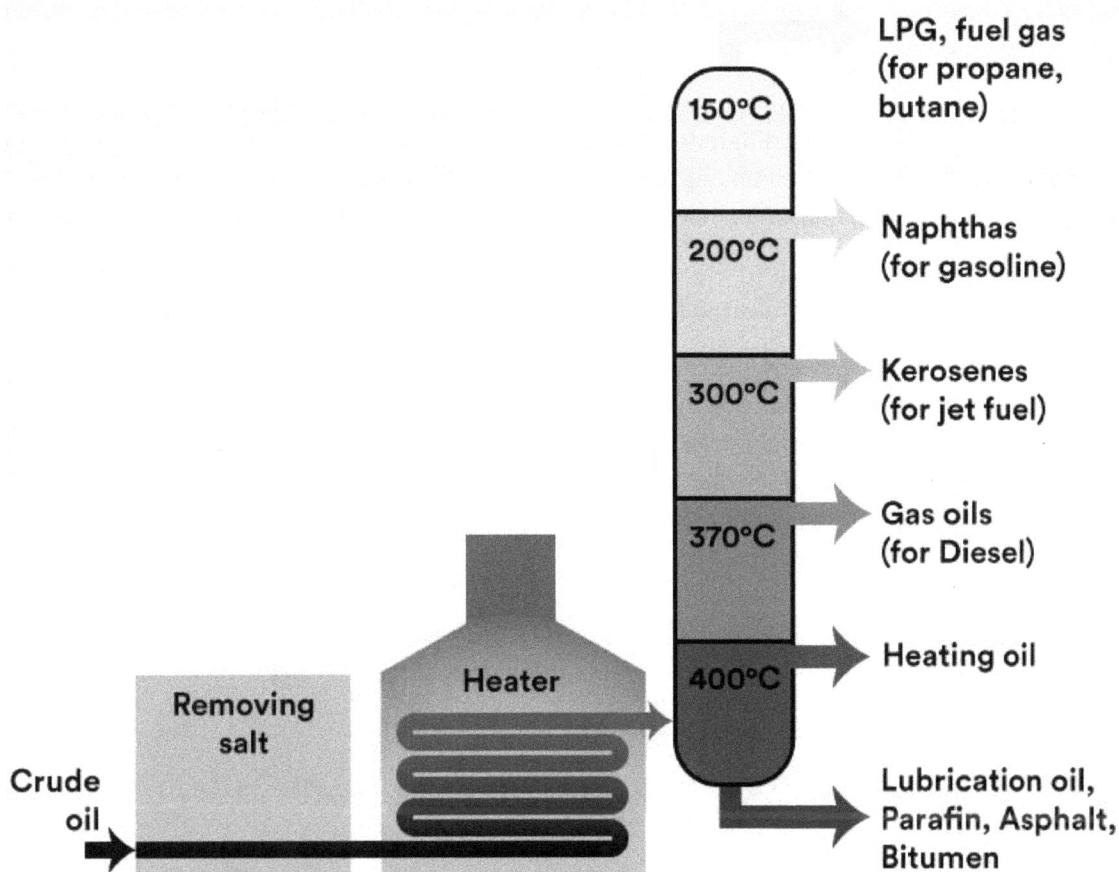

Figure 18: Crude oil distillation unit (working principle). FrankvEck, CC BY-SA 4.0, via Wikimedia Commons.

1. Removal of Impurities

The initial step in refining crude oil involves the elimination of various impurities such as sulphur, salts, water, and non-hydrocarbon substances. These impurities pose significant challenges to the refining process and the quality of the final products. Common methods for removing impurities include chemical treatment, filtration, and heating. For instance, certain secondary refining processes like hydrofining are specifically designed to reduce sulphur and other elements from heavy oil fractions, thereby increasing the quality of light motor fuels produced from these sources [271]. The effective removal of impurities is crucial for improving overall efficiency and reducing the environmental impact of refining operations [272].

2. Distillation

Once the impurities are removed, the crude oil is subjected to distillation. This process involves heating the oil in a distillation column, which operates by exploiting the different boiling points of hydrocarbons. As the crude oil is heated, lighter fractions, such as gasoline components, vaporize and ascend to the upper sections of the column, while heavier fractions, including diesel and asphalt, condense at lower levels.

Crude oil is heated in a furnace to temperatures around 350–400°C and fed into a distillation column. As the oil vaporizes, different hydrocarbon components condense at various heights in the column depending on their boiling points. Lighter fractions like liquefied petroleum gas (LPG) and gasoline rise to the top, while heavier products like diesel, kerosene, and residual oils settle lower down.

Each of these fractions becomes the feedstock for further processing or is directly blended into final fuel products.

The fractions collected at varying heights exemplify the stratification of crude oil into its components, which enables the selective harvesting of different products for subsequent applications[271, 272]. This distillation method serves as a fundamental technique in oil processing, reflecting both historical and modern advancements in refining processes [272].

Figure 19: Richmond Chevron Refinery with backdrop of San Pablo Bay. Bastique, CC BY-SA 4.0, via Wikimedia Commons.

3. Conversion Processes

Not all components of crude oil are suitable in their distilled form. Some of the heavier, less valuable fractions are upgraded through conversion processes to yield more useful products. Following distillation, some fractions undergo further processing through various conversion methods. Catalytic cracking and hydrocracking are significant processes that break down larger hydrocarbon molecules into smaller, lighter ones, thereby enhancing the yield of gasoline and other valuable products [271, 273]. Catalytic reforming is another pivotal technique that alters the structure of hydrocarbons to enhance characteristics like the octane rating, which is critical for high-performance fuels [273]. These conversion processes often utilize sophisticated catalysts, such as zeolites, which significantly improve efficiency and product quality in refining operations [274].

Key conversion methods include:

- **Cracking (Catalytic or Thermal Cracking):** Breaks long hydrocarbon chains into shorter, more valuable ones like gasoline and diesel.

- **Hydrocracking:** Uses hydrogen under high pressure and temperature to break down heavy molecules and remove impurities, producing high-quality fuels.

- **Coking:** Converts heavy residual oils into lighter products and a solid carbon-rich material called petroleum coke.

These processes increase the yield of high-demand fuels such as gasoline and jet fuel from every barrel of crude.

Cracking is a vital chemical process in petroleum refining, where large and complex hydrocarbon molecules are broken down into smaller, more valuable components. This process helps meet the demand for high-performance fuels and petrochemical feedstocks. There are two main types of cracking: steam cracking and catalytic cracking, each tailored to produce different types of products under specific conditions.

Steam cracking is conducted at very high temperatures, typically between 1050 and 1150 K, and is designed to produce high yields of alkenes such as ethene and propene, which are essential building blocks in the chemical industry. Feedstocks can include ethane, propane, butane from natural gas, or heavier fractions like naphtha and gas oil from crude oil. The process involves preheating the feedstock, mixing it with steam to reduce the risk of carbon formation, and then passing it rapidly through tubular reactors to prevent unwanted side reactions. The output is rapidly quenched to stabilize the desired products. The yield depends on both the type of feedstock and the severity of the operating conditions. For instance, using ethane yields a higher proportion of ethene, while naphtha produces more raw pyrolysis gasoline (RPG) and a broader range of hydrocarbons.

Catalytic cracking, particularly fluid catalytic cracking (FCC), uses a powdered zeolite catalyst to break down heavier hydrocarbons like gas oil into lighter products such as branched and cyclic alkanes, which are ideal for gasoline. This process is carried out at lower temperatures (700–800 K) compared to steam cracking and relies on a fluidized bed reactor. The catalyst is continuously regenerated by burning off coke deposits. The products include high-octane gasoline, light olefins, and a heavy residue used as fuel. FCC is highly flexible, allowing adjustments in temperature and catalyst type to target specific outputs. A variant, hydrocracking, uses hydrogen and a platinum catalyst at high pressures (up to 80 atm) to produce saturated alkanes and cycloalkanes. This is especially valuable for producing clean, high-quality fuels.

Isomerisation involves rearranging straight-chain hydrocarbons into branched isomers to improve fuel performance, particularly the octane number. For example, normal butane can be converted into isobutane, which is more suitable for blending into high-octane gasoline. The reaction is carried out at mild temperatures (around 300 K) over an aluminium chloride catalyst, and separation of the branched and straight chains is done using distillation or molecular sieves.

Reforming, or catalytic reforming, restructures hydrocarbons to produce compounds with higher octane ratings, such as aromatics and branched alkanes. Typically, straight-chain alkanes from the naphtha fraction are mixed with hydrogen and passed over a platinum or rhenium catalyst on alumina at about 700 K and 30 atm. Reforming also produces hydrogen as a byproduct, which is reused in hydrotreating and hydrocracking processes. The resulting reformate is rich in benzene, toluene, and xylene—key inputs for petrochemical manufacturing.

Alkylation is the process of combining smaller alkanes and alkenes to produce high-octane, branched-chain hydrocarbons like isooctane. Conducted with acid catalysts (sulfuric or hydrofluoric acid) at low temperatures, alkylation is essential for blending high-performance gasoline. Waste acid from the process is often recycled through sulfur recovery systems to regenerate usable acid.

Dealkylation is the removal of alkyl groups, typically from aromatic compounds. For example, methylbenzene (toluene) can be converted into benzene by removing a methyl group using hydrogen and a catalyst at high temperatures and pressures. This reaction helps increase the production of benzene, a more valuable feedstock in the chemical industry.

Disproportionation is a reaction where a single molecule transforms into two different products. An example is the conversion of methylbenzene into benzene and dimethylbenzenes using zeolite catalysts under moderate conditions. This process is useful for selectively generating valuable chemical feedstocks like 1,4-dimethylbenzene.

Polymerization involves the chemical bonding of small hydrocarbon molecules (monomers) into larger chains called polymers. Alkenes such as ethene and propene, typically produced through cracking, are used to manufacture plastics like polyethylene and polypropylene. Refineries often include polymerization units to produce oligomers for surfactants or larger polymers used in packaging, construction, and textiles.

Overall, these chemical processes are central to maximizing the value of crude oil by producing high-performance fuels and feedstocks for petrochemicals. Each process plays a strategic role in shaping the output slate of a modern oil refinery, ensuring that both energy and industrial chemical needs are met efficiently.

4. Blending and Treatment

Once the fractions have been converted, they are blended to produce specific oils, including gasoline, diesel, jet fuel, and heating oil. The blending process is essential for ensuring that the final products meet industry standards for stability and performance. Additionally, further treatments may be necessary to enhance the quality of the blends, which can include methods to reduce viscosity and improve oxidation resistance [275]. Such treatments ensure that the products can withstand various operational conditions and have desirable characteristics in their final applications.

To meet environmental regulations and engine requirements, the separated and converted fuel components are treated to remove impurities like sulphur, nitrogen, water, and metals. This is done through:

- **Hydrotreating:** Reacts the fuel fractions with hydrogen to remove sulphur and other contaminants.

- **Desalting:** Removes salts and water from the crude before distillation.

- **Blending:** Refined components are combined with additives (e.g., anti-knock agents, detergents) to produce finished fuels with desired performance characteristics.

At this stage, fuels such as unleaded gasoline, ultra-low sulphur diesel, and aviation kerosene (jet fuel) are ready for distribution.

5. End Products

After refining, the final products are transported for use in several applications, including transportation fuels, heating, and as feedstocks for the petrochemical industry. The refined fuels are essential for powering vehicles, generating energy, and serving as raw materials for producing various chemicals and materials [276]. Continued advancements in refining technologies are pivotal for meeting the growing global demand for these products while simultaneously addressing environmental concerns [271].

6. Storage and Distribution

Once the refined fuels are produced, they are stored in large tanks and transported to end users via pipelines, tankers, rail, and road trucks. Distribution infrastructure ensures the final fuel reaches gas stations, airports, industrial facilities, and households.

Resources and Costs

Refining crude oil into usable fuels is a resource-intensive and capital-heavy process requiring substantial physical inputs, infrastructure, specialized human expertise, and large-scale investment. The feasibility and economics of refining vary significantly depending on geographic location, access to crude supply, energy costs, environmental regulations, and the quality of existing infrastructure. Each stage in the refining process demands specific resources and incurs considerable costs, with notable differences in practices and expenditure across global regions.

The first step, separation or fractional distillation, is a physical process that requires a continuous supply of crude oil, typically delivered by pipelines, tankers, or rail. This crude is heated using energy derived mainly from natural gas or refinery fuel gas to temperatures between 350 and 400°C. Distillation columns—large towers with internal trays made of heat-resistant alloys—separate the vaporized hydrocarbons based on boiling points. Cooling systems and condensers help manage vapour condensation at various column heights. The process also relies on trained engineers and operators for safe and efficient operation. Constructing a simple atmospheric distillation unit can cost between $500 million and $1 billion depending on capacity. Energy costs are substantial, accounting for 30% to 60% of operational expenditure, with distillation alone consuming up to 2% of the energy content in the crude barrel. For example, the Reliance Industries refinery in Jamnagar, India, the largest globally, processes over 1.2 million barrels per day using highly optimized, energy-efficient systems.

In the conversion phase, heavier hydrocarbon fractions are transformed into more valuable, lighter fuels. This requires high-temperature, high-pressure reactors built from corrosion-resistant materials. Processes such as catalytic cracking, hydrocracking, and coking are used, with hydrogen being a critical input for hydrocracking and hydrotreating—often produced on-site via steam methane reforming. These operations depend on catalysts like zeolites and metal compounds, which must be replaced periodically, and significant amounts of water and steam for thermal regulation and cleaning. The cost of building hydrocracking units alone ranges from $1.5 to $3 billion. High-performance catalysts cost between $10,000 and $30,000 per tonne, and hydrogen production costs add further expense, especially when not integrated into the refinery gas stream. The Motiva Port Arthur refinery in Texas, which invested over $10 billion in upgrades, including hydrocrackers, represents North America's largest and most advanced facility.

Treatment and finishing processes follow, aiming to purify and enhance the fuels to meet environmental and engine performance standards. This stage includes hydrotreating with hydrogen to remove sulphur and other contaminants, as well as desalting to eliminate corrosive salts from crude oil. Additives are blended into final fuel products to improve performance, reduce emissions, and enhance stability. Hydrotreating units alone may cost between $300 million and $1 billion, and stringent environmental regulations such as Euro VI and EPA Tier 3 standards increase compliance costs through continuous emissions monitoring, sulphur recovery systems, and wastewater treatment. Refineries in the European Union, particularly in Germany and the Netherlands, are leaders in producing ultra-low sulphur fuels, but also face higher per-barrel processing costs due to regulatory compliance.

The final stage involves storage and distribution. Refined fuels are stored in steel tanks ranging from 10,000 to 50,000 cubic meters in capacity and transported through extensive pipeline networks, marine tankers, fuel trucks, and railcars. Building storage tanks typically costs between $2 and $5 million each, while pipeline infrastructure can cost $1 million to $5 million per mile depending on geography and environmental permitting. For example, Saudi Aramco maintains a vast infrastructure system including the 1,200-kilometre East–West pipeline that connects crude and refined product flows from oil fields to Red Sea ports, supporting both domestic distribution and global exports.

Beyond the main production phases, additional operational and environmental costs must be considered. Refineries employ large workforces of skilled staff and contractors, which adds to fixed labour costs. The refining process also consumes vast quantities of water, necessitating robust wastewater treatment systems. Environmental regulations impose further expenses, such as those arising from carbon pricing and flaring restrictions. For instance, under the European Union's Emissions Trading Scheme (EU ETS), refiners must purchase emissions permits, which can significantly increase operating expenditures.

Transforming crude oil into market-ready fuels is a complex and capital-intensive operation that demands careful management of energy, materials, and environmental factors. Refineries in regions like the Middle East benefit from low energy and crude costs, while those in Europe and North America face higher costs due to strict regulations and advanced product requirements. Despite these regional differences, all refineries must balance efficiency, environmental compliance, fuel quality, and market dynamics to ensure long-term profitability and sustainability.

The economic evaluation of refining crude oil into usable fuels involves a dual consideration of significant costs versus valuable outputs. Refining operations demand substantial capital investments, with estimates indicating that constructing and maintaining refining infrastructure can exceed billions of dollars. The intricate engineering systems required for refining necessitate advanced technology, high-quality materials, and a skilled workforce, all of which contribute to the ongoing operational expenses associated with the refining process [277, 278]. Energy consumption is a crucial factor; for example, fractional distillation can consume significant energy per barrel of crude oil, while processes such as hydrocracking necessitate further energy inputs and active catalysts, thereby increasing operational costs [278].

In addition to direct costs associated with infrastructure and energy, environmental compliance is increasingly significant, particularly in regions like the European Union and California, which impose strict emissions regulations that include desulfurization to reduce sulphur content in fuels [279, 280]. These regulatory frameworks impose additional financial burdens that affect the cost per barrel of fuel produced, especially in high-cost regions where energy inputs are variable and regulatory pressures are more pronounced.

Despite these costs, the benefits of refining crude oil are considerable and multi-faceted. Refineries play an essential role in maximizing economic output from crude oil by transforming it into a variety of high-demand products such as gasoline, diesel, jet fuel, and petrochemicals, which are critical across various sectors, including transportation and industry [278]. The economies of scale realized in modern megarefineries, such as those operated by Reliance in India and Aramco in Saudi Arabia, enable these entities to take advantage of cost efficiencies and reliance on domestic crude sources, enhancing profitability within lower-cost operational environments [281].

Moreover, refined fuels benefit from a well-established global logistics network, which facilitates easier distribution and storage compared to crude oil or emerging fuel alternatives such as hydrogen [280]. This logistical advantage underscores the continued high demand for refined fuels, particularly in developing nations where energy needs are expanding rapidly. Therefore, the geographical placement of refineries in proximity to crude supplies, along with favourable regulatory landscapes, often dictates operational success and profitability. The integration of refining operations into global supply chains underpins the ongoing relevance of these energy sources within the global economy [282].

Volumes

As of recent global data, it is reported that approximately 4.3 to 4.5 billion tonnes (or about 90 to 95 million barrels per day) of crude oil are processed annually worldwide, translating to roughly 33 to 35 billion barrels per year [283]. This massive scale of crude oil processing reflects a combination of high global demand for fuels and various petroleum products, the impacts of industrial and economic development, the lack of comprehensive alternatives, and the strategic and economic significance of oil.

1. **High Global Demand for Fuels and Products**: Crude oil serves as the primary feedstock for fuels essential for transportation, including gasoline, diesel, and jet fuel, and contributes significantly to non-fuel products such as petrochemicals, asphalt, and lubricants [284]. The diversity of products derived from crude oil underscores the foundational role it plays in modern economies, which are heavily dependent on energy derived from petroleum [285].

2. **Industrial and Economic Development**: Developing nations are experiencing increasing energy consumption driven by urbanization, expanding middle classes, and the rapid growth of industrial activities. These factors contribute to the steady increase of refined petroleum products [283]. The ongoing industrialization in countries like China and India has been a key driver of global oil demand, as these nations continue to develop their infrastructure and economies. Daily oil demand rose significantly from around 60 million barrels in the early 2000s to approximately 84 million barrels per day in 2020, reflecting the growing energy appetite among emerging markets [283].

3. **Lack of Complete Alternatives**: Despite advances in renewable energy and electric vehicles, the global energy climate remains highly reliant on oil due to several persistent barriers. Electric vehicle adoption varies regionally, and comprehensive infrastructures for alternatives, such as hydrogen and biofuels, remain underdeveloped, particularly in aviation and maritime transport [284]. The dependency on oil in heavy industries, along with the efficiency and performance standards vehicles and machinery demand, further reinforces the continued reliance on crude oil as a primary energy source [283].

4. **Strategic and Economic Value**: The refining of crude oil operates not merely to satisfy domestic energy needs but also serves crucial economic and strategic roles in

international trade. Countries rich in oil resources, such as Saudi Arabia and Russia, capitalize on crude oil as a major export commodity to bolster their economies while ensuring energy security [284]. Oil refining integrates deeply into the fabric of global trade, with a significant flow of crude being shipped internationally to be processed in diverse refineries [283].

The processing of over 4 billion tonnes of crude oil annually is a complex yet critical part of the global energy system, reflecting the diverse and essential uses of oil. With ongoing investments in alternative energy sources and technologies, it remains vital to monitor how these developments could reshape crude oil's central position in global markets.

Refineries and Chemical Processing

Oil refineries are industrial process plants that transform crude oil into usable petroleum products such as gasoline, diesel, jet fuel, heating oil, liquefied petroleum gas (LPG), asphalt, lubricants, and petrochemical feedstocks like ethylene and propylene. These facilities form a crucial part of the downstream petroleum industry, converting raw hydrocarbons into fuels and materials essential for transportation, industry, heating, and power generation.

Globally, refineries operate in nearly every major economy. As of 2020, the combined capacity of all crude oil refineries was approximately 101.2 million barrels per day. The largest refinery complex in the world is the Jamnagar Refinery in Gujarat, India, owned by Reliance Industries, which processes over 1.2 million barrels per day. Other notable refineries include the Paraguaná Refinery Complex in Venezuela and the Ulsan Refinery in South Korea. The U.S., with numerous refineries across Texas, Louisiana, and California, remains a major refining hub, although no large new refineries have been built since the 1980s due to permitting challenges and environmental regulations. Instead, existing refineries have been upgraded and expanded.

Refineries are highly complex and resemble large chemical plants, with extensive networks of pipes, towers, and processing units. The refining process starts with the fractional distillation of crude oil, separating it into various hydrocarbon fractions based on boiling points. Heavier components are further processed using methods like cracking (breaking down larger molecules), reforming (restructuring molecules), and treating (removing impurities). Advanced techniques also produce petrochemicals directly from crude oil, bypassing traditional intermediates like naphtha.

Historically, oil refining dates back over a millennium, with early developments in China and the Middle East. The modern refining industry began in the mid-19th century with the invention of kerosene production and grew rapidly in the 20th century with the rise of automobiles and aviation. Technological innovations during and after World War II significantly advanced refining capabilities.

Today, refineries are optimized for efficiency and product yield. They also include environmental safeguards to comply with air and water quality standards. Units such as sulphur recovery systems, wastewater treatment, and emissions controls are now standard. Many refineries also co-generate electricity and utilize waste heat, enhancing energy efficiency.

Refineries are categorized by their configuration and complexity. Simple refineries perform basic distillation, while complex ones include conversion and treatment units that maximize output of high-demand products like gasoline and jet fuel. The configuration depends on the type of crude processed and the market demand in a given region.

Oil refineries produce over 6,000 by-products beyond fuels. These include plastics, fertilizers, synthetic rubber, detergents, and even pharmaceuticals. Refined sulphur, paraffin wax, petroleum coke, and petrochemical building blocks like BTX (benzene, toluene, xylene) are also major outputs.

Despite growing pressure for decarbonization, oil refineries remain vital to the global economy. They are adapting by integrating renewable feedstocks (e.g., bio-crude, used cooking oil) and exploring carbon capture technologies. Some are transitioning toward "refineries of the future," co-producing hydrogen, synthetic fuels, and electricity alongside traditional fuels.

In summary, refineries are essential infrastructure in the global energy system. Their ability to adapt to changing technologies, environmental policies, and market demands will shape the future of energy and materials supply in the coming decades.

The chemical processes in an oil refinery are highly integrated and complex, designed to convert crude oil into various refined products that meet specific quality, regulatory, and performance standards. These processes are organized into several major systems, beginning with the preparation of the crude and extending through separation, conversion, treatment, and final product finishing.

The refining process begins with the desalter unit, which removes water-soluble impurities such as salts and other contaminants from crude oil. These salts can corrode equipment and poison catalysts in later stages, so this early washing step is critical. Larger refineries with more than 100,000 barrels per stream day (bpsd) typically also include a pre-flash or pre-distillation unit, which partially separates lighter components before the main distillation, improving energy efficiency.

At the heart of the refinery lies the crude oil distillation unit, which uses atmospheric pressure to separate the crude into fractions based on boiling points. Heavier components that remain after this step are sent to the vacuum distillation unit, where the pressure is lowered to allow further distillation without reaching thermal degradation temperatures.

Next, the various fractions undergo hydrotreating, beginning with the naphtha hydrotreater. This process uses hydrogen to remove sulphur and nitrogen compounds before the naphtha is sent to the catalytic reformer, which rearranges molecular structures into higher-octane aromatics for gasoline. The reforming process releases hydrogen, which is recycled into other refinery units, such as the distillate hydrotreater, which desulfurizes diesel and kerosene streams under higher pressures.

Fuel, Energy and Net Zero

To convert heavier oils into more valuable lighter products, refineries employ units like the fluid catalytic cracker (FCC), which uses a solid catalyst to "crack" large hydrocarbon molecules into smaller, more volatile compounds like gasoline and olefins. The hydrocracker, by contrast, uses hydrogen and high pressure to achieve similar results with better control over the end product's quality and fewer impurities.

Desulfurization is also achieved in the Merox unit, which oxidizes sulphur compounds in LPG, kerosene, or jet fuel into less harmful forms. Older techniques like doctor sweetening and caustic washing are less common today but still used in specific applications.

For very heavy residues that resist catalytic or hydrocracking, coking units such as delayed cokers convert them into lighter fuels and a solid carbon-rich by-product known as petroleum coke. Meanwhile, alkylation units use acid catalysts (sulfuric or hydrofluoric acid) to combine isobutane and olefins into alkylate, a high-octane blending component essential for gasoline production.

Other chemical processes include dimerization, which combines olefins into larger, branched molecules suitable for gasoline. However, these highly reactive products must be quickly stabilized or blended to avoid gum formation. Isomerization helps improve octane numbers by converting linear alkanes into branched forms, particularly valuable for blending and as feedstock for alkylation.

Refineries often produce their own hydrogen for hydrotreating and hydrocracking through steam reforming, which converts methane (natural gas) and water into hydrogen and carbon monoxide. The refinery infrastructure includes liquefied gas storage vessels, usually spherical or cylindrical under pressure, to store propane and similar fuels.

Hydrogen sulphide extracted from desulfurization processes is managed through an amine gas treater, followed by a Claus unit, which recovers elemental sulphur. This sulphur, a by-product of hydroprocessing, is widely used in agriculture and industry. Wastewater containing hydrogen sulphide is treated in sour water strippers, which release the gas for sulphur recovery.

Supporting systems include cooling towers, boiler plants for steam generation, and instrument air systems for pneumatic control devices. Wastewater treatment systems use API separators and biological treatment units such as activated sludge tanks to ensure environmental compliance and water reuse.

For the production of lubricants and specialized fuels, solvent refining with cresol or furfural is employed to remove undesirable aromatics. Solvent dewaxing eliminates waxy materials like petrolatum to ensure fluidity of products such as motor oils at low temperatures.

Finally, refined products and feedstocks are stored in large vertical storage tanks equipped with vapor control systems and surrounded by containment berms for spill protection. These tanks store crude oil, intermediate products, and finished fuels prior to distribution by pipeline, tanker, rail, or truck.

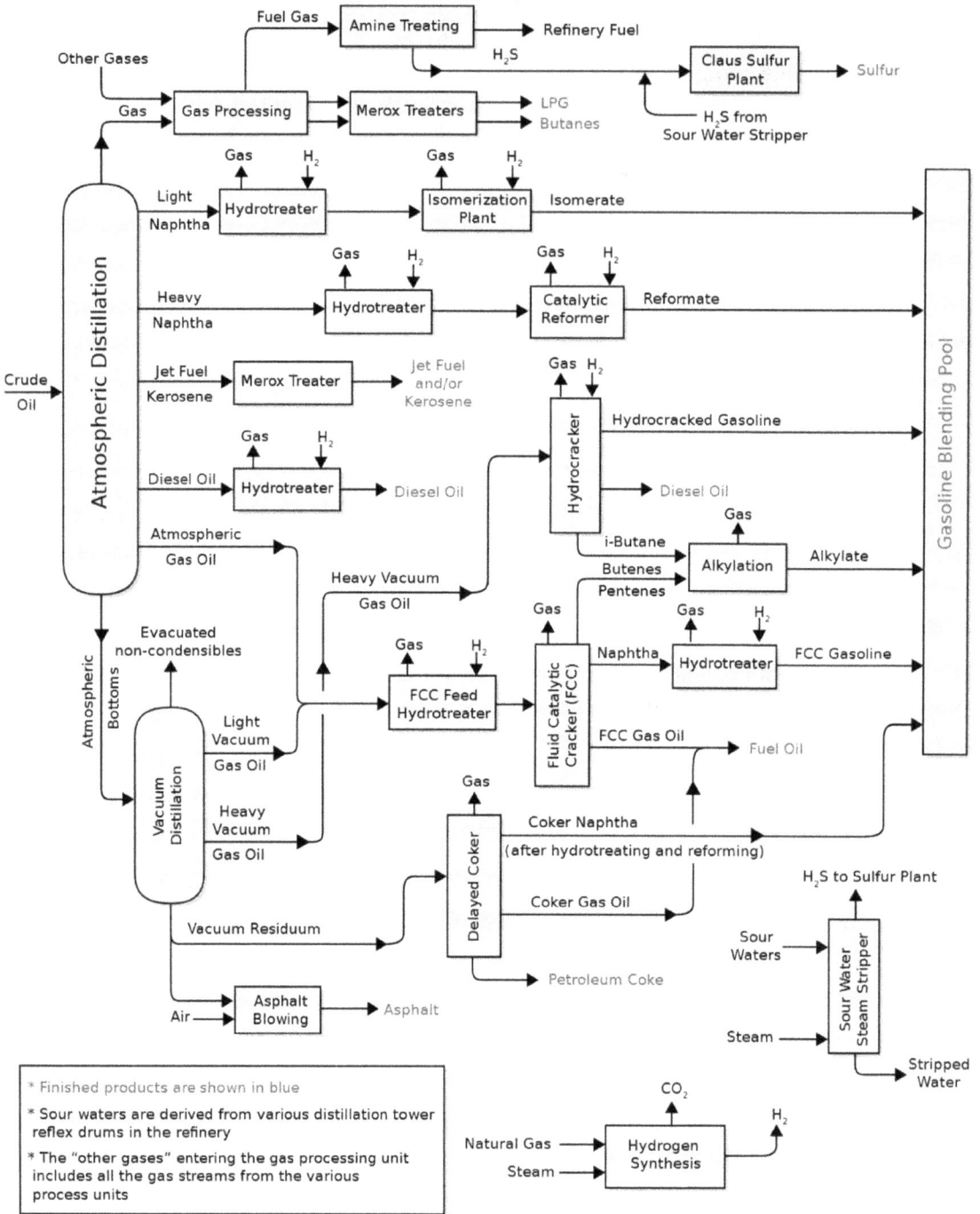

Figure 20: Schematic process flow diagram of the processes used in a typical oil refinery. Mbeychokderivative work: Begoon, CC BY-SA 3.0, via Wikimedia Commons.

In summary, the multitude of chemical processes in a refinery work in concert to maximize the value extracted from every barrel of crude oil. They enable the production of clean-burning, high-performance fuels and petrochemicals, while complying with increasingly stringent environmental standards.

Emissions and Environmental Impact

The environmental impact of oil refining is a significant concern due to its multi-faceted nature, contributing to air emissions, water pollution, waste generation, and high energy consumption. Refineries, as energy- and emissions-intensive industrial facilities, face considerable challenges in managing and reducing these environmental impacts, which have essential regulatory, technical, and financial implications.

Air quality degradation from oil refining is primarily due to emissions of greenhouse gases (GHGs) and various hazardous air pollutants. The International Energy Agency (IEA) reports that oil refineries emitted over 1 billion tonnes of CO_2 in 2022, contributing to approximately 5% of global industrial emissions [286]. Major sources of CO_2 emissions include the combustion of fossil fuels for heat and steam, steam methane reforming for hydrogen production, and flaring of excess gases, which further contributes to volatile organic compounds (VOCs) and particulate matter emissions [286, 287]. VOCs, released during storage and transfer processes, are precursors to ground-level ozone, exacerbating urban air quality issues [288]. Additionally, refineries release sulphur oxides (SO_x) and nitrogen oxides (NO_x) during combustion processes, which can lead to acid rain and respiratory health issues, thus creating significant public health concerns [272].

Particulate matter (PM), emitted during combustion and flaring, poses serious health risks when inhaled and is a significant contributor to deteriorating air quality [288]. These emissions reflect the extensive operational processes and inefficiencies that modern refining technologies still grapple with, despite advancements in emission control technologies [287, 289].

The refining process consumes large volumes of water, which, coupled with wastewater generation, can lead to significant water pollution. Wastewater from refineries may contain oil and grease, heavy metals such as lead and nickel, and other hazardous substances that can disrupt aquatic ecosystems and harm water quality when discharged improperly [290]. The treatment processes, such as API oil-water separators and biological treatments, aim to mitigate these effects, but challenges remain in fully eliminating harmful discharges [272].

Refineries also generate substantial amounts of solid and hazardous waste, including sludges from wastewater treatment processes, spent catalysts containing hazardous metals, and petroleum coke [291]. These wastes require careful management to avoid soil contamination and groundwater pollution, leading to potential health hazards in surrounding communities [292]. Effective hazardous waste management practices are critical; without such measures, improper disposal can exacerbate environmental degradation and pose long-term risks to human health [292].

Refining operations are notably energy-intensive, accounting for a significant portion of operational costs [289]. Distillation columns, hydrocrackers, and steam systems are identified as major energy consumers within the refinery [293]. However, innovations in energy efficiency, such as waste heat recovery systems, have shown promise in mitigating energy consumption and lowering environmental footprints [294]. For example, the adaptation of organic Rankine cycle systems can harness waste heat, substantially reducing overall energy needs [294, 295].

Flaring, necessary for safety during process upsets, inadvertently produces CO_2 and VOCs, highlighting the need for better management practices, such as flare gas recovery systems [287, 289]. Fugitive emissions from equipment leaks contribute significantly to GHG emissions and require rigorous leak detection and repair programs to minimize their impact on climate change [289].

The operation of refineries is governed by stringent environmental regulations, such as the U.S. Clean Air Act and the EU Industrial Emissions Directive, which mandate compliance through continuous emissions monitoring and pollutant reduction targets [296]. These regulations are essential for guiding the industry toward a more sustainable operational model and ensuring that refineries invest in cleaner technologies [289].

Efforts to mitigate the environmental impact of refining include the implementation of sulphur recovery units, hydrogen recovery systems, and advanced process control technologies to optimize energy and water use [296]. Many refineries are also exploring the transition to "green refineries" that integrate renewable energy sources to further reduce their carbon footprint [289, 296].

Oil refineries are indispensable to modern industrial society, but they pose substantial environmental challenges. Air emissions, water usage, hazardous waste, and energy demand all contribute to the environmental footprint. While modern refineries are more efficient and cleaner than those of the past, the refining industry remains under pressure to decarbonize, reduce pollutants, and improve sustainability. Continuous innovation, strict regulatory enforcement, and investment in clean technologies are essential for reducing the environmental impact of refining operations.

Part 3

Fuel Use Around the World

Chapter 7

Transportation

Having examined how fuels are extracted, refined, and categorised, the book now turns to one of their most vital applications: transportation. Chapter 7 explores the central role fuels play in powering the global movement of people and goods—across land, sea, and air. Transportation is not only one of the largest consumers of fuel worldwide, but also a major contributor to greenhouse gas emissions. This chapter delves into the types of fuels used in road vehicles, aviation, rail, and maritime sectors, comparing their efficiencies, environmental impacts, and ongoing innovations. It also discusses the growing push for decarbonisation in the transport sector through the introduction of electric vehicles, alternative fuels like biofuels and hydrogen, and shifts in policy and infrastructure. As the world moves toward net-zero goals, transforming how transportation is fuelled will be a key part of the energy transition journey outlined throughout this book.

Transport as a Major User of Fuel

Transportation is a vital part of modern life, providing access to goods, services, employment, and education. It includes multiple modes—such as road, rail, maritime, aviation, and pipelines—which collectively support passenger mobility and freight logistics. Today, the global transportation sector is predominantly reliant on oil, with over 90% of all transport energy derived from petroleum products. Transportation alone accounts for nearly two-thirds of global oil consumption and approximately 16% of global greenhouse gas (GHG) emissions. The impacts of transport-related emissions are not just environmental; they also have profound public health and social equity implications. Low-income and marginalized communities tend to experience higher exposure to air pollutants from transport, reinforcing environmental injustice [297].

The largest share of energy use in transportation comes from short-distance travel, particularly by cars, light trucks, and motorcycles. These vehicles use mainly gasoline—about 97% of the global light vehicle fleet relies on it—and are highly inefficient, with less than 1% of the fuel energy used to move the driver. This inefficiency makes the electrification of personal vehicles a key priority for global decarbonization efforts. Transitioning to electric vehicles (EVs), supported by renewable electricity, offers a major opportunity to reduce both emissions and energy waste. Global EV sales are rising, especially in countries like Norway and Iceland, and the worldwide EV fleet has grown by over 600% between 2018 and 2023. However, EV penetration still remains relatively low globally, with EVs making up only 3% of all registered cars in 2023 [297].

On the other hand, long-distance travel via aviation, maritime shipping, and long-haul trucking is more difficult to decarbonize. These modes require high energy densities and infrastructure that is currently built around fossil fuels. Aviation and shipping, in particular, rely almost exclusively on fossil fuels, including jet kerosene, heavy fuel oil, marine diesel, and gas oil. Even rail, which is partly electrified in some regions, remains about 52% fossil-fuelled globally—mainly via diesel-powered locomotives. Trucks and buses, which are responsible for a large share of freight transport, still run 96% on fossil fuels, with diesel dominating [297].

Globally, transportation represents 28% of total final energy consumption, with the U.S. slightly higher at 30% of primary energy consumption. In the U.S., light-duty vehicles make up more than half of transportation energy use. The global transport sector is also energy-intensive, with passenger cars and airplanes among the most inefficient modes. For freight, air cargo is the most energy-intensive option, while rail and maritime shipping are more energy efficient [297].

Efforts to decarbonize transportation include improving urban design to reduce the need for travel, shifting freight and passenger loads to more efficient modes like rail or public transit, encouraging behavioural changes (e.g., biking or walking), and investing in alternative fuels and clean vehicle technologies. Electrification is central to this shift, especially for road vehicles, supported by the development of smart grids, charging infrastructure, and battery innovation. For aviation and maritime transport, sustainable fuels such as green hydrogen, methanol, and sustainable aviation fuels (SAFs) offer some promise, though they are still under development and currently expensive [297].

Policies play a crucial role in enabling this transition. Measures like carbon pricing, fuel efficiency standards, public investments in clean infrastructure, and incentives for EV purchases can accelerate decarbonization. However, significant barriers remain. Fossil fuels are often underpriced and do not reflect their true environmental and social costs. The lack of global coordination on fuel standards, the need for major infrastructure overhauls, and concerns about battery production—including material sourcing, cost, and environmental impacts—are major obstacles. Additionally, some transport modes like long-haul aviation and heavy trucking are especially hard to electrify due to weight and range constraints [297].

Overall, the transport sector is a major contributor to climate change and environmental degradation but also represents one of the most promising areas for clean energy transition. Decarbonizing transport requires integrated strategies that span vehicle technologies, urban planning, fuel innovation, and global policy alignment. If successful, these efforts can reduce pollution, improve public health, support energy security, and promote a more equitable and sustainable mobility future [297].

Road Vehicles: Gasoline and Diesel

The road transportation sector is a significant user of fossil fuels, particularly gasoline and diesel, powering billions of internal combustion engine (ICE) vehicles worldwide. This fuel mix displays substantial regional variations in consumption patterns based on local infrastructure, economic conditions, and regulatory frameworks.

Global Fuel Use Patterns

Gasoline generally serves as the main fuel for light-duty vehicles across many regions, especially in North America and parts of Asia and Latin America. It is preferred due to its higher energy density and favourable combustion properties for spark-ignition engines [298]. Diesel, conversely, is the fuel of choice for heavy-duty vehicles due to its superior fuel efficiency and energy content, making it ideal for freight transportation. According to the International Energy Agency (IEA), road transport accounts for approximately 63% of global oil demand in the transport sector, with 58% being diesel and 42% gasoline [299].

Regional Trends and Differences

In the United States, gasoline predominates in the light-duty vehicle market, while diesel fuels heavy-duty applications. The historical dominance of diesel in regions such as Europe has been influenced by tax incentives and fuel efficiency advantages. However, these trends have recently reversed due to regulatory crackdowns on emissions following incidents like the Dieselgate scandal [300].

The Dieselgate scandal, or the Volkswagen emissions scandal, erupted in September 2015 when the U.S. Environmental Protection Agency (EPA) unveiled that Volkswagen Group had embedded "defeat devices" in the software of its diesel vehicles. These devices were designed to disable emissions controls under normal driving conditions, allowing emissions of nitrogen oxides (NOx) to breach permissible levels by as much as 40 times the legal limit established in the United States. Volkswagen had marketed these vehicles under the "Clean Diesel" label, promoting them as environmentally friendly products aligned with contemporary ecological values [301, 302].

The breach of trust and deceptive practices by Volkswagen extended globally, impacting approximately 11 million vehicles worldwide. This included various brands within the Volkswagen Group, such as Audi, Porsche, Škoda, and SEAT. The scandal predominantly affected cars sold in Europe and North America, particularly those equipped with the EA189 diesel engines [303, 304]. The ramifications of Dieselgate have been significant, resulting in legal and financial consequences for the company, which, by 2023, had incurred costs exceeding $35 billion in penalties, settlements, and vehicle buybacks across multiple jurisdictions [305]. High-profile arrests followed, including that of former CEO Martin Winterkorn, who was indicted in the U.S. and charged in Germany [306].

Studies have quantified the health implications associated with the excess NOx emissions from Volkswagen vehicles. Research indicated that these emissions may have resulted in fatalities estimated to range from 5 to 50 in the United States, along with substantial economic burdens due to health-related costs linked to the pollution [307, 308]. The scandal not only devastated Volkswagen's reputation but also significantly eroded consumer confidence in diesel vehicles, inciting broader scrutiny of the automotive industry regarding compliance with emissions standards [309]. The aftermath led to reforms in vehicle emissions testing, notably with the introduction of the Worldwide Harmonized Light Vehicles Test Procedure (WLTP) in the European Union, which aimed to ensure stricter adherence to emission regulations [310].

Moreover, the incident highlighted systemic issues concerning corporate governance, ethical compliance, consumer protection, and regulatory frameworks globally [311, 312]. The scandal prompted crucial discussions regarding the need for robust regulatory enforcement and improved corporate accountability in light of escalating environmental concerns [309, 313]. The dialogue arising from Dieselgate underscores vulnerabilities in regulatory systems and corporate cultures that prioritize profit over ethical standards, marking a significant turning point for sustainable transportation practices [314, 315].

The Volkswagen Dieselgate scandal exemplifies the complex interplay between corporate ethics, consumer trust, environmental responsibility, and regulatory compliance. It serves as a stark reminder of the far-reaching consequences that corporate misconduct can have on public health, environmental sustainability, and market competition integrity [316, 317]. The scandal has notably impacted the automotive industry's trajectory, leading to an increased emphasis on electric vehicles and alternative propulsion technologies as viable solutions to combat environmental degradation.

In the European Union, diesel and gasoline usage characterize the energy landscape of road transport. Data from 2022 show that diesel accounted for 65.4% of the energy consumed, while motor gasoline represented 25.2%, leaving alternative fuels, including renewables and biofuels, with a minor contribution of about 6.4% [318]. The reliance on petroleum fuels can be attributed to established transport infrastructures and consumer preferences, despite ongoing efforts to shift toward sustainable practices such as electrification [319].

The Asia-Pacific region, particularly within APEC economies, shows even higher proportions of energy consumption from road transport, estimated at 86%. Oil products dominate the energy mix, making up approximately 89% of total energy use in the transport sector [320, 321]. With rapid economic growth, this region reflects urgent policy action needs to mitigate environmental impacts consequential to fossil fuel dependence [318, 322].

In China, gasoline fuels the majority of passenger vehicles, while diesel is widely used in commercial fleets. The country has also begun to embrace electric vehicles (EVs), which may reshape future fuel consumption patterns [323].

India exhibits a distinct scenario where diesel is extensively used in commercial transport, particularly among trucks and buses, although policy shifts are gradually promoting gasoline usage among personal vehicles [324]. Meanwhile, Latin America and Africa show varied patterns; for example, countries like Brazil incorporate ethanol in gasoline, while others rely heavily on diesel for public transport [325]. Southeast Asia features a mix of gasoline for personal vehicles and diesel for freight transport, with emerging biofuel policies indicating a potential future shift in fuel reliance [326].

In Turkey, as of 2017, diesel fuels represented the largest segment of fuel consumption in road transport, accounting for 77.79% of the total, followed by auto gas (LPG) at 13.35% and gasoline at 8.86% [327]. These patterns emphasize a strong focus on diesel fuel in transportation, indicating the necessity for energy efficiency improvements and diversification of fuel sources for sustainable development.

Environmental Impact and Transition

The environmental ramifications of gasoline and diesel usage are significant, particularly regarding greenhouse gas emissions and local air pollutants, including NOx and particulate matter, which are notably higher in diesel engines [328]. Strategies to mitigate these impacts are gaining traction, including the adoption of cleaner combustion technologies and promoting other energy alternatives. Dual-fuel systems that blend gasoline and diesel aim to capitalize on the benefits of both fuels while reducing emissions [329]. Moreover, the global push for cleaner technologies, such as electric vehicles and biofuels, may significantly alter the landscape of road transport, as seen in various innovations and regulatory changes [330].

Heavy Vehicle Road Transport

Diesel is the preferred fuel choice for trucks around the world due to a combination of energy efficiency, torque characteristics, durability, and economic factors that make it well-suited for the heavy demands of freight transport.

One of the main reasons is diesel's high energy density. Diesel fuel contains more energy per litre than gasoline, which means trucks can travel longer distances on a single tank. This makes it more efficient for long-haul transport, reducing refuelling frequency and downtime. Diesel engines also have better fuel economy, typically offering 20–30% more miles per gallon compared to equivalent gasoline engines, which is crucial for businesses trying to minimise operational costs in logistics and distribution.

Another key advantage is torque output. Diesel engines generate more low-end torque than gasoline engines, which is essential for moving heavy loads. High torque at low engine speeds improves pulling power, making diesel ideal for applications such as towing trailers, climbing hills, or starting from a stop under heavy load—common requirements for freight and cargo transport.

Figure 21: An MGM Kenworth C509 prime mover with a 60-metre A-quad combination in Port Hedland, Western Australia. SquiddyFish, CC BY 4.0, via Wikimedia Commons.

Durability and engine longevity are also major factors. Diesel engines are built to withstand higher compression ratios and more rugged conditions, giving them a longer lifespan than gasoline engines. Many diesel truck engines can run for over a million kilometres before requiring a major overhaul, which makes them cost-effective in the long run despite higher initial purchase costs.

Economically, diesel infrastructure is well-established, with fuelling stations, maintenance knowledge, and supply chains all geared toward supporting diesel-powered vehicles. In addition, diesel engines tend to require less frequent maintenance than gasoline engines, further contributing to their popularity in commercial transport fleets.

In short, diesel remains the dominant fuel for trucks because it combines superior energy efficiency, load-handling capability, long-term reliability, and cost-effectiveness—qualities that align well with the operational demands of the global freight and logistics industry.

Converting the global truck fleet from diesel to alternative fuels such as electricity, hydrogen, natural gas, or synthetic/e-fuels represents a significant logistical and financial challenge, likely costing trillions of dollars worldwide. The total cost depends greatly on the type of fuel chosen, whether the trucks are replaced entirely or retrofitted, the extent of infrastructure upgrades required, and regional economic conditions.

Electric trucks are currently two to three times more expensive than diesel trucks. A heavy-duty Class 8 electric truck like the Tesla Semi is estimated to cost between $180,000 and $200,000, while a diesel equivalent might cost $120,000 to $150,000. With more than 50 million commercial trucks operating globally across light, medium, and heavy-duty categories, converting the entire fleet to electric would cost approximately $7.5 trillion. This estimate includes vehicles and charging equipment. Additional infrastructure such as charging stations, electrical grid enhancements, and depot rewiring could require another $500 billion to $1 trillion. Beyond cost, electric trucks face challenges such as limited range, lengthy recharge times, heavy battery weights, and increased strain on electricity grids.

Hydrogen fuel cell trucks are even more expensive, currently priced at $300,000 or more. If economies of scale bring future costs down to $250,000 per vehicle, converting 50 million trucks would still amount to around $12.5 trillion. Unlike electricity, hydrogen infrastructure is still in its infancy. Building an adequate global network of hydrogen refuelling stations, along with production facilities and storage systems, could cost an additional $1 to $2 trillion. High fuel costs, complex storage needs, and a lack of widespread refuelling infrastructure further complicate the transition to hydrogen-powered trucks.

Natural gas-powered trucks, using compressed natural gas (CNG) or liquefied natural gas (LNG), are relatively more affordable than electric or hydrogen alternatives. These trucks typically cost $30,000 to $50,000 more than diesel models. Retrofitting existing vehicles may cost between $20,000 and $40,000 per truck, but newer replacements are often preferred. On average, converting the fleet at $50,000 per unit would total about $2.5 trillion. Infrastructure development for natural gas refuelling is moderately advanced, with an estimated global build-out cost of $100 to $300 billion. However, environmental concerns over methane leakage and reliance on fossil fuels (unless using biogas) present limitations.

Fuel, Energy and Net Zero

Synthetic and e-fuels offer the advantage of compatibility with existing diesel engines, requiring only minimal modifications to vehicles. The challenge lies in the high cost of production. Currently, e-fuels cost between $3 and $5 per litre, compared to about $1 per litre for conventional diesel. Trucking consumes roughly 17 to 20 million barrels of diesel per day, or 3.2 to 3.6 billion litres. A $2 per litre price premium could result in an additional $2 billion per day in operating costs, translating to an annual burden of approximately $730 billion. Although e-fuels require little change to current distribution infrastructure, their widespread use would demand vast amounts of renewable electricity, posing significant feasibility and sustainability concerns.

Each of these alternatives brings a unique set of financial, technical, and environmental considerations. While synthetic fuels offer the lowest transition cost in terms of vehicle compatibility and infrastructure, they come with a high ongoing fuel cost. Electric and hydrogen vehicles promise lower emissions but require substantial investment in both vehicles and energy systems. Natural gas presents a middle ground but faces scrutiny over its environmental footprint. Ultimately, a global transition away from diesel trucks will likely require a combination of these technologies tailored to specific regional conditions and long-term climate goals.

Table 2: Summary Table (approximate, for 50 million trucks).

Fuel Type	Vehicle Conversion Cost	Infrastructure Cost	Annual Fuel Cost Premium	Key Challenge
Electric (BEV)	$7.5 trillion	$0.5–1 trillion	Lower operating cost	Range, charging, grid
Hydrogen	$12.5 trillion	$1–2 trillion	High (current prices)	Fuel cost, infra
Natural Gas	$2.5 trillion	$0.1–0.3 trillion	Slightly lower	Methane leaks
Synthetic Fuels	Minimal	Minimal	$730 billion/year	Fuel production cost

The total cost of a global transition away from diesel trucks depends on the chosen fuel pathway, but could easily exceed $10 trillion over several decades. This cost must be weighed against the economic, environmental, and health benefits, including reduced greenhouse gas emissions, lower urban air pollution, improved energy security, and alignment with climate goals such as net-zero by 2050.

Governments and industry leaders are exploring mixed-fuel strategies—deploying battery-electric trucks for urban and regional routes, hydrogen for long-haul freight, and synthetic fuels where replacement is impractical.

Outlook

Despite advances in electric vehicles and alternative energy sources, gasoline and diesel will likely continue to dominate the global vehicle fleet in the near term, especially in developing countries where economic growth remains strongly linked to fossil fuel consumption [331]. The IEA forecasts that oil demand from road transport may peak in developed nations before 2030, while emerging markets may see sustained growth without stringent climate policies [332].

The diverse global landscape of vehicle fuel usage is shaped by a combination of historical trends, regional preferences, and evolving technological advancements. While efforts to transition toward more sustainable energy sources are underway, the significance of gasoline and diesel in current transportation remains paramount.

Aviation Fuels

Aviation fuels, specifically formulated for aircraft, are primarily petroleum-based and designed to meet rigorous safety and performance specifications that differ significantly from those used in ground vehicles. These specialized fuels must exhibit unique properties and include specific additives to ensure efficient operation under the demanding conditions encountered at high altitudes and temperatures. The properties required for aviation fuels include high thermal stability, appropriate flash and freezing points, as well as detonation resistance, which are crucial for ensuring reliable engine performance and safety [333-335].

Among the key categories of aviation fuels, Jet Fuel and Aviation Gasoline (Avgas) are prominent. Jet Fuel, mainly kerosene-based, is utilized in turbine-engine aircraft such as turboprops and jets, with common varieties being Jet A, Jet A-1, and Jet B [335, 336]. The physical characteristics of Jet A-1 make it particularly suitable for high-performance engines, reflecting a high density and energy content per unit volume which enhances engine performance [337, 338]. Conversely, Aviation Gasoline, or Avgas, is typically used in piston-engine aircraft, especially smaller models. The high-octane ratings of Avgas provide the required performance for general aviation applications including training and recreational flying [333, 339].

Jet fuel is the most widely used type of aviation fuel, powering commercial airliners, military jets, and many private aircraft with turbine engines.

Jet A and Jet A-1 are the most common commercial jet fuels:

- Jet A is used primarily in the United States and has a freezing point of −40°C.

- Jet A-1 is used internationally and has a slightly lower freezing point of −47°C, making it more suitable for long-haul international flights.

- Both are kerosene-based fuels refined from crude oil and contain additives to prevent icing and static buildup.

JP fuels are military-grade jet fuels:

- JP-5 is a high-flash-point kerosene-based fuel used on aircraft carriers.

- JP-8 is similar to Jet A-1 but includes corrosion inhibitors and anti-icing additives for military aircraft.

Avgas is used in piston-engine aircraft, typically smaller private planes and training aircraft. It is a high-octane fuel designed for engines that require precise combustion characteristics:

- Avgas 100LL (Low Lead) is the most common type, despite still containing small amounts of tetraethyl lead to increase octane rating and prevent knocking.

- Avgas is more volatile and flammable than jet fuel and requires specialized storage and handling.

In recent years, Sustainable Aviation Fuels (SAF) have garnered significant attention as an innovative solution aimed at reducing emissions associated with traditional aviation fuels [340]. SAF can be produced from renewable feedstocks through various methods, including biofuel production from biomass [341]. The International Civil Aviation Organization (ICAO) has underscored the importance of SAFs in fostering sustainable practices within the aviation industry, highlighting their potential for decreased greenhouse gas emissions [342, 343].

Sustainable aviation fuels (SAFs) are a growing class of alternative jet fuels made from non-petroleum sources such as:

- Biomass (e.g., algae, used cooking oil, agricultural residues)

- Municipal solid waste

- Synthetic fuels created from captured CO_2 and hydrogen using renewable electricity (power-to-liquid fuels)

SAFs are chemically similar to conventional jet fuel and can be blended with Jet A-1 up to 50% without requiring modifications to aircraft or fuelling infrastructure. Their use reduces lifecycle carbon emissions significantly, sometimes by over 80%, depending on the feedstock and production method.

Emerging aviation fuels are being explored to reduce the carbon footprint of the aviation industry and offer sustainable alternatives to fossil-derived jet fuel. Among these, biofuels represent the most immediate and scalable solution. Produced through processes like biomass-to-liquid (BtL), sustainable aviation fuel (SAF) can be derived from a variety of feedstocks, including used cooking oil, agricultural waste, and non-food energy crops. Some straight vegetable oils have also been evaluated for use in aviation engines, although they are less common. One of the key advantages of SAF is its compatibility with existing aircraft technology. As long as the fuel meets strict specifications for lubricity, density, and chemical composition, it can often be used as a drop-in fuel with little or no modification to current aircraft engines and fuel systems. This compatibility extends to its ability to adequately swell elastomer seals in aircraft fuel lines, which is essential for safe and reliable operation.

Blended with conventional jet fuel or used independently under controlled conditions, SAF can deliver significantly lower emissions of particulate matter and greenhouse gases (GHGs). The lifecycle emissions of SAF can be up to 80% lower than conventional fossil-based fuels, depending on the feedstock and production process. Despite these benefits, widespread adoption remains limited due to economic, political, and technical challenges. SAF is still substantially more expensive than conventional jet fuel, often costing two to five times as much. Additionally, global supply chains and production capacity are still developing, and supportive policies, incentives, or mandates are not yet universally implemented.

Compressed natural gas (CNG) and liquified natural gas (LNG) are also being studied as future aviation fuel sources. While natural gas is more commonly associated with ground transport and power generation, its potential for aviation has been explored through experimental platforms. The Tupolev Tu-155, a Soviet-era aircraft, was a notable example of using LNG as a testbed fuel. More recently, Boeing's Subsonic Ultra Green Aircraft Research (SUGAR) team proposed the "SUGAR Freeze" concept under NASA's N+4 program, which examined natural gas use for next-generation aircraft. However, natural gas in both CNG and LNG form presents serious drawbacks in aviation due to its relatively low energy density, especially by volume. Even in liquified form, natural gas offers significantly less energy per unit volume than conventional jet fuel, making it less efficient for long-distance flights unless major breakthroughs in tank design and airframe integration are achieved.

Hydrogen, particularly when produced through electrolysis powered by renewable energy sources, offers a nearly carbon-free alternative. Hydrogen can be used in two primary ways: combusted directly in jet engines or converted into electricity through fuel cells to power electric propulsion systems. When used in fuel cells, hydrogen produces only water as a byproduct, whereas direct combustion does result in some nitrogen oxide (NOx) emissions due to high-temperature reactions with air. One of hydrogen's greatest advantages is its high energy content by weight, but its extremely low energy density by volume remains a significant design challenge. Hydrogen must be stored either as a cryogenic liquid below 20K or as a highly pressurized gas, typically at 250–350 bar. The storage tanks required to contain it safely at such pressures are currently much heavier than the hydrogen itself, offsetting many of its potential weight advantages. Furthermore, cryogenic handling and the integration of bulky tanks into aircraft structures demand new airframe designs, such as blended wing body configurations, which could accommodate larger internal volumes without significantly increasing aerodynamic drag.

While hydrogen shows long-term promise, it is not expected to enter commercial aviation use at scale for at least a couple of decades. Current research and prototype development are accelerating, especially since 2020, but the industry timeline for adoption remains lengthy. Therefore, in the near-to-mid term, the most feasible and impactful alternatives to fossil-based jet fuel are aviation biofuels and synthetic "e-jet" fuels, produced from green hydrogen and captured carbon dioxide. These fuels, collectively termed Sustainable Aviation Fuel (SAF), present the most viable pathway for decarbonizing aviation over the next 10–30 years while longer-term hydrogen solutions continue to mature.

Emerging aviation technologies include:

- Hydrogen as a combustion fuel or feedstock for fuel cells. It offers zero carbon emissions when burned or used in fuel cells, but it requires new aircraft designs and infrastructure due to its storage challenges.

- Electric propulsion, powered by batteries or hybrid systems, is being developed for short-haul and regional aircraft. Energy density of current batteries limits their use to smaller planes over shorter distances.

The specifications for aviation fuels are tightly regulated and are necessity-driven, given that varying flight conditions necessitate optimal fuel composition. Apart from the standard qualities like low volatility and high flashpoint for kerosene used in jet fuels, jet fuels also need to maintain cleanliness and withstand extreme operational environments [334, 335, 337]. Distributing and handling aviation fuels demands a robust supply chain to ensure the maintenance of cleanliness and adherence to safety protocols from the refinery to the aircraft [334, 344]. This involves stringent quality control measures to prevent contamination that could result in engine failures or performance issues [334, 345].

Global Consumption and Impact

Aviation fuel accounts for approximately 6–8% of global oil consumption and contributes around 2–3% of global CO_2 emissions. The sector's emissions are growing due to increasing global air travel, making decarbonization a priority. Transitioning to SAFs and alternative propulsion systems is central to international aviation climate strategies such as those outlined by ICAO (International Civil Aviation Organization) and IATA (International Air Transport Association).

The global patterns of aviation fuel use reflect a convergence of factors stemming from the demands of both commercial and military aviation. Notably, jet fuel, primarily Jet A and Jet A-1, continues to dominate global aircraft operations, accounting for over 85% of aviation fuel usage as of the early 2020s. This prevalence of jet fuel aligns with the significant consumption trends observed in long-haul international flights, which incur higher fuel usage per trip compared to shorter flights due to greater distances and heavier payloads [346]. Since the pandemic, commercial airlines experienced a sharp drop in fuel consumption, with approximately 300 billion litters consumed in 2019. Recovery is anticipated as global air traffic rebounds, though exact projections vary [347, 348].

On a regional scale, fuel consumption varies greatly. North America remains one of the largest consumers of aviation fuel due to its extensive air travel networks and high per capita travel rate [349]. Europe also maintains a significant share; however, its aviation fuel usage is shifting towards sustainable alternatives due to stringent regulatory frameworks aimed at reducing environmental impacts [350]. The Asia-Pacific region, led by nations like China and India, is identified as the fastest-growing area for aviation fuel demand, supported by rising economic conditions and infrastructural investments geared towards fostering air travel [351]. Conversely, regions like Africa and South America generally exhibit lower consumption rates, which are likely to increase as their economies develop and aeronautical infrastructures improve [349].

In military aviation, fuel usage accounts for approximately 5–10% of the total aviation fuel market, with entities such as the U.S. Department of Defense being significant consumers of jet fuel [349]. Conversely, general aviation, which predominantly utilizes Avgas—the fuel for smaller piston-engine aircraft—represents about 2-3% of total fuel consumption, underscoring the larger market share captured by commercial operations [352].

Seasonality and operational dynamics also influence aviation fuel consumption. Fuel usage peaks during summer and holiday travel seasons, with an increase in flight volumes reported, although advancements in aircraft design and materials have led to gradual improvements in fuel efficiency [353]. The average fuel consumption for commercial aviation is estimated at about 3–3.5 litres per 100 passenger-kilometres, depending on the type of aircraft and operational route [354].

A critical trend in aviation fuel use is the gradual introduction of Sustainable Aviation Fuels (SAFs), which currently comprise less than 0.2% of global aviation fuel consumption [355]. The International Air Transport Association (IATA) has set targets to increase SAF utilization as part of broader sustainability goals, supported by significant policy initiatives such as the EU's "RefuelEU Aviation" and tax incentives in the U.S. Inflation Reduction Act [351, 356]. The development of SAFs aims to mitigate the aviation sector's carbon footprint, which contributes notably to global $CO2$ emissions [356].

The landscape of global aviation fuel use is characterized by a blend of commercial and military requirements, heavily influenced by regional characteristics, economic dynamics, and environmental policies. The anticipated growth of aviation fuel consumption, along with the industry's pivot towards Sustainable Aviation Fuels, highlights the dual challenge of meeting energy needs while striving to reduce ecological impacts.

Converting the global aviation industry from fossil-based jet fuels to sustainable alternative fuels such as sustainable aviation fuels (SAFs), synthetic or e-fuels, biofuels, or hydrogen represents one of the most ambitious and technically complex undertakings in the decarbonisation of global transport. This transition would require massive investments in fuel production, aircraft redesign, airport infrastructure, and regulatory adaptation, with total estimated costs running into trillions of U.S. dollars over several decades.

The production of alternative aviation fuels is one of the most capital-intensive components of the transition. Sustainable aviation fuels (SAFs) currently cost between $2.50 and $6.00 per litre, which is significantly higher than the $0.70 to $1.00 per litre typically paid for fossil-based jet fuel. Today, SAFs account for less than 0.2% of the total global aviation fuel demand. To replace jet fuel entirely—estimated at around 300 billion litres per year pre-COVID—SAF production would need to increase approximately 1500-fold. To reach this scale by 2050, it is estimated that $1.5

trillion to $2.5 trillion in capital investment would be required to build the necessary production capacity.

Synthetic or e-fuels, made by combining green hydrogen and captured carbon dioxide through Fischer–Tropsch synthesis, offer another pathway. These fuels currently cost between $3 and $6 per litre, though costs may fall to $1.50 to $2.00 by 2040 as technologies mature and scale. However, their production requires an immense amount of renewable electricity—between 5,000 and 10,000 terawatt-hours annually on a global scale. Building out the infrastructure needed for electrolysis, carbon capture, and synthesis could cost an additional $3 to $4 trillion.

From the aircraft perspective, SAFs offer an advantage because they can be blended with conventional jet fuel and used in existing aircraft engines with no need for significant modifications. This makes SAF adoption relatively cost-effective, with only minor additional expenses for fuel certification, performance testing, and updates to airline management systems. In contrast, hydrogen-powered aircraft would require a complete redesign of the airframe, propulsion systems, and fuel tanks. The rollout of hydrogen aviation would also necessitate major airport infrastructure changes to support hydrogen production, liquefaction, storage, and refuelling. Hydrogen aircraft are expected to cost 20% to 50% more than conventional designs, and retrofitting or replacing the global commercial fleet could cost between $3 and $5 trillion over 30 years. Hydrogen-specific infrastructure at airports would add a further $1 to $2 trillion.

Upgrading airport and distribution infrastructure is another key cost consideration. For SAFs, this includes developing blending facilities, new storage tanks, and pipeline upgrades at major aviation hubs, which could collectively cost between $300 billion and $500 billion worldwide. Transitioning to hydrogen, however, would require building entirely new systems for cryogenic storage, handling, and fuelling, with estimated costs of $1 to $2 trillion depending on the extent of deployment.

Governments around the world will need to play a crucial role through policy incentives and carbon pricing to help bridge the cost gap between fossil fuels and sustainable alternatives. Financial support mechanisms such as subsidies, tax credits, and grant programs are likely to run into hundreds of billions of dollars globally over the coming decades. Carbon pricing systems like the European Union Emissions Trading Scheme (EU ETS) will also help drive competitiveness by increasing the cost of emissions-intensive fuels, but alone may not be sufficient to fund the transition. Many airlines will still require targeted financial assistance to manage the economic risks of adopting new fuel technologies.

In summary, the shift to sustainable aviation fuels and other non-fossil propulsion systems will demand coordinated global investment, innovation, and regulatory alignment. While SAFs offer a relatively low-barrier, near-term option, the longer-term potential of hydrogen and synthetic fuels will require far more systemic transformation—both technically and financially.

Table 3: Total Global Cost Estimate (by 2050).

Component	Estimated Cost (USD)
SAF/Synthetic fuel production	$1.5 – $4.0 trillion
Aircraft redesign (if hydrogen)	$3.0 – $5.0 trillion
Airport infrastructure	$0.3 – $2.0 trillion
Policy and carbon incentive costs	$0.5 – $1.0 trillion
Total	**$5.3 – $12.0 trillion**

Outlook

Global aviation fuel demand is expected to rise steadily through 2050, especially in Asia, Africa, and Latin America. While fuel efficiency and SAFs will play key roles in reducing emissions, overall consumption will remain substantial due to expanding air travel markets. Transitioning the aviation sector to low-carbon fuels will require major investment in fuel technology, infrastructure, and international cooperation.

Shipping and Marine Fuels

Shipping and marine fuels are a critical component of the global transportation and trade network, powering everything from small coastal vessels to massive container ships and oil tankers. Marine fuels—commonly referred to as "bunker fuels"—are primarily derived from crude oil and include a range of products such as heavy fuel oil (HFO), marine gas oil (MGO), and marine diesel oil (MDO). As the shipping industry moves toward decarbonization and compliance with environmental regulations, there is increasing attention on cleaner alternatives, including liquefied natural gas (LNG), methanol, ammonia, biofuels, and hydrogen-based fuels. The shipping industry is currently experiencing a pivotal shift towards more sustainable marine fuels, primarily due to mounting regulatory pressures and environmental concerns. Traditionally, heavy fuel oil (HFO), often referred to as bunker fuel, has been the dominant energy source for maritime propulsion. However, the International Maritime Organization (IMO) has implemented stringent regulations that necessitate a transition to low-sulphur fuels, with the sulphur content capped at 0.50% as of 2020 [263, 357]. These regulatory frameworks underline the industry's critical need to reduce emissions associated with ship operations.

Historically, the majority of ocean-going vessels have relied on heavy fuel oil (HFO), a residual product from crude oil refining. HFO is thick, tar-like, and relatively inexpensive, but it contains high levels of sulphur and other pollutants such as nitrogen oxides (NOx) and particulate matter. Marine diesel oil and marine gas oil are more refined and cleaner-burning but also more expensive. These fuels are used more commonly in smaller vessels, auxiliary engines, and within regulated emission control areas (ECAs), where stricter environmental standards apply.

Alternative fuels are gaining prominence as the shipping sector seeks to meet sustainability goals and decrease the environmental impact of maritime operations. Liquefied natural gas (LNG) is emerging as a leading transitional fuel due to its significantly lower emissions of nitrogen oxides (NOx), sulphur oxides (SOx), and a reduction in CO2 emissions when compared to traditional fuels [358, 359]. Moreover, biofuels derived from organic materials, such as biodiesel and hydrotreated vegetable oils (HVO), are being explored for their potential to reduce greenhouse gas emissions. Studies indicate that biofuels can play an essential role in replacing fossil fuels and achieving long-term sustainability targets within the sector [359, 360].

Ammonia and hydrogen represent the frontier of zero-emission fuels, garnering increased attention for their potential to reduce carbon footprints. Nonetheless, challenges persist regarding their storage, infrastructure development, and the conversion technology required to utilize these fuels effectively on a large scale [361, 362]. Methanol is also being investigated as a feasible alternative due to its compatibility with existing fuel systems and its environmental benefits [363, 364].

Alternative and Emerging Marine Fuels

To meet environmental targets, various alternative fuels are being tested and deployed:

- **Liquefied Natural Gas (LNG):** LNG is one of the most widely adopted alternatives today, offering a reduction in CO_2 emissions by about 20%, and near elimination of SOx and particulate emissions. However, methane slip—a phenomenon where unburned methane escapes into the atmosphere—poses a concern due to methane's high global warming potential.
- **Methanol:** Methanol is gaining attention as a marine fuel due to its ease of storage and lower emissions profile. It can be made from fossil sources (gray methanol), biomass (bio-methanol), or renewable electricity and CO_2 (e-methanol). It burns cleaner than oil-based fuels and is easier to handle than LNG, although its energy density is lower.
- **Ammonia:** Ammonia is a zero-carbon fuel when produced from renewable sources. It contains no carbon atoms and therefore emits no CO_2 when combusted. However, burning ammonia produces nitrogen oxides and requires careful handling due to its toxicity. Research and pilot projects are ongoing to assess its viability in large-scale shipping.
- **Biofuels:** Bio-based marine fuels are derived from organic matter, such as waste oils, algae, or lignocellulosic biomass. They can be used as drop-in replacements in existing engines and infrastructure with little modification. Biofuels offer lifecycle emission reductions but face scalability and sustainability challenges, particularly concerning feedstock availability and land use.
- **Hydrogen:** Hydrogen can be used in combustion engines or fuel cells, producing only water as a byproduct. It offers excellent emissions performance but has a low volumetric

energy density and requires cryogenic or high-pressure storage. As with ammonia, hydrogen-powered vessels require extensive redesign and safety adaptations.

The focus on sustainability is further amplified by the IMO's targets, which aim for a 40% reduction in carbon intensity relative to 2008 levels by 2030 and a 70% reduction by 2050 [359, 365]. This pressure not only promotes alternative fuel adoption but also fosters innovation within the industry, leading to substantial opportunities for technological advancements and efficiency improvements [362, 366]. However, the path toward widespread alternative fuel adoption is fraught with challenges, including the necessity of creating supportive infrastructure and the economic implications of transitioning from conventional fuels [357, 367, 368].

The ongoing transition within the maritime industry towards more sustainable marine fuels is driven by stringent regulations and significant environmental challenges. While traditional heavy fuel oil continues to dominate, rising alternatives like LNG, biofuels, ammonia, hydrogen, and methanol are redefining the landscape of maritime fuel use. The industry's commitment to reducing greenhouse gas emissions is reshaping its operational frameworks, albeit accompanied by notable challenges in infrastructure, economics, and technology development.

Regulation and Environmental Concerns

The International Maritime Organization (IMO) has played a key role in regulating marine fuel emissions. In January 2020, the IMO implemented a global sulphur cap that reduced the permissible sulphur content in marine fuel from 3.5% to 0.5%. In Emission Control Areas such as parts of North America, Northern Europe, and the Mediterranean, even stricter limits apply—requiring sulphur content below 0.1%. These changes have led to a shift toward very low sulphur fuel oils (VLSFO), marine gas oils, and scrubber installations on ships to clean exhaust emissions.

The shipping industry is also a significant contributor to global greenhouse gas emissions, accounting for about 2–3% of global CO_2 emissions. This has spurred a broader push toward decarbonization, including IMO targets to reduce total annual GHG emissions from international shipping by at least 50% by 2050 (compared to 2008 levels), with increasing calls for net-zero emissions by the end of the century.

Global Consumption of Shipping and Marine Fuels

Global shipping constitutes a fundamental component of international trade, with around 90% of global goods transported via maritime routes. This vast industry heavily relies on marine fuels, consuming over 200 million tonnes annually as reported by the International Maritime Organization (IMO) [359, 369]. The principal marine fuels employed include heavy fuel oil (HFO), very low sulphur fuel oil (VLSFO), marine diesel oil (MDO), and marine gas oil (MGO) [358]. Despite increasing interest in alternative fuels such as liquefied natural gas (LNG), methanol, and biofuels, these options currently account for a minor portion of total fuel consumption in the maritime sector [370, 371].

The IMO's Data Collection System (DCS) provides crucial insights into fuel usage by different ship categories. In 2023, container ships, bulk carriers, and tankers represented the majority of fuel consumption [372]. Notably, these types of vessels predominantly utilize HFO and VLSFO. The latter has seen a surge in usage following the implementation of the IMO 2020 sulphur cap, which restricts the sulphur content in marine fuels to a maximum of 0.5% [373]. This regulatory shift has been pivotal in addressing the environmental repercussions associated with high sulphur emissions from traditional fuel oils.

The environmental impact of marine fuel consumption is substantial. In 2022, international shipping was responsible for approximately 2% of global energy-related CO_2 emissions, alongside significant contributions to air pollutants including nitrogen oxides (NO_x) and sulphur oxides (SO_x), which pose serious risks to human health and the environment [374]. The maritime shipping industry is reported to produce around 3% of global CO_2 emissions, 15% of NO_x emissions, and 16% of SO_x emissions, underscoring the urgency for more sustainable practices within the sector [359, 369].

The path towards decarbonization is crucial for the maritime sector, with the IMO establishing ambitious targets, aiming for a reduction of at least 50% in greenhouse gas (GHG) emissions by 2050 compared to 2008 levels [375]. The adoption of alternative fuels is essential in this context, as they present potential pathways to mitigate the harmful emissions traditionally associated with marine fuels [376]. However, the transition remains complex and requires continued investment in innovative technologies and regulatory frameworks to enhance the efficiency and sustainability of maritime shipping operations.

Infrastructure and Costs

Transitioning to alternative marine fuels involves major infrastructure changes. Ports must be equipped with refuelling systems for LNG, methanol, hydrogen, or ammonia. Shipbuilders and retrofitters face complex technical requirements, and operators must manage higher fuel and capital costs. For example, LNG-capable ships are more expensive to build and require special cryogenic tanks. The cost of alternative fuels remains a barrier, with prices for green ammonia or e-methanol far exceeding those of conventional fuels.

Future Outlook

The decarbonization of marine fuels is a long-term challenge but also a growing opportunity. Regulatory frameworks such as the EU's inclusion of maritime emissions in its Emissions Trading Scheme (ETS), carbon taxes, and fuel mandates are accelerating change. The shipping industry is also exploring hybrid propulsion systems, battery-electric coastal vessels, and wind-assisted technologies to reduce fuel consumption.

While heavy fuel oil still dominates the marine fuel market, the landscape is rapidly evolving. A combination of regulatory pressure, technological innovation, and environmental urgency is pushing the global shipping industry toward cleaner, more sustainable fuels. The future of marine propulsion will likely involve a diverse mix of fuels and technologies tailored to vessel types, routes, and regional regulations.

Rail and Alternative Powertrains

The energy landscape of rail transport is influenced by a variety of factors such as geography, infrastructure, and technological advancements. Traditionally, diesel fuel has been the primary power source for rail transport, especially in contexts where electrification is either economically unfeasible or geographically challenging. However, the rail sector is increasingly embracing electrification and alternative powertrains to reduce emissions and enhance energy efficiency.

Figure 22: Aurizon diesel locomotive 2322D in Queensland, 2018. Kgbo, CC BY-SA 4.0, via Wikimedia Commons.

Fuel, Energy and Net Zero

Historically, diesel locomotives have been instrumental in rail systems, primarily utilized in freight operations across North America, Australia, and certain parts of Asia, as well as in regional passenger services, especially in rural regions. Specifically, diesel-electric locomotives generate electricity through their diesel engines, which subsequently drives electric traction motors, enabling high levels of efficiency and significant torque. Nonetheless, this reliance on diesel contributes to greenhouse gas emissions and various local pollutants, including nitrogen oxides (NOx) and particulate matter (PM) [377]. The aging diesel fleet continues to operate extensively, accounting for approximately 85% of global railway mileage, particularly in freight operations where diesel traction remains prevalent [378].

In contrast, rail systems in regions such as Europe, Japan, and China predominantly feature electrified tracks, underpinned by overhead wires or third-rail systems. These electrified systems support high-speed rail transportation, urban transit networks, and mainline routes that favour energy efficiency. Moreover, electric trains offer environmental advantages, emitting zero tailpipe pollutants and benefiting from renewable energy sources, such as solar or wind-generated electricity that may power their operations [379]. This transition not only boosts operational efficiency but also aligns with global sustainability initiatives targeting reduced carbon footprints [380].

In response to both the limitations of diesel engines and the challenges of full-scale electrification, several innovative powertrains are emerging in rail transport. Hydrogen fuel cell trains, such as Alstom's Coradia iLint, and battery-electric trains represent promising technologies capable of operating on non-electrified routes while producing minimal emissions—primarily water vapor for hydrogen trains [381]. Meanwhile, hybrid trains, which integrate diesel engines with batteries, are being employed to reduce fuel consumption and emissions during short runs or shunting operations, showing more favourable emissions and efficiency profiles compared to conventional diesel-only locomotives [382].

Overall, the ongoing shift towards electrification and alternative propulsion systems within the rail industry is driven by national climate objectives and progressive emissions regulations. Technological improvements in battery design, hydrogen production, and the development of lightweight materials contribute to this transformation. For example, developments in battery-electric vehicles are seen as interim solutions that may pave the way for wider electrification [383]. Countries like Germany and China lead these advancements, while some regions, notably the U.S. and Australia, continue to rely heavily on diesel, potentially hindering substantial progress in emissions reduction [380, 384].

Fuel Consumption

Railway fuel consumption is influenced by various operational and technical factors, resulting in significant variability across different contexts. A benchmark from the research literature indicates that diesel locomotives consume approximately 9.4 litres of fuel per kilometre travelled, which provides a baseline for assessing fuel efficiency in railway operations [385].

However, as noted in various studies, actual fuel consumption varies based on factors such as train type, speed, load, track conditions, and operational practices [386].

The type of train plays a crucial role in determining fuel consumption. Diesel locomotives, relying solely on internal combustion engines, are generally less fuel-efficient compared to electric locomotives that draw power from overhead wires or a third rail. Electric trains not only consume less fuel but also operate with lower emissions, thus presenting a cleaner profile [387, 388]. Hybrid locomotives blend the advantages of diesel engines with battery systems and regenerative braking, showing improved efficiency compared to traditional diesel models [382, 389]. This adaptability can enhance operational performance, especially in varied rail environments [389].

Figure 23: Freight train of TRA pulled by the R167 diesel locomotive, between Erjie and Yilan. jason199567, CC BY-SA 2.0, via Wikimedia Commons.

Speed is another critical parameter affecting fuel consumption. Research indicates that increased speeds lead to higher aerodynamic drag, necessitating more energy for propulsion, which consequently escalates fuel consumption [390]. Conversely, operating trains at moderate speeds is associated with better fuel economy, as this reduces power demands and enhances overall system efficiency [391, 392]. Additionally, the load that a train carries directly impacts its energy requirements; heavier trains require more energy to accelerate or maintain momentum, thereby increasing fuel usage [392, 393].

Track conditions further complicate fuel consumption metrics. Well-maintained rail infrastructure reduces friction and resistance, facilitating smoother train operations and improving fuel efficiency. In contrast, poorly maintained tracks contribute to higher levels of vibration and resistive forces, leading to increased energy demands [394, 395]. Operational practices, including driver behaviour, idling times, and control system implementations, also influence fuel efficiency. Evidence indicates that disciplined driving habits, such as gradual acceleration and strategic braking, can significantly lower fuel consumption [396, 397].

Fuel consumption in the railway sector is typically quantified in litres per kilometre, allowing for straightforward comparisons of energy use across different locomotives and services. However, for comprehensive environmental assessments, metrics like kilograms of CO_2 equivalent per passenger-kilometre are employed [387, 398]. In freight transport, energy usage is often expressed in joules per tonne-kilometre, facilitating comparisons across various modes of transportation [392, 399].

Railway fuel consumption is determined by a complex interplay of mechanical, environmental, and operational factors. While benchmarks such as 9.4 L/km for diesel locomotives provide a useful reference point, actual consumption can significantly deviate based on operational context. Innovations in hybrid technology, increasing electrification of railway networks, and enhanced operational strategies continue to offer promising avenues for optimizing fuel use and mitigating environmental impacts within the rail industry [398, 400].

Trends and Future Outlook

The rail industry is experiencing a gradual shift towards cleaner energy sources. The use of renewable energy to power rail networks is projected to double by 2030, reaching about 25% of energy consumption . Technological advancements, such as the development of hydrogen fuel cells and battery-electric trains, are also contributing to this transition.

The rail transport sector operates in a highly competitive environment where even small improvements in fuel efficiency or fluctuations in fuel costs can significantly impact profitability and competitiveness. Many energy-efficiency measures not only save fuel but also offer additional benefits, such as increased productivity and reduced maintenance costs. One key strategy is weight reduction. Traditionally, freight cars have been made of steel, but newer designs explore the use of lighter materials such as aluminium, composites, or plastics. Innovations in design, such as replacing mechanical systems with electronic fly-by-wire controls, can also help reduce the weight of rolling stock [401].

Double stacking is another efficiency measure where two high-cube containers are placed on top of one another, allowing trains to carry up to 40% more freight without requiring additional power. This method can ease congestion and improve network capacity. However, it is important to consider that increased weight and aerodynamic drag might offset some of the energy savings. In

Australia, double stacking is limited by lower axle-load restrictions and loading gauge constraints on certain routes. Speed and throttle management also play an important role in reducing energy use. Operating at lower speeds and minimising throttle application reduces drag and limits unnecessary braking, which helps conserve energy. These operating parameters are influenced by gradients, loading times, instructions from train controllers, and interactions with other trains [401].

Improving logistics through new electronic controls allows trains to run closer together, increasing track utilisation and throughput. Efficient freight handling—such as container switching across a horizontal platform instead of moving entire railcars—can drastically cut the time needed to reconfigure trains. This method can allow for an entire train to be reconstituted in under 15 minutes. Technologies such as driver assistance software, GPS receivers, and portable loggers further support fuel-efficient operations by advising drivers on optimal power use, such as gradual acceleration or coasting before deceleration [401].

Aerodynamic design is crucial for reducing energy loss from air resistance. Locomotives and railcars with improved streamlining, reduced gaps between cars, and covered wheel sets can lower drag significantly. In some cases, drag losses on intermodal cars may account for up to 30% of total energy consumption. Solutions such as streamlining the sides and underfloors, using air bags to close gaps, and installing wheel covers can reduce aerodynamic drag without needing to replace existing rolling stock. Software tools are also available to evaluate the aerodynamic efficiency of entire trains and optimise the arrangement of wagons [401].

Wheel and rail friction is another key factor. Excessive friction increases energy use and wear, especially on curved tracks. Lubrication systems, applied either from wayside equipment or onboard devices, can reduce energy use and noise, and prolong track and wheel life. Braking technologies have also evolved, with electronically controlled pneumatic (ECP) braking allowing simultaneous braking across all cars in a train. This improves safety and enables the use of longer trains with heavier loads, especially beneficial on undulating terrain where inertia helps conserve power on inclines [401].

Anti-idling devices are another fuel-saving solution. These devices automatically shut down the engine during periods of inactivity and restart it when needed. Though regulations in some areas (such as Australian mainlines) restrict engine shutdowns, these devices can still be used effectively on private tracks. In terms of propulsion technologies, switching from DC to AC traction motors in locomotives offers better traction and lower maintenance due to fewer locomotives being needed for the same task. AC motors also enable faster service turnaround [401].

Hybrid locomotives combine batteries and diesel generators, using regenerative braking to store energy and improving overall fuel efficiency. This technology could be especially beneficial for freight routes involving long downhill segments, such as Australian mining railways heading to the coast. Research into hybrid applications continues, especially for line-haul operations [401].

Innovation is also driving solar and hydrogen-powered trains. The world's first 100% solar-powered train began operating in Byron Bay, New South Wales, in 2017 [401]. In Europe, companies are trialling solar panels on railway sleepers to generate electricity for overhead lines and local communities. Hydrogen fuel cell trains have already entered service in Germany, offering a clean alternative to diesel. These trains can run up to 1000 km on a single fill and emit only steam and water. The UK is retrofitting older trains with hydrogen systems to replace diesel units on non-electrified routes [401].

Hydrogen is particularly suited for regional routes due to its long range and flexible deployment. Although the upfront cost is high, proponents suggest the lower operating costs could recover the investment within a decade. AI-powered control systems also contribute to efficiency by learning optimal braking and power patterns in real time, coordinating locomotive performance, energy supply infrastructure, and operational strategies [401].

Figure 24: The world's first hydrogen train, the Alstom Coradia iLint, pictured at the Baie-Saint-Paul station along the Train de Charlevoix route during its North American demonstration service period between Montmorency Falls near Quebec City and Baie-Saint-Paul. Scarlett Kang, CC BY-SA 4.0, via Wikimedia Commons.

Finally, mode shifting—encouraging the movement of freight from road to rail—can substantially reduce transport-related energy use and emissions. With greater investments in intermodal infrastructure and digital integration, the rail sector has significant potential to offer cleaner and more efficient freight solutions across national and regional supply chains [401].

Transport Fuels and Economics

The relationship between a country's economy and the types of fuels used in its transportation systems is influenced by various economic factors such as income levels, industrial structure, energy infrastructure, government subsidies, and access to capital. High-income economies often have greater financial capacity to invest in cleaner and advanced fuels. For instance, in Norway, around 80% of new car sales are electric, with this transition supported by significant government incentives such as tax breaks, toll exemptions, and a largely hydroelectric power grid [402]. Germany has similarly implemented measures, including subsidies for electric vehicle (EV) adoption, showcasing how economic strength facilitates the shift from fossil fuels to more sustainable energy forms [402].

In contrast, low- and middle-income economies tend to prioritize affordability and reliability, which often results in a reliance on traditional fossil fuels such as diesel and petrol. Many regions in Africa and South Asia struggle with inadequate electrical grid capacity, making the development of widespread EV charging infrastructure challenging. In India, diesel continues to be favoured for long-haul freight and public buses due to its economic advantages, despite ongoing initiatives to promote compressed natural gas (CNG) to address urban air pollution [403]. Economic constraints and infrastructural limitations mean that investments in cleaner technologies, which typically require significant upfront costs, frequently remain unfeasible [403].

Oil-producing countries like Saudi Arabia and Russia display strong dependence on gasoline and diesel, supported by domestic production and government subsidies that lower fuel prices. Such subsidies have historically impeded the transition to cleaner fuels by diminishing the incentive for efficiency [403]. However, Saudi Arabia's recent diversification efforts indicate a push towards hydrogen and synthetic fuels to modernize its transport sector and reduce dependence on fossil fuels [403].

Resource-scarce but economically advanced nations like Japan and South Korea are pursuing alternative fuels as part of their energy security and climate strategies. Japan's focus on hydrogen fuel cell vehicles aims to reduce oil dependency, while South Korea combines EV incentives with national investments in battery technology, illustrating a proactive approach to achieving long-term sustainability amidst global energy challenges [403].

Furthermore, countries with robust logistics sectors, such as the United States and China, exemplify how economic interests drive fuel choice policies. The U.S. has traditionally relied on diesel and gasoline due to established infrastructure, yet current trends reflect a marked shift towards electrification, bolstered by increased subsidies and tax credits for urban transit and light-duty vehicles [402]. Meanwhile, China has become the largest EV market globally, employing a combination of industrial policies, consumer subsidies, and emissions regulations to rapidly deploy EVs while also exploring hydrogen technologies for heavy transport applications [404, 405].

Chapter 8

Electricity Generation

Having explored the sources and processes that convert raw fuels into usable forms of energy, the book now turns to one of the most critical pillars of modern civilisation: electricity generation. Electricity is not a fuel itself but rather a highly versatile energy carrier that underpins nearly every aspect of contemporary life—from lighting homes and powering industries to enabling digital communication and driving innovation. Chapter 8 builds on the foundational understanding of fuels developed in earlier chapters, particularly fossil fuels, renewable alternatives, and bioenergy, to examine how these energy sources are transformed into electricity. This chapter will explore the engineering principles, infrastructure, and technologies behind electricity generation, highlighting both historical and emerging systems. It also sets the stage for later discussions about energy storage, grid reliability, and decarbonisation, all of which are central to achieving a sustainable and net-zero energy future.

Fuel for Power Plants (Coal, Gas, Oil, Biomass)

Fuel used in power plants is central to how countries generate electricity, and it varies based on factors such as resource availability, technology, environmental policies, and economics. The four primary fuels used in thermal power generation are coal, natural gas, oil, and biomass. Each has distinct characteristics in terms of energy content, efficiency, environmental impact, infrastructure needs, and global distribution.

Coal has historically been the dominant fuel for electricity generation, especially in industrialized and emerging economies. It is an abundant and widely distributed fossil fuel, extracted from mines and transported by rail, barge, or conveyor to power stations.

In a coal-fired power plant, coal is pulverized into fine powder, then burned in a furnace to heat water in a boiler, producing steam. This steam drives a turbine connected to a generator, converting thermal energy into electricity. Modern coal plants can be subcritical, supercritical, or ultra-supercritical, with increasing levels of efficiency and pressure.

Coal provides a high energy output but has significant environmental drawbacks. It emits large amounts of carbon dioxide (CO_2)—a major greenhouse gas—along with particulate matter, sulphur dioxide (SO_2), and nitrogen oxides (NO_x), which contribute to air pollution and acid rain. Technologies like flue gas desulphurization, electrostatic precipitators, and carbon capture and storage (CCS) are used to reduce emissions.

Coal remains the primary source of electricity in countries such as India, China, South Africa, and parts of the United States, but many governments are phasing it out due to climate concerns.

Natural gas is increasingly used in power generation due to its lower emissions, higher thermal efficiency, and flexibility in balancing variable renewable energy sources. It is composed primarily of methane and is extracted through drilling, often with hydraulic fracturing (fracking), then transported by pipelines or liquefied for shipping.

Gas power plants come in two main types:

- **Open-cycle gas turbines (OCGT):** These are quick to start and suitable for peaking power, but less efficient.

- **Combined-cycle gas turbines (CCGT):** These use a gas turbine and a steam turbine in tandem, capturing waste heat to generate additional electricity and achieving efficiencies of up to 60%.

Natural gas emits about 50–60% less CO_2 than coal per unit of electricity generated. It also produces less SO_2 and particulates, although methane leakage during extraction and transport can reduce its climate advantage if not controlled.

Countries like the United States, Qatar, and Russia have large natural gas reserves and rely heavily on gas for electricity. It is also the preferred backup fuel in many regions with high shares of wind and solar energy.

Oil is less commonly used for power generation today due to high cost, volatile prices, and emissions intensity. However, it still plays a role in regions without access to alternative fuels or where energy security requires diversified sources.

Oil-fired power plants typically use diesel, heavy fuel oil (HFO), or residual fuel oils in boilers or turbines. These are used mostly for peak load, backup, or off-grid power in remote locations, islands, or developing nations.

Oil-fired generation is highly polluting, producing large quantities of CO_2, NO_x, and particulates. Due to the environmental impact and high fuel prices, most developed countries have phased out or significantly reduced oil use in electricity generation.

Nations like Saudi Arabia, Venezuela, and some Caribbean islands still maintain oil-fired plants, often linked to subsidized domestic oil production.

Biomass power generation uses organic materials such as wood chips, agricultural residues, food waste, and energy crops. When burned in a boiler, these materials generate heat, which produces steam to run turbines for electricity.

Biomass can be used in dedicated biomass plants or co-fired with coal in traditional coal plants to reduce emissions. It is considered renewable because the CO_2 released during combustion is partially offset by the carbon absorbed by the plants during their growth phase. However, sustainability depends on feedstock type, land use changes, and supply chain emissions.

Advanced biomass systems, such as anaerobic digestion, gasification, and pyrolysis, convert organic material into gas or liquid fuels that can be burned more efficiently.

Biomass plays an important role in countries like Sweden, Brazil, Finland, and the UK, especially where forestry or agricultural industries provide abundant feedstock. It is also promoted in developing countries for rural electrification, where waste can be turned into energy.

Table 4: Summary - Fuel used in power plants.

Fuel Type	Advantages	Challenges	Common in
Coal	Abundant, reliable, cheap	High CO_2 and pollutant emissions	China, India, South Africa
Natural Gas	Cleaner than coal, efficient	Methane leakage, price fluctuation	USA, EU, Middle East
Oil	High energy density, flexible	Expensive, polluting	Remote/off-grid regions
Biomass	Renewable, waste-to-energy	Land use, sustainability issues	Scandinavia, UK, Brazil

Coal for Energy Production

Coal is used to produce energy primarily through combustion in coal-fired power plants, where the chemical energy stored in coal is converted into electricity. This transformation involves several stages that turn raw coal into usable electrical power.

The process begins with the delivery of coal to the power station, typically via rail, ship, or conveyor belt. Once onsite, the coal is stored in large silos or stockpiles. Before combustion, the coal is crushed into a fine powder using a pulverising mill. Pulverising increases the surface area of the coal, allowing it to burn more efficiently and completely.

The powdered coal is then injected into a boiler furnace, where it mixes with heated air and combusts at very high temperatures, typically between 1,400°C and 1,700°C. The resulting heat energy is used to convert water circulating within tubes inside the boiler into high-pressure steam.

As the water absorbs this heat, it becomes steam and is collected in a steam drum. This high-pressure steam, often around 3,500 psi and 600°C, is directed into a turbine. The steam flows through multiple stages of turbine blades—high-pressure, intermediate-pressure, and low-pressure—causing the turbine shaft to rotate. This rotation is transferred to a generator, where mechanical energy is converted into electrical energy through electromagnetic induction.

Inside the generator, a rotor spins within a stator to produce electricity. The rotor's magnetic field induces an electric current in the stator coils. This electrical output is then transformed to higher voltages using transformers and distributed through power lines to homes, businesses, and industries.

Figure 25: Photo of a coal-fired power plant in Shuozhou, Shanxi, China. Kleineolive, CC BY 3.0, via Wikimedia Commons.

After exiting the turbine, the steam is directed into a condenser, where it is cooled using water from nearby natural sources or cooling towers. The steam condenses back into liquid water and is pumped back into the boiler to repeat the cycle, improving overall plant efficiency and conserving water resources.

Burning coal, however, generates significant pollutants, including sulphur dioxide (SO_2), nitrogen oxides (NO_x), carbon dioxide (CO_2), and particulate matter. To manage these emissions, power plants use various pollution control technologies. Electrostatic precipitators and baghouse filters capture fly ash particles. Flue gas desulfurization units, known as scrubbers, remove sulphur dioxide. Selective catalytic reduction (SCR) systems reduce NO_x emissions. Carbon capture and storage (CCS) technologies, though still under development, are designed to trap and store CO_2 emissions to limit their environmental impact.

Coal-fired power plants have played a major role in the industrialisation of many economies by providing a reliable energy source. However, due to their high emissions, these plants are increasingly being phased out or modernised. Many nations are now turning toward cleaner and more sustainable energy sources such as natural gas, nuclear power, wind, and solar to meet climate goals and reduce air pollution.

In 2022, global coal consumption surged to approximately 8.3 billion tonnes, marking a record high. Of this, around 36%—about 3 billion tonnes—was utilized for electricity generation. The significant demand for coal is driven by its characteristics of abundance and affordability, as well as existing infrastructures that facilitate its use in many countries [406]. Globally, coal accounts for nearly 30% of the world's energy supply, underscoring its role as a dominant energy source, particularly in industrial applications such as steel and cement production, which are challenging for decarbonization efforts [407].

The economic dynamics of coal consumption reflect its competitive pricing relative to other energy resources. In regions lacking subsidies for alternatives like natural gas or renewables, coal often remains the most viable option [408]. Countries, particularly in Asia, continue to rely heavily on coal due to affordability and accessibility. This reliance is evident as coal usage spiked in regions such as China, where it constitutes a substantial share of the energy mix [409]. Additionally, developed nations have also reverted to coal, especially in light of energy supply instability driven by geopolitical tensions and rising costs [410].

Nonetheless, coal combustion is the leading contributor to carbon dioxide emissions, accounting for approximately 30% of global CO_2 emissions as of 2020 [407]. Countries transitioning to cleaner energy sources face the difficult balance of maintaining energy affordability while addressing environmental impacts [411]. For instance, nations like China and India, despite pledging to reduce coal consumption, find it challenging to disengage from coal due to energy security concerns and ongoing economic demands [412]. Transition strategies to diminish coal dependence are underway, focusing on enhancing renewable energy infrastructure, improving energy efficiency, and developing carbon capture technologies [413].

Despite the environmental implications of coal use and ongoing global discussions on achieving carbon neutrality, coal remains a core component of the energy landscape, particularly in developing nations [411]. The dual challenge of ensuring energy security while mitigating carbon emissions continues to be a critical point of contention as many countries navigate their energy policies in a rapidly changing global context [411].

As of early 2025, the global landscape of coal-fired power generation consists of approximately 2,500 coal-fired power stations operating worldwide, which include over 7,000 individual generating units. These facilities collectively produce about one-third of the world's electricity, making them essential to energy supply but also significant contributors to greenhouse gas emissions [414, 415]. The reliance on coal-fired plants remains a critical issue, particularly as they produce not only carbon dioxide but also various air pollutants that impact public health [416, 417].

China leads globally with a staggering number of operational coal power plants—approximately 3,168—which constitutes over half of the worldwide coal-fired capacity [418]. Following China, India has about 256 operational plants, while the United States is home to roughly 408 coal-fired power stations [419]. Other countries such as Russia, Japan, and South Africa also hold notable positions in global coal power generation [418]. This distribution highlights the ongoing reliance on coal as a primary energy source, especially in developing regions where economic constraints drive the continued construction and operation of coal facilities [420].

Despite the growth in coal-powered infrastructure, there is a notable trend toward transitioning away from coal due to environmental concerns and the increasing availability of renewable energy sources. In 2024, global coal power capacity witnessed a 2% increase, largely fuelled by the addition of 70 gigawatts of new coal plants in China [420]. This contrasts sharply with the experiences of other nations such as the United Kingdom, which decommissioned its last coal-fired power station by September 2024, formally marking the end of coal-powered electricity generation in the country [418, 419].

The global coal scenario embodies a dichotomy: while some regions are accentuating coal capacity to satisfy energy requirements, others are actively phasing out coal in favour of sustainable alternatives. This multidirectional trend is influenced by a variety of economic, political, and environmental factors that dictate each region's energy strategy, illustrating the complexities of transitioning to cleaner energy sources amid persistent coal investments in certain economies [415, 421, 422].

The dual challenges of expanding coal infrastructure in some nations while managing the shift to renewable energy underscore the intricate balance required to meet global energy demands while mitigating environmental impacts. This interplay is expected to shape the future of energy generation significantly in the coming years [423].

Natural Gas for Energy Production

Natural gas is widely used for energy production due to its high efficiency, clean-burning characteristics, and versatility across industrial, residential, and electricity generation applications. In power generation, natural gas is primarily combusted in specialized power plants, where its chemical energy is converted into electrical or thermal energy. The specific technologies and methods used to harness energy from natural gas depend on the type of plant involved.

There are two main types of natural gas-fired power plants: open-cycle gas turbines (OCGTs) and combined-cycle gas turbines (CCGTs). Open-cycle plants operate by combusting natural gas in a chamber to produce high-temperature, high-pressure gases. These gases are directed over turbine blades, causing the turbine shaft to spin and drive an electricity generator. These plants are valued for their quick start-up times, making them suitable for meeting peak demand, though they operate with a lower thermal efficiency of about 30–40%. In contrast, combined-cycle plants enhance overall efficiency (up to 60%) by capturing the waste heat from the gas turbine's exhaust. This heat is used to generate steam in a heat recovery steam generator (HRSG), which then powers a secondary steam turbine, producing additional electricity.

Figure 26: Kawasaki natural gas power station which is owned by Kawasaki Natural Gas Power Generation Ltd. (KNGPG), headquartered in Kawasaki ward, Kawasaki City. インターネット川崎ガイド, CC BY 3.0, via Wikimedia Commons.

The electricity generation process begins with the delivery of natural gas via pipelines to the power plant. After mixing with air, it is combusted in a turbine chamber. The resulting hot gases spin turbine blades connected to a generator, which converts mechanical energy into electricity through electromagnetic induction. In CCGT systems, the hot exhaust gases are routed to an HRSG, which produces steam to drive a steam turbine for additional power generation. The steam is then cooled, condensed, and recycled within the system, enhancing efficiency and reducing water waste.

Natural gas is also utilized in cogeneration or combined heat and power (CHP) plants, where both electricity and useful thermal energy are produced from a single fuel source. Rather than losing excess heat during electricity generation, CHP systems recover it for industrial processing, water heating, or district heating. These systems can reach overall energy efficiencies of 70–80%, making them especially valuable in urban and industrial settings.

Beyond power plants, natural gas plays a significant role in industrial and residential heating. It fuels industrial boilers used for process heating and chemical manufacturing. In homes and businesses, natural gas powers water heaters, ovens, stoves, and space heaters, offering a reliable and cost-effective energy solution.

Environmentally, natural gas is considered cleaner than coal and oil. It emits 50–60% less carbon dioxide than coal and 20–30% less than oil when burned, and it also generates lower levels of nitrogen oxides, sulphur dioxide, and particulate matter. However, methane—the primary component of natural gas—is a highly potent greenhouse gas. If it leaks during extraction, transportation, or distribution, it can significantly offset the climate benefits of using natural gas instead of other fossil fuels.

Natural gas remains a highly flexible and relatively efficient energy source, particularly when deployed in combined-cycle and cogeneration systems. It continues to play a central role in the global energy mix, supporting electricity generation, heating, and industrial processes. While it presents some environmental advantages over coal and oil, minimizing methane leaks and improving infrastructure are critical to ensuring its role as a cleaner transitional fuel in the shift toward more sustainable energy systems.

As of 2022, the global landscape of natural gas-fired power generation includes approximately 4,500 operational power stations housing around 9,300 individual generating units across 129 nations. This widespread adoption is particularly notable in the United States, which leads with about 987 natural gas power plants, followed closely by countries such as Russia and China, where the numbers stand at approximately 177 and 163 stations, respectively [424]. Other nations with significant natural gas infrastructure include Germany, India, Japan, the United Kingdom, Australia, Canada, and Saudi Arabia, reflecting a diverse international investment in this energy source to meet growing electricity needs [425, 426].

Natural gas-fired power plants account for nearly 23% of global electricity production, playing a crucial role in the energy mix. Their popularity stems largely from their relatively lower carbon emissions compared to traditional coal and oil sources [415, 427]. For instance, research indicates that electricity generated from natural gas can yield climate benefits, particularly when compared with coal and oil-based generation, showcasing a marked reduction in carbon footprints [428]. However, despite these advantages, the expansion of natural gas infrastructure raises significant concerns with respect to long-term climate goals. Methane emissions associated with natural gas production and transportation can undermine its perceived environmental benefits. Studies have demonstrated that leakage rates of 2%-5% can significantly alter the climate impact balance, making natural gas's overall contribution to climate change comparable to that of coal plants over a 20-year span [427, 429]. As such, the discourse around natural gas transitions towards a more comprehensive appraisal, emphasizing the potential risks these practices pose to global climate objectives [426].

The juxtaposition of natural gas's role as a 'bridge fuel' towards a renewable energy future remains contentious. While its ability to reduce reliance on more polluting fossil fuels is well-documented, emerging literature posits that the narrative surrounding natural gas as a transitional energy source is increasingly viewed with scepticism amid the urgent need for a pivot to wholly renewable technologies. This viewpoint underscores the potential hazards of entrenching natural gas infrastructures, which could lock in carbon emissions and delay the transition to fully renewable energy systems [426, 430]. The imperative to critically evaluate the long-term sustainability of natural gas reliance is echoed throughout recent academic contributions, driving the conversation towards innovative solutions for carbon neutrality in the energy sector [431-433].

Oil for Energy Production

Oil is used to produce energy primarily through combustion in power plants and engines, where its chemical energy is converted into thermal, mechanical, or electrical energy. Although oil is not as widely used for electricity generation as coal or natural gas, it plays a significant role in transportation and in regions with limited access to alternative fuels. The process of using oil to generate energy involves various stages and applications, depending on whether it is for power generation, industrial use, or transportation.

In oil-fired power generation, refined petroleum products such as heavy fuel oil (HFO), diesel, or kerosene are burned to produce electricity. These power plants operate similarly to coal or natural gas plants but are less common due to higher fuel costs and emissions. Crude oil is first refined and then transported to power stations, where it is stored in large tanks. The oil is pumped into a combustion chamber and ignited with air, generating high temperatures that convert water into steam. The high-pressure steam spins a turbine connected to a generator, which produces electricity. After passing through the turbine, the steam is cooled, condensed back into water, and recycled into the system.

The most widespread use of oil-derived fuels is in internal combustion engines (ICEs), powering vehicles, ships, and small-scale generators. In gasoline and diesel engines, fuel is combusted to expand gases and move pistons, generating mechanical energy to drive vehicles. In aviation, kerosene-based jet fuels are used in turbine engines, where compressed air and fuel create high-speed exhaust for thrust. Portable generators, often used in remote or emergency situations, also rely on diesel or gasoline to generate electricity.

Industrially, oil is used to produce heat for manufacturing, chemical processes, and food production. Oil-fired furnaces and boilers generate steam for sterilization or direct heat for material processing. Combined heat and power (CHP) systems also use oil to simultaneously generate electricity and usable heat, enhancing overall energy efficiency in industrial operations.

Oil as an energy source offers several advantages. It has a high energy density, making it efficient for transportation and storage. It is also flexible and reliable, especially for backup generation and in off-grid areas. However, there are notable challenges. Burning oil emits significant greenhouse gases and pollutants, contributing to climate change and air quality issues. Oil prices are volatile and influenced by global markets, and as a finite resource, its extraction and use pose environmental and geopolitical risks.

Due to these drawbacks, oil's role in power generation has been declining. Today, it accounts for roughly 3% of global electricity generation and is primarily used in oil-rich or remote regions with limited access to other fuels. Nevertheless, oil remains a key energy source in global transportation and industry. As countries shift toward cleaner alternatives like solar, wind, and natural gas, oil is being phased out in many power systems to reduce emissions and environmental impact.

Figure 27: Oil refinery and power plant Holthausen near Lingen, Germany. Hartmut Schmidt Heidelberg, CC BY 4.0, via Wikimedia Commons.

As of 2025, there isn't a definitive global count of oil-fired power stations due to variations in reporting standards and the dynamic nature of the energy sector. However, several comprehensive databases provide insights into the prevalence and distribution of these facilities worldwide.

Fuel, Energy and Net Zero

The Global Energy Observatory (GEO) maintains a list of oil-fired power plants, offering detailed information on their locations and capacities. Similarly, the Global Oil and Gas Plant Tracker (GOGPT) by Global Energy Monitor catalogues oil and gas-fired power plants with capacities of 50 megawatts (MW) or more (20 MW or more in the European Union and the United Kingdom), including their operational status.

While exact numbers fluctuate due to new constructions, retirements, and conversions, these resources indicate that oil-fired power stations remain operational in various countries, particularly in regions where alternative energy sources are less accessible. Notably, countries like Saudi Arabia, Iran, Russia, China, and the United States have significant oil-fired electricity generation capacities.

It's important to note that the global reliance on oil for electricity generation has been declining, driven by concerns over greenhouse gas emissions and the increasing competitiveness of renewable energy sources. Consequently, many oil-fired power plants are being decommissioned or repurposed to align with evolving energy policies and environmental goals.

Biomass for Energy Production

Biomass is used to produce energy through a variety of processes that convert organic materials—such as wood, agricultural residues, animal waste, and dedicated energy crops—into heat, electricity, or biofuels. It is considered a renewable energy source because the carbon dioxide (CO_2) released during combustion is approximately offset by the CO_2 absorbed by plants during their growth. The method of converting biomass to energy varies depending on the form of biomass and the technology applied.

Biomass sources are diverse and include wood and wood waste (like sawdust and forest residues), agricultural residues (such as straw and corn stover), animal manure, energy crops (including switchgrass and fast-growing trees), the biodegradable portion of municipal solid waste, and organic industrial waste from food processing and breweries. These materials can be used in different ways to generate energy.

One of the most common methods is direct combustion, where biomass is burned in boilers to produce steam, which then drives turbines to generate electricity. This process is frequently used in standalone biomass power plants and in combined heat and power (CHP) systems, with efficiency ranging from 20–30%, or higher in CHP setups. Another method is co-firing, where biomass is mixed with coal in existing power plants. This approach reduces greenhouse gas emissions and leverages existing infrastructure, although it often requires biomass to be pre-processed for compatibility with coal systems.

Anaerobic digestion is a biological process in which microorganisms break down organic matter in the absence of oxygen, producing biogas—mainly methane and carbon dioxide. This process is suitable for inputs like manure, sewage, and food waste. The biogas can be used in engines or turbines to generate electricity, or upgraded to biomethane for use in pipelines or as vehicle fuel. The process also produces digestate, a nutrient-rich by-product that can be used as fertilizer.

Gasification involves heating biomass in a low-oxygen environment to produce syngas, a mixture of carbon monoxide, hydrogen, and methane. This syngas can be cleaned and used in gas turbines or engines, or as a feedstock for chemicals and fuels. Gasification offers higher efficiency than direct combustion but requires more advanced technology. Similarly, pyrolysis decomposes biomass at high temperatures without oxygen to produce bio-oil, biochar, and syngas. The bio-oil can be refined into liquid fuels, while biochar has uses in agriculture and carbon sequestration.

Figure 28: SICET, Biomass power plant in Ospitale di Cadore, Italy. Tiia Monto, CC BY-SA 3.0, via Wikimedia Commons.

Biomass power generation systems may consist of boilers to produce steam from solid biomass and turbines and generators to convert thermal energy into electricity. CHP systems provide both electricity and useful heat for buildings, industrial processes, or district heating networks, making efficient use of the fuel.

Environmental considerations are important in biomass energy use. While often described as carbon-neutral, the actual impact depends on responsible sourcing and land-use practices. Emissions from combustion—such as particulates, nitrogen oxides, and volatile organic compounds—can be managed through scrubbers, filters, and catalytic converters. Large-scale biomass production must be balanced with food security and forest conservation to avoid negative consequences.

The advantages of biomass include its renewable nature, widespread availability, use of waste products, support for rural economies, and compatibility with existing infrastructure. However, it also presents challenges such as lower energy density compared to fossil fuels, the need for extensive storage and handling facilities, and the requirement for emissions control to meet air quality standards. Its sustainability depends heavily on how the biomass is sourced and managed.

Biomass is a versatile and renewable energy source capable of producing electricity, heat, and fuels. Its use spans combustion, gasification, anaerobic digestion, and pyrolysis technologies. While it provides several environmental and economic benefits, its long-term viability hinges on sustainable practices and advanced technological applications. Biomass continues to play a significant role in global efforts to transition to cleaner energy systems, especially in regions rich in agricultural and forestry resources.

As of early 2024, approximately 4,971 active biomass power plants are operating globally, with a combined installed electrical capacity of about 83.8 gigawatts (GW) [434-436]. The biomass energy sector continues to experience growth, with projections anticipating that the number of biomass power plants could rise to around 5,980 by 2033, achieving an approximate total capacity of 96.8 GW [434]. This anticipated increase reflects the growing global emphasis on renewable energy sources as societies strive to reduce fossil fuel dependence and transition to cleaner energy technologies, prompted by climate concerns and policy frameworks aimed at promoting sustainable energy practices [434, 435].

Figure 29: Gonoike Biomass Power Plant, Kamisu City, Ibaraki, Japan. Σ64, CC BY 4.0, via Wikimedia Commons.

Regions with abundant organic resources, particularly the United States and several European nations, have taken significant strides in biomass electricity production. The U.S. remains a leader in this area, primarily utilizing wood waste and agricultural residues for energy generation [435, 437]. European countries like Germany, Sweden, and the United Kingdom have also heavily invested in biomass facilities as part of their renewable energy strategies, thus enhancing energy independence and contributing to the European Union's climate goals [436, 437].

The role of biomass power plants in diversifying the global energy mix is notable. Unlike more variable renewable sources such as wind and solar, biomass can provide consistent, dispatchable power, making it a critical component in balancing energy supply and demand [437, 438]. This flexibility allows biomass energy to complement other renewable sources, thereby enhancing overall grid reliability and stability [436, 437]. Notably, the integration of biomass power generation can contribute to a reduction in greenhouse gas emissions, facilitating a sustainable transition away from fossil fuels [439, 440].

As the biomass sector evolves, it is not only important for reducing reliance on fossil fuels but also for supporting the energy transition needed to effectively address climate change. The projected increase in both the number of biomass plants and their capacities underscores the significance of policy support and technological advancements in realizing the full potential of biomass as a renewable energy source [434, 439].

Energy Production through Wind and Solar Electricity Generation

The transition from fossil fuels to renewable sources such as solar and wind is technically feasible and increasingly supported by advancements in technology and a growing global commitment to decarbonization. However, this transition involves navigating a complex landscape of challenges that include infrastructure reconfiguration, economic considerations, system reliability, and comprehensive policy reform.

Renewable energy sources, particularly solar and wind, are globally abundant and strategically advantageous due to their widespread availability. Many regions possess the capacity to generate more electricity from these renewables than they consume, as demonstrated in studies related to the integration of renewable energies into existing grid systems [441, 442]. Technological advancements have led to significant cost reductions in solar photovoltaics (PV) and wind turbines, making them competitive with traditional fossil fuels. Recent literature emphasizes that the generation of solar and wind energy incurs minimal direct greenhouse gas emissions, which is crucial for meeting climate change mitigation goals [443, 444].

Nevertheless, the intermittent nature of solar and wind resources presents a challenge. Unlike fossil fuel plants that can generate electricity on demand, renewables depend on environmental conditions and time of day, necessitating supplementary systems to maintain reliability and dispatchability [445, 446]. Current strategies to address this intermittency involve integrating large-scale energy storage solutions, such as batteries and pumped hydro systems, alongside advanced grid management technologies [446]. These enhancements are essential for moving away from fossil-based baseload generation and ensuring a consistent energy supply even when renewable output is low.

The transition also raises significant questions regarding infrastructure and land use. Utility-scale solar and wind projects require considerable land, potentially conflicting with agricultural land use, ecological preservation efforts, and the rights of local communities, including Indigenous populations [447, 448]. As renewable energy projects proliferate, there is an urgent need for policies that ensure sustainable land use practices, promoting synergy between energy development and community concerns [443].

The economic dimensions of this transition are critical. The fossil fuel industry supports millions of jobs globally, and regions dependent on coal, oil, and natural gas face significant economic disruption as renewable technologies gain traction [449, 450]. Managed transition strategies, including carbon pricing mechanisms and subsidies for clean energy development, are necessary to facilitate a just transition for workers and communities affected by the decline of fossil fuel industries [443, 451].

As the global energy landscape evolves, integrating complementary technologies such as natural gas with carbon capture and storage (CCS), nuclear energy, and green hydrogen can play a vital role in maintaining grid stability and reducing emissions, particularly in sectors where renewables cannot fully meet demand [444, 447]. This transition to a low-carbon energy system requires a strategic blend of multiple technologies and infrastructure investments, illustrating the need for cohesive energy policy frameworks that foster collaboration among stakeholders in the energy sector [452, 453].

While the technical feasibility of replacing fossil fuel power plants with solar and wind energy exists, the practical implementation of such a transition is fraught with multifaceted challenges. It requires thorough planning that addresses infrastructure needs, economic impacts, land use conflicts, and robust policy support frameworks. The path forward depends not only on technological innovation but also on committed and coordinated efforts across multiple sectors to create a reliable, inclusive, and sustainable energy future.

Comparing Fuel Sources by Region and the Global Energy Mix

The global energy landscape is undergoing a significant transformation as regions adapt their electricity generation strategies to align with resource availability, economic considerations, and policy commitments. While fossil fuels—primarily coal and natural gas—still account for about 61% of global electricity generation as of 2023, there is a clear shift toward cleaner alternatives. Coal remains the largest single source at 35%, followed by natural gas. Meanwhile, renewables such as hydro, wind, and solar, along with nuclear energy, collectively contribute around 39% of global electricity production, setting a new record for low-emission generation.

The global energy mix is undergoing a complex and critical transformation. Historically dominated by fossil fuels, particularly coal, oil, and natural gas, the world is slowly but increasingly shifting towards low-carbon energy sources like renewables and nuclear. Despite the growth in clean energy, fossil fuels still make up the majority of global electricity generation, with coal remaining the largest single contributor. According to *Our World in Data* and the International Energy Agency (IEA), coal and natural gas collectively account for over half of electricity generation, while oil contributes a smaller share. On the other hand, renewables—led by hydropower, wind, and solar—along with nuclear energy, now represent nearly 39% of global electricity generation, marking steady progress toward decarbonisation [454].

In Asia, China leads the world in renewable energy deployment while also being the largest consumer of coal. Remarkably, China reached its 2030 wind and solar targets six years ahead of schedule, although coal continues to form a significant part of its power mix. India, facing surging electricity demand, also relies heavily on coal but is rapidly expanding its solar capacity. In Southeast Asia, countries like Indonesia and Vietnam are striving to balance economic growth with energy needs by continuing coal use alongside renewable energy development.

Africa presents a diverse picture. Countries such as Ethiopia and the Democratic Republic of Congo rely heavily on hydroelectric power, whereas South Africa depends primarily on coal. Nevertheless, renewable energy—particularly solar—is gaining momentum across the continent, aiming to address energy access challenges and sustainability goals.

Fuel, Energy and Net Zero

In the Americas, the United States predominantly uses natural gas for power generation, although wind and solar are rapidly growing, and coal use has dropped to record lows. Canada maintains a largely clean energy profile, with hydroelectricity dominating and additional growth in wind and solar. Latin American countries like Brazil and Chile are also leading renewable investment, with Brazil leveraging hydro resources and Chile utilizing its solar-rich Atacama Desert.

Europe is at the forefront of the global energy transition. The European Union has committed to ambitious renewable energy targets, and countries like Germany and Spain have made substantial investments in wind and solar technologies. However, this transition comes with challenges, including ensuring grid reliability and addressing energy security concerns during periods of fluctuating supply.

In Oceania, Australia is experiencing a boom in renewable energy, especially rooftop solar, yet coal still accounts for a significant share of electricity generation, posing obstacles to emission reductions. New Zealand, on the other hand, is a global leader in clean electricity, relying predominantly on hydroelectric and geothermal sources.

Looking ahead, the global trajectory points toward a steady transition from fossil fuels to renewable energy. This shift is driven by technological innovation, supportive policy frameworks, and growing environmental awareness. While fossil fuels currently dominate, the continued decline in renewable energy costs and improvements in efficiency suggest that the next decade will be transformative for electricity generation worldwide.

Regional Patterns and Differences

In the Asia-Pacific region, energy consumption is dominated by coal, particularly in China and India. China, despite its continued reliance on coal, has made remarkable strides in renewable energy, becoming a global leader in solar and wind capacity. India, facing fast-growing energy demand, also depends heavily on coal but is rapidly scaling its solar infrastructure. Southeast Asian nations such as Indonesia and Vietnam maintain high coal use while beginning to embrace renewables [454].

Europe shows a more advanced transition, with countries like Germany, Spain, and the Nordic states significantly investing in wind and solar energy. Though natural gas and coal are still part of the mix, their roles are diminishing. Countries such as Sweden and Norway, with abundant hydro and nuclear capacity, are among the cleanest energy producers globally [454].

North America displays a similar pattern. In the United States, natural gas is the dominant source of power, but renewables—especially wind and solar—are rapidly expanding. Coal use has significantly declined. Canada benefits from vast hydropower resources and has a comparatively clean electricity mix [454].

Africa presents a diverse and emerging energy landscape. While hydroelectric power plays a key role in countries like Ethiopia and the Democratic Republic of Congo, South Africa still relies heavily on coal. Solar projects are increasingly being deployed across the continent to address energy access and climate goals, but structural and financial challenges remain [454].

In Latin America and the Caribbean, countries like Brazil and Chile are making significant progress. Brazil's electricity is largely hydro-powered, and Chile is leveraging its world-class solar resources in the Atacama Desert. These regions are moving away from traditional liquid fuels toward a more balanced energy mix [454].

Key Global Trends

Several major trends are shaping the global energy transition. First, renewables—especially wind and solar—are experiencing the fastest growth. This expansion is driven by falling technology costs, policy incentives, and environmental pressure. Second, coal's share in power generation is declining globally, although it remains dominant in some countries due to affordability and availability. Third, natural gas is serving as a transitional fuel in many countries, offering lower emissions than coal while complementing intermittent renewables. However, it is still a fossil fuel and contributes to emissions [454].

There is also substantial regional variation in energy strategy, influenced by factors like natural resource availability, policy frameworks, economic development, and infrastructure. For instance, countries rich in hydropower (e.g., Norway, Canada) or solar potential (e.g., Australia, Chile) have a natural advantage in transitioning toward renewables[454].

Global energy use has evolved slowly over centuries. Prior to the Industrial Revolution, traditional biomass—mainly wood and charcoal—dominated global energy supply. The rise of coal in the 19th century, followed by oil and gas in the 20th century, powered rapid industrial and economic growth. Nuclear energy emerged in the 1960s, while modern renewables such as wind and solar only began to grow significantly in the last few decades [454].

Data from the *Energy Institute* shows that although the share of low-carbon energy has grown significantly—tripling since the 1960s—it still only makes up about one-sixth of global primary energy consumption. Hydropower and nuclear remain the largest low-carbon contributors, but wind and solar are catching up fast. The challenge now is not just to grow low-carbon energy to meet new demand, but to replace existing fossil fuel use outright [454].

Despite this progress, the pace of decarbonisation is not fast enough to meet international climate goals. Total fossil fuel consumption continues to rise in absolute terms due to growing global energy demand. This underscores a critical reality: the climate doesn't respond to percentages, but to absolute emissions. To address this, countries must not only scale up low-carbon energy faster but also begin to rapidly phase down fossil fuel production and consumption [454].

Wealthier countries have a vital role to play—not only by cutting their own emissions but also by helping lower-income nations develop clean energy systems through investment, technology transfer, and fair trade. This global cooperation is essential, especially as poorer countries face the dual challenge of economic development and climate responsibility.

The global energy mix is changing—but not quickly enough. Coal and natural gas remain dominant, although their shares are falling in some regions. Renewables and nuclear are rising but must expand much faster to replace fossil fuels and mitigate climate change. The path forward will require strong political will, substantial investment, technological innovation, and international collaboration to ensure a fair and effective energy transition for all [454].

Energy Mix

The global energy landscape is at a turning point, reflecting a significant operational shift as countries realign their electricity generation frameworks in response to resource availability, economic conditions, and evolving environmental policies. As of 2023, fossil fuels—namely coal and natural gas—remain pivotal, accounting for approximately 61% of global electricity generation, with coal alone contributing around 35% [455, 456]. Nonetheless, a notable rise in low-emission energy sources has been observed, where the combined share of hydropower, wind, solar, and nuclear power has reached almost 39% [457, 458]. This transition is crucial as it is driven by growing technological efficiencies and a mounting imperative to mitigate climate change [455, 459].

Historically, the reliance on fossil fuels facilitated rapid industrial growth throughout the 20th century [460, 461]. The current momentum toward a low-carbon future is catalysed by declining costs associated with renewables and broad policy support aimed at reducing carbon footprints [462]. This progression represents a significant change but highlights the ongoing challenge of fossil fuel dominance, which complicates global decarbonization efforts [463, 464]. The literature indicates that while renewables and nuclear energy are advancing, their capacity to fully supplant fossil fuel use remains constrained due to entrenched energy systems [465, 466].

A closer examination of regional dynamics reveals pronounced variations in energy transitions. In Asia, for instance, China, the largest coal consumer, is also spearheading investments in renewable infrastructure, having made significant advancements in solar and wind capacity [456, 461]. Despite this, coal remains essential to satisfy extensive energy demands, with parallels evident in India's reliance on coal, albeit complemented by a surge in solar energy contributions [467, 468]. Southeast Asian nations, such as Indonesia and Vietnam, are similarly positioning their energy strategies around coal while striving toward renewable deployment.

Africa presents a dual energy narrative, with countries like Ethiopia generating a substantial portion of their energy from hydropower. In contrast, South Africa's energy sector remains heavily coal-dependent [461, 469]. The adoption of solar energy technologies is expanding, particularly to improve electricity access for underserved communities, although significant barriers remain [456, 470].

In the Americas, a gradual transition to greener energy systems is evident. The United States now sees natural gas leading electricity generation, accompanied by a rapid expansion in wind and solar energy, signalling a decline in coal usage [455, 460]. Canada capitalizes on its hydroelectric resources, while nations in Latin America, particularly Brazil and Chile, are investing extensively in renewable infrastructure, leveraging existing hydropower and solar potential [456, 469].

The European Union is at the forefront of this energy transition, where substantial investments in renewable capacities are supported by stringent policy frameworks. Nations like Germany and Spain exemplify this shift; however, they still face challenges related to energy security and grid management amidst these changes [466, 471]. Furthermore, countries such as Norway and Sweden maintain high shares of hydropower and nuclear energy, showcasing some of the cleanest energy systems globally [464, 472].

Australia represents an intriguing case in Oceania, rapidly expanding its renewable energy base while simultaneously remaining reliant on coal [458, 473]. In contrast, New Zealand extensively utilizes hydro and geothermal power for its electricity needs, positioning it favourably on a global scale [455, 467].

The COVID-19 pandemic and subsequent economic recovery efforts have also played significant roles, with fossil fuels remaining entrenched in energy consumption patterns due to affordability and availability [460, 469]. Despite renewables claiming a higher percentage of the overall energy mix, total fossil fuel consumption has not declined in absolute terms, underscoring a paradox where increases in renewable capacity do not translate into proportional emission reductions [456, 474].

To effectively combat climate change, the global focus should not only be on expanding renewable energy resources but also on displacing existing fossil fuel-based energy production [455, 459]. This necessitates enhanced technological and financial investments as well as robust international cooperation to align efforts toward a low-carbon future [471, 474]. Such a commitment is particularly essential from wealthier nations, which must set an example in reducing emissions and facilitating the clean energy transition in less affluent regions [463, 469].

There are several compelling arguments for why some countries cannot or should not immediately stop using coal and natural gas for energy production. Key among these arguments are economic realities, energy security, technological constraints, and socio-political considerations.

First, in many regions, particularly those with limited renewable energy infrastructure or unstable electricity grids, coal and natural gas are essential for providing reliable and dispatchable power. Unlike solar and wind energy, which are intermittent and dependent on weather conditions, fossil fuels offer a stable baseload electricity supply that can be ramped up quickly to meet peak demand. For instance, countries such as Poland and South Africa rely heavily on their domestic coal production to ensure energy independence and grid reliability, mitigating the risks of relying solely on imports or less stable energy sources [475].

The economic implications of fossil fuel industries further complicate the issue. In many developing nations, such as Indonesia and India, coal and natural gas industries are significant contributors to livelihoods, providing jobs and supporting local economies through royalties and taxes. A rapid transition away from these resources could result in significant economic upheaval and job losses. For example, studies have documented how the economically vital coal sector in Australia and other nations faces dire consequences from abrupt energy policy shifts [462]. The economic stability provided by fossil fuels reinforces their current utility, particularly in regions reliant on these industries for tax revenues and employment [476].

Technologically, transitioning to renewable energy sources takes time and substantial investment. Many countries lack the necessary technology to fully realize renewables at scale. A successful energy transition requires advanced energy storage systems to counteract the intermittency of green energy and a modernized electrical grid capable of integrating decentralized power sources [477]. Developing economies, such as Nigeria and Pakistan, currently face technical and financial obstacles that complicate rapid transitions to clean energy infrastructures [478].

Also important to this discourse are socio-political considerations. The historical responsibility for greenhouse gas emissions is often cited by developing nations advocating for a fairer transition model. They argue that wealthier nations, which are primarily responsible for past emissions, should take the lead in decarbonizing while allowing developing countries to utilize fossil fuels to foster economic growth and poverty alleviation [479]. A stark example of this is seen in Bangladesh and Vietnam, where coal remains a practical and affordable solution to expand access to electricity for underserved populations, and natural gas is often viewed as a bridging fuel during this transition [480].

Lastly, geopolitical factors play a significant role in discussions around fossil fuel reliance. Nations rich in fossil fuel resources may leverage these assets for geopolitical influence, as seen with countries like Russia and Qatar, where oil and gas exports are critical tools in foreign policy and economic strategy [481]. The phasing out of fossil fuels could substantially disrupt their economic stability and international standing, complicating the energy transition discourse [479].

While the necessity to transition to renewable energy is clear for addressing climate change, a blanket approach suggesting all countries should immediately phase out fossil fuels disregards their unique circumstances. A nuanced and flexible approach must be adopted—one recognizing each country's development stage, energy requirements, and socio-political contexts [482]. Promoting international cooperation to facilitate clean energy investments and develop fair policy frameworks will be essential for achieving a just and effective energy transition.

Based on the arguments presented—including energy security, economic dependency, infrastructure readiness, socio-political factors, and the need for a just transition—each country requires a tailored fuel mix that balances its current realities with future sustainability goals. The ideal energy mix for power generation should consider a country's resources, policy environment, and development priorities.

As examples, Australia is well-positioned to embrace renewables due to its abundant solar and wind resources. The ideal fuel mix for Australia includes 50–60% renewables, particularly solar and wind, supported by 20–30% natural gas as a transitional and dispatchable source. Energy storage and hydroelectric generation could make up another 10–20%, while coal use should be reduced to minimal levels. As coal still plays a significant role in the national grid, continued investment in grid upgrades and storage systems is necessary to enable a phased coal exit.

The United Kingdom has a strong foundation in offshore wind and advanced grid capabilities. Its ideal mix would consist of 60–70% renewables, supported by 20–25% nuclear power to provide reliable baseload electricity. Around 5–10% of the mix could be made up of natural gas, ideally with carbon capture and storage (CCS) to mitigate emissions. Less than 5% would come from other backup or peaking sources. With coal usage already in steep decline, the UK is well placed to further decarbonise its power system.

Germany's energy transition strategy, Energiewende, places a high priority on renewables and energy efficiency. Its ideal mix would include 60–65% from wind, solar, and biomass, with 10–15% natural gas, preferably blended with green hydrogen or equipped with CCS. Imports and energy storage technologies would account for another 15–20%, and coal use would be minimal and short-term. Following the phase-out of nuclear, Germany needs robust backup systems and cross-border energy trade to maintain grid stability during the transition.

India faces rising electricity demand and continues to rely heavily on coal for affordable energy access. A balanced and realistic fuel mix for India would include 40–50% renewables such as solar, wind, hydro, and biomass. Coal, particularly from domestic sources, would remain at 30–35% in the short to medium term but should be gradually reduced. Natural gas could contribute 10–15%, and less than 10% could come from nuclear and emerging storage technologies. To achieve a cleaner future, India will require significant international funding and technical support for grid modernisation.

China, the world leader in renewable energy capacity, still depends heavily on coal, particularly for industrial power needs. An optimal fuel mix would include 45–55% renewables—mainly hydro, solar, and wind—complemented by 25–30% coal, which should steadily decline. Nuclear power should account for 10–15%, with 5–10% from natural gas and energy storage. China's centralized energy planning and manufacturing dominance enable rapid transformation, though balancing industrial growth with decarbonisation remains a challenge.

The United States has diverse energy resources and a federal structure that allows states to lead in clean energy adoption. An ideal mix would consist of 45–55% renewables, including wind, solar, hydro, and geothermal. Natural gas would serve as a transition fuel at 20–25%, while nuclear power would provide 10–15% of the baseload supply. Energy storage and other emerging technologies would account for another 5–10%, with coal phased out to less than 5% nationally. Federal policy support and infrastructure investment can accelerate this shift.

Canada already boasts one of the cleanest electricity profiles globally, driven by hydropower. Its optimal energy mix would include 60–70% hydropower and other renewables, 15–20% nuclear energy (mainly in Ontario and New Brunswick), and 10–15% natural gas. Fossil fuels should be minimal in the rest of the country. Canada should continue to expand renewables and invest in electrification of other sectors while preserving its low-carbon advantage.

Fuel, Energy and Net Zero

Saudi Arabia, traditionally reliant on oil and gas, is beginning to diversify its energy sources. The ideal fuel mix for Saudi Arabia includes 30–40% solar and wind, taking advantage of its high solar irradiance. Natural gas would still contribute 40–50%, offering dispatchable power. Around 10–15% could come from hydrogen and emerging low-carbon technologies. Oil use should be minimized for domestic power, preserving it for export and economic value. Saudi Arabia's Vision 2030 strategy supports this transition, focusing on green hydrogen and clean energy innovation.

Table 5: Summary Ideal Fuel Mix by Country (Examples).

Country	Ideal Fuel Mix	Rationale
Australia	50–60% Renewables (solar, wind) 20–30% Natural Gas 10–20% Storage and Hydro <10% Coal	Abundant solar and wind. Rapid rooftop solar growth. Grid still coal-dependent. Natural gas and storage needed for transition.
United Kingdom	60–70% Renewables (offshore wind, solar) 20–25% Nuclear 5–10% Natural Gas (with CCS) <5% Other/Backup	Advanced grid, declining coal use, strong offshore wind. Nuclear for baseload. Gas with CCS as flexible support.
Germany	60–65% Renewables (wind, solar, biomass) 10–15% Natural Gas (with CCS or green hydrogen) 15–20% Imports and Storage <5% Coal	Energiewende policy focus. Needs imports and gas for stability post-nuclear. Phasing out coal.
India	40–50% Renewables (solar, wind, hydro, biomass) 30–35% Coal 10–15% Natural Gas <10% Nuclear and Storage	Rising demand. Domestic coal vital. Expanding solar. Needs global support and modern grid.
China	45–55% Renewables (hydro, solar, wind) 25–30% Coal 10–15% Nuclear 5–10% Gas and Storage	Global leader in renewables. Still industrial coal reliance. Central planning enables rapid scaling.
United States	45–55% Renewables (wind, solar, hydro, geothermal) 20–25% Natural Gas 10–15% Nuclear 5–10% Storage and Other <5% Coal	Diverse resources. State-driven clean energy growth. Gas as bridge fuel. Coal declining.
Canada	60–70% Hydropower and Renewables	Clean grid dominated by hydro. Focus on

Country	Ideal Fuel Mix	Rationale
	15–20% Nuclear 10–15% Natural Gas Minimal Fossil Fuels Elsewhere	electrification and minimal coal/oil.
Saudi Arabia	30–40% Solar and Wind 40–50% Natural Gas 10–15% Hydrogen and Emerging Tech Minimal Oil	Vision 2030 supports renewables. Oil reserved for export. Hydrogen development underway.

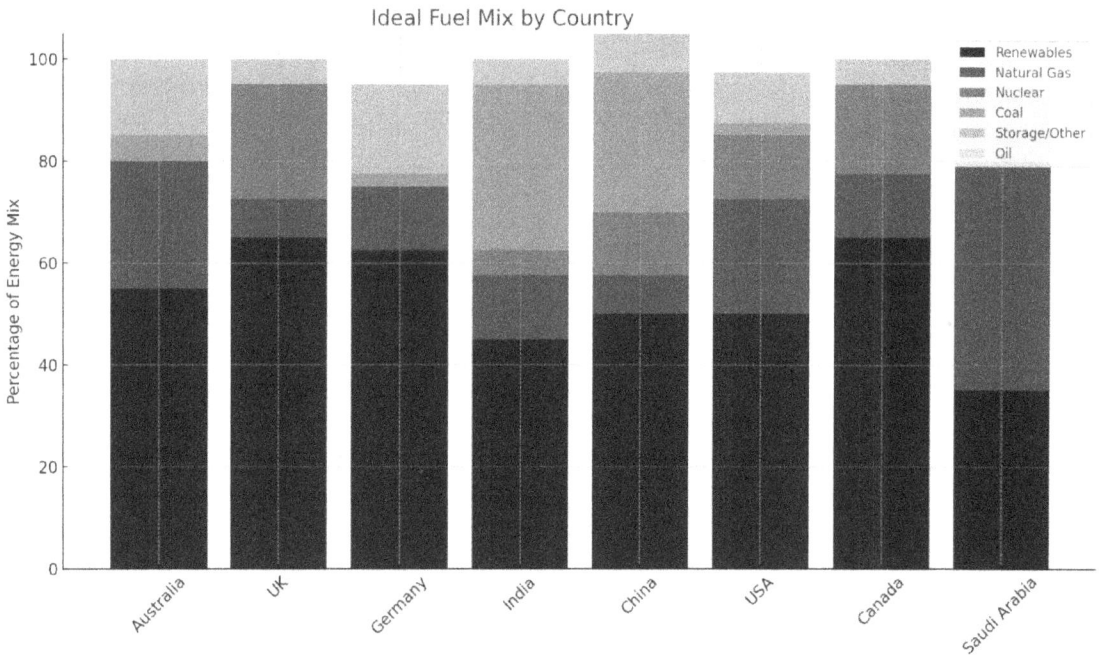

Figure 30: Ideal Fuel Mix by Country (Examples).

Each country's ideal energy mix must reflect its own economic, environmental, and social realities. Wealthier nations can lead the way with aggressive decarbonisation, while developing economies require a more gradual, supported transition. A fair and effective global energy transition depends on international cooperation, policy innovation, and inclusive investment strategies that respect regional diversity and development needs.

Role in Energy Grids

The modern energy grid is a dynamic and interconnected system that relies on various fuels and technologies to generate and deliver electricity efficiently, reliably, and sustainably. Each fuel type plays a unique role in supporting the grid's operation, and their relevance often depends on geographic, economic, and technological factors. Understanding the role of each energy source is essential to shaping effective policies and ensuring a balanced energy transition.

Coal has traditionally been used as the backbone of electricity generation in many countries. It provides a stable form of baseload power, meaning it runs continuously to meet the minimum level of electricity demand at all times. Coal-fired power plants are most efficient when operating steadily over long periods but are less flexible in adjusting output quickly. Although coal is gradually being phased out in many developed countries due to its high greenhouse gas emissions, it remains important in energy systems that lack reliable alternatives or where coal is a domestically abundant resource, such as in India, China, and South Africa.

Baseload power refers to the minimum continuous electricity demand required to maintain operations of essential services and activities throughout a 24-hour period. Traditionally, this demand has been met by centralized generation facilities that can operate continuously, such as coal, nuclear, and hydroelectric power plants. These conventional baseload plants are engineered to maintain steady output with minimal interruptions, a key feature that supports the inherent stability of power grids [483]. While conventional sources like coal and nuclear energy deliver reliable baseload power, they pose environmental and sustainability challenges, prompting a re-evaluation of the energy mix toward lower-carbon alternatives [484, 485].

Renewable energy sources, such as solar and wind, present challenges in fulfilling baseload capacity due to their intermittent nature. For instance, solar power generation is limited to daylight hours and can be influenced by weather conditions, while wind energy output varies significantly with atmospheric conditions [486]. This variability impedes the ability of renewables to guarantee the consistent generation necessary for baseload power, leading to potential instabilities in the grid if not managed appropriately [483]. As a result, a significant reliance on these sources may lead to frequency and voltage fluctuations, necessitating the incorporation of energy storage solutions to address the variability and ensure a steady power supply when renewable generation dips [487, 488].

In response to these challenges, technological innovations are emerging to enhance the reliability of renewable energy for baseload needs. The development of sophisticated energy storage systems, alongside diversified energy portfolios that integrate wind, solar, geothermal, and hydropower technologies, is critical to mitigating the instability typically associated with renewable energy generation [489, 490]. Geothermal energy, in particular, stands out as a dependable baseload source due to its capacity for continuous generation without the need for extensive energy storage systems [485, 491]. This reliability makes geothermal a unique asset in future energy strategies, potentially enabling a transition toward a low-carbon grid capable of consistently meeting baseload demands [492].

While traditional renewable resources such as solar and wind struggle to provide continuous baseload power due to their inherent variability, a multifaceted approach that includes robust energy storage solutions, advancements in smart grid technology, and diversification of renewable resources can effectively help fulfill baseload requirements. The inclusion of geothermal energy offers a promising pathway toward achieving a more stable and sustainable energy mix that enhances grid resilience while minimizing environmental impact [493, 494].

Natural gas plays a crucial role in modern energy grids due to its flexibility and responsiveness. Gas-fired power plants, especially combined-cycle units, can be ramped up or down quickly, making them ideal for meeting peak electricity demand or backing up intermittent renewable sources like wind and solar. In many countries, natural gas serves as a transitional fuel, helping reduce emissions from coal while supporting grid reliability during the expansion of renewable capacity. Its cleaner combustion relative to coal makes it an attractive option, although it still emits carbon dioxide and methane, a potent greenhouse gas.

Peak electricity demand is a critical concept in understanding energy consumption patterns and grid management. It refers to the maximum amount of electricity consumed during a specified period—be it daily, seasonally, or annually—often coinciding with extreme weather conditions, such as hot summer afternoons or cold winter evenings when heating and lighting needs surge [495]. Utilities and grid operators must meticulously prepare to meet these peak demands to prevent blackouts or service interruptions, thereby highlighting the importance of maintaining surplus capacity during these critical moments [496].

The characteristics of peak demand reveal that each peak event is typically of short duration but possesses high intensity, necessitating rapid increases in electricity generation [497]. Peak periods tend to show predictable timing, influenced by daily and seasonal patterns, such as heightened electrical use in late afternoons during hot months or evenings of cold winters [498]. Furthermore, electricity infrastructure undergoes significant stress during these peak times, requiring systems to be robust enough to adapt quickly and flexibly to demand [499].

To manage peak demand effectively, the electricity grid relies on "peaking power plants," predominantly natural gas turbines. These facilities can be turned on and off with ease and can ramp up generation swiftly to address sudden spikes in demand, albeit at a potentially higher cost and reduced efficiency compared to baseload plants [500]. However, the reliance on these fossil-fuel-based sources presents challenges as nations strive for cleaner energy alternatives.

Renewable energy sources, such as solar and wind, are becoming increasingly significant in the energy mix. However, they struggle to meet peak demand reliably due to inherent intermittency and variability in power production [501]. The generation from solar panels typically diminishes at sunset when electricity demand can peak, creating a 'duck curve' phenomenon [502]. Furthermore, while energy storage systems, such as batteries, present a potential solution for balancing supply and demand, the technological constraints—especially in terms of scalability, affordability, and long-duration storage—pose ongoing challenges [503].

Several strategies are being pursued to enhance the compatibility of renewables with peak demand requirements. These include the implementation of battery storage systems and larger utility-scale systems [503]. Additionally, demand response (DR) programs have emerged as effective mechanisms to shift consumer demand from peak to off-peak periods, assisting in load management while alleviating stress on the grid [496]. By employing these measures alongside hybrid systems that incorporate fast-ramping natural gas or hydroelectric power, utilities aim for a more reliable energy supply that accommodates both renewable integration and peak demand management [504].

Figure 31: A large battery storage facility by the local electricity company in Walenstadt, Switzerland. Kecko from Eastern Switzerland, CC BY 2.0, via Wikimedia Commons.

Oil, while historically significant in power generation, now plays a limited role in most modern electricity grids due to its high cost and carbon emissions. However, it remains a vital backup source, particularly in remote areas, islands, or developing countries where alternative fuels are unavailable or unreliable. Oil generators provide emergency power during grid failures or fuel shortages and are also used in military operations and off-grid installations.

Nuclear power offers another form of low-carbon baseload electricity. It operates continuously at high capacity and is not subject to the variability of weather like renewables. Nuclear plants are capital-intensive and take years to build, but once operational, they provide stable and predictable energy for decades. In countries like France, the UK, and Canada, nuclear energy plays a central role in ensuring grid stability while contributing significantly to climate goals through zero-emission generation.

Figure 32: Tihange Nuclear Power Station (Huy, Belgium). Trougnouf (Benoit Brummer), CC BY 4.0, via Wikimedia Commons.

Low-carbon baseload electricity is a crucial concept in the transition to a sustainable energy system, emphasizing reliable, continuous power generation with minimal greenhouse gas (GHG) emissions. Baseload power refers to the constant minimum electricity demand on a grid that must be continuously met, irrespective of factors such as time of day or seasonal variations [483]. This reliability is paramount for essential services like healthcare and transportation, requiring systems that are not only always available but also predictable and economical [483].

Several technologies are categorized as low-carbon baseload providers. Nuclear power, for instance, is renowned for its ability to operate continuously and emit virtually zero GHGs during operation [505]. Extensive studies highlight that integrating nuclear with renewable sources can enhance overall energy efficiency and reduce land use, essential for maximizing low-carbon energy outputs [506]. Large-scale hydropower exemplifies low-carbon baseload electricity, offering substantial emissions reductions when water flows are managed sustainably. However, the impacts of climate change on hydropower reliability and generation capacity remain a concern [507].

Additionally, geothermal energy presents a reliable and low-emission energy source. Innovative geothermal systems can provide continuous power, thus contributing significantly to the reliability of the grid [508]. Biomass, particularly when integrated with carbon capture technologies, can be classified as low-carbon baseload under sustainable management practices; however, its effectiveness depends on the specific management practices employed [509]. Fossil fuel sources integrated with Carbon Capture and Storage (CCS) also present a method to reduce emissions from traditional baseload plants, though they still depend on non-renewable inputs [510].

Although renewables like solar and wind are celebrated for their low-carbon attributes, they currently lack the consistent dispatchability required for baseload electricity. Their inherent intermittency—a product of variable weather conditions—makes them incapable of reliably meeting baseload demand without substantial backup systems [511]. The increase in renewable capacity necessitates baseload generation sources to stabilize the grid, as renewables alone cannot provide the steady power needed for consistent demand [512]. Advanced energy storage solutions are being researched to counter these limitations, but current technologies cannot adequately address long-duration storage needs [511, 513].

Thus, while renewables are integral to a clean energy future, they cannot yet serve as reliable low-carbon baseload sources. Until technological advancements enable effective long-duration energy storage, low-carbon baseload technologies like nuclear, hydropower, and enhanced geothermal systems remain pivotal in delivering uninterrupted, low-emission electricity around the clock, ensuring systemic stability in power grids [507].

Hydropower is one of the most established renewable energy sources, especially in countries with abundant river systems. It provides both baseload and dispatchable power, depending on the size of reservoirs and water flow management. Large-scale hydropower plants can store energy and release it during high-demand periods, offering natural energy storage. Hydropower is also vital for grid balancing and frequency regulation, making it a reliable partner for intermittent renewables. In Canada, Norway, and Brazil, hydropower is a cornerstone of the energy mix.

Grid balancing and frequency regulation play crucial roles in ensuring the stability and reliability of power systems. These functions are essential because electricity supply must be continuously matched with demand in real-time. Slight discrepancies can lead to significant consequences, including power outages and potential damage to infrastructure [514, 515].

Grid balancing is the practice of maintaining a constant equilibrium between electricity generation and consumption. It is essential for preventing voltage and frequency variations, which can arise when supply exceeds demand or vice versa [516]. The mechanisms employed by system operators for effective grid balancing include the use of baseload generation, which provides a stable supply of electricity, and flexible generation sources such as natural gas turbines and hydroelectric power, which can quickly adjust output to match the demand [517]. Modern strategies also incorporate demand-side management (DSM), which allows consumers to reduce their electricity consumption during peak times through incentives [518, 519]. Furthermore, energy storage systems, including batteries, are deployed to absorb surplus energy or deliver it when needed, thus enhancing grid stability [514, 520].

Frequency regulation is another critical component of grid management, involving the maintenance of the grid's operational frequency within predefined limits, typically around 50 Hz or 60 Hz, depending on the region [521]. When demand outstrips supply, grid frequency drops, while an oversupply will cause frequency to rise. This fluctuation can lead to equipment damage and system failures if not swiftly corrected [522]. Power stations are typically equipped with automatic control systems that can adapt output in seconds or milliseconds, thus enabling them to respond to instantaneous changes in frequency [523]. The importance of frequency stability cannot be overstated, as it safeguards critical infrastructure such as transformers and industrial machinery from potential disruptions caused by frequency deviations [524].

While the integration of renewable energy is crucial for reducing carbon emissions, it presents distinct challenges concerning grid balancing and frequency regulation. The variability inherent in renewable sources, particularly in wind and solar, results in unpredictable output, complicating efforts to maintain balance [525]. Additionally, many renewable technologies lack mechanical inertia—an essential feature of traditional generators such as coal, nuclear, and hydroelectric plants, which naturally stabilize grid frequency [526]. Renewables like solar photovoltaics and wind turbines primarily utilize power electronics for grid connection, which traditionally do not contribute to inertia [527].

Moreover, the inability of renewables to be dispatched reliably on demand limits their effectiveness in providing necessary reserves for grid regulation [528]. This delayed response capability requires supplementary systems, such as fast-acting battery storage or hybrid power plants that combine renewable generation with conventional sources, to provide the necessary flexibility [529].

To address these challenges, modern energy systems are increasingly employing advanced technologies. Innovations such as grid-forming inverters are being developed to mimic the inertia provided by traditional generators, enhancing stability [530]. Battery storage technologies offer rapid frequency response options, which are becoming integral in high-penetration renewable setups [514]. Additionally, hybrid power plants that integrate renewables with gas or storage solutions create a more resilient energy mix that can adapt to varying grid demands [531].

As these technologies continue to evolve, there remains a pressing need for a comprehensive approach that includes sophisticated grid management software and robust demand response programs. Together, these measures will enhance the resilience and efficiency of future electricity systems, especially as reliance on renewable sources intensifies [515, 525].

Effective grid balancing and frequency regulation are vital for maintaining a reliable electricity system. While renewable energy sources contribute significantly to sustainable generation, their integration poses challenges that necessitate advanced technological solutions and strategic planning for optimal performance in real-time electricity management.

Solar and wind energy are the fastest-growing renewable sources globally due to falling costs and supportive policies. They generate *clean electricity* without fuel inputs or emissions, making them crucial for decarbonisation. However, their output is *intermittent and weather-dependent*, which poses challenges for grid stability. Therefore, their expansion must be supported by energy storage, demand-side management, and flexible backup power sources. Distributed solar, such as rooftop panels, also contributes to *decentralised generation*, reducing the need for long transmission lines.

Biomass energy, derived from organic materials like wood, crop waste, and animal manure, is a form of dispatchable renewable power. It can be burned or converted into gas to generate electricity in a way that mimics fossil fuels but with lower net emissions if sustainably sourced. Biomass can provide baseload electricity in rural or agricultural regions and plays a role in waste-to-energy systems. It is particularly useful in areas with large forestry or agricultural sectors.

Energy storage technologies, such as lithium-ion batteries and pumped hydro storage, are becoming indispensable in modern grids. These systems do not generate electricity themselves but store excess power when supply exceeds demand—especially from solar and wind—and release it when needed. This makes storage vital for integrating renewables, managing peak loads, and improving grid resilience. As the cost of battery technology declines, storage is expected to play an increasingly central role in grid management.

Pumped hydro storage is the most widely used and mature form of large-scale energy storage in the world. It plays a crucial role in supporting electricity grids by storing excess energy and releasing it when demand rises. This helps to balance supply and demand, integrate variable renewable energy sources like wind and solar, and ensure grid stability.

Richard Skiba

Figure 33: Llyn Peris as seen from the Dinorwig to Nan Peris slate trail. Llyn Paris is used as the lower reservoir for the Hydro-electric power station built into the mountainside. Water from an upper reservoir (Marchyn Mawr) generates electricity during daytime peaks but at night when the demand lessens the same turbines pump the water up to upper reservoir for use the next day. Denis Egan, CC BY 2.0, via Wikimedia Commons.

Pumped hydro works like a giant battery by using gravity and water movement to store and generate electricity. It involves two reservoirs positioned at different elevations. During periods of low electricity demand or surplus renewable generation—such as on sunny or windy days—electric pumps use the excess electricity to move water from the lower reservoir to the upper one. This process effectively converts electrical energy into gravitational potential energy. Later, when electricity demand increases or renewable output drops, the stored water is released from the upper reservoir. As it flows back down to the lower reservoir, it passes through turbines that spin and drive generators, producing electricity that is fed back into the grid.

The round-trip efficiency of pumped hydro storage typically ranges between 70% and 85%, meaning most of the energy used to pump water uphill can be recovered. Depending on system design, pumped hydro can store energy from a few hours up to several days, making it suitable for both short-term balancing and longer-duration energy needs. These facilities can range in size from a few megawatts to several gigawatts, with storage capacities reaching into the thousands of megawatt-hours.

Fuel, Energy and Net Zero

Globally, there are several well-known examples of pumped hydro projects. In Australia, the Snowy 2.0 expansion is expected to add 2,000 MW of generating capacity and 350,000 MWh of storage. The Bath County facility in the United States is the largest of its kind, generating up to 3,003 MW. In the United Kingdom, Dinorwig—also known as "Electric Mountain"—provides rapid-response power to help stabilize the grid.

Pumped hydro offers a range of advantages. It is well-suited to managing national electricity systems with high shares of renewable energy. The infrastructure is long-lasting, often exceeding 50 years with proper maintenance. It has a fast response time and can ramp from zero to full output in less than a minute, making it ideal for meeting peak demand and regulating frequency. Additionally, its operational costs are relatively low once the system is built.

However, there are limitations to pumped hydro. It requires specific geographical conditions, including suitable terrain for constructing two reservoirs at different elevations. Building such infrastructure involves high upfront capital investment and can take years of planning and approval. Environmental impacts are also a concern, as construction can disrupt water ecosystems and land use, especially in sensitive areas.

As renewable energy sources like wind and solar become more prevalent, pumped hydro is increasingly important. It enables excess generation to be stored—such as solar power produced at midday—and used later during evening demand peaks. It helps smooth power fluctuations to maintain a stable grid and can replace fossil-fuel peaking plants with clean, dispatchable energy.

Hydrogen, especially green hydrogen produced via electrolysis using renewable electricity, is an emerging energy carrier with the potential to revolutionize electricity systems. It can be stored and used to generate power in fuel cells or turbines, offering long-duration and seasonal storage that complements the short-term flexibility of batteries. Hydrogen can also be blended into natural gas pipelines or used to decarbonize industrial and transport sectors. Although still in early development, countries like Germany, Australia, and Saudi Arabia are investing heavily in hydrogen infrastructure.

Investing in hydrogen infrastructure offers a range of potential benefits for countries aiming to transition towards cleaner energy systems. Countries with varying energy goals, economic conditions, and natural resources can tailor hydrogen investments to their unique contexts. For instance, green hydrogen, produced via electrolysis using renewable energy, plays a pivotal role in decarbonizing hard-to-abate sectors, such as steel manufacturing, aviation, and long-distance transportation, where electrification alone may not suffice. It provides a viable pathway for these sectors to achieve net-zero emissions while enhancing energy storage capabilities and improving grid reliability by utilizing surplus renewable electricity [532, 533].

The global hydrogen landscape is accelerating, with significant backing from major economies like the European Union, Japan, and South Korea, all of which have developed national hydrogen strategies supported by both public funding and private investments. This collective momentum can create an environment conducive to establishing robust hydrogen infrastructures, fostering expertise, and encouraging strategic partnerships, thereby positioning countries as leaders in the emerging hydrogen economy [534, 535].

However, the challenges associated with hydrogen infrastructure cannot be understated. Currently, green hydrogen production costs are higher than fossil fuel alternatives, making the economic feasibility of hydrogen projects a pressing concern. Investments are necessary not only for hydrogen production but also for building the requisite infrastructure, including electrolysers and transportation networks. The relatively low energy efficiency of hydrogen, characterized by conversion losses, raises questions about its competitiveness compared to direct electrification methods [536].

Moreover, the complexity in transporting and storing hydrogen, due to its low energy density, necessitates advanced technologies for compression, liquefaction, or conversion into hydrogen carriers, all of which can elevate costs and present technical hurdles [537, 538]. Environmental considerations also complicate the hydrogen narrative, especially concerning blue hydrogen, which is derived from natural gas using carbon capture technologies. Blue hydrogen's reliance on fossil fuels and issues related to methane leakage can compromise its perceived climate benefits, highlighting the importance of pursuing truly renewable hydrogen production methods [539, 540].

Investing in hydrogen infrastructure can be an astute strategy for countries aiming to meet energy transition goals. The approach must be prudent and tailored, focusing on sectors where hydrogen offers distinct advantages. Strategic investments, supportive regulatory frameworks, and international collaboration can enable nations to maximize the potential of hydrogen as a cornerstone of sustainable energy systems. As development progresses, hydrogen can serve as a complementary solution to electrification, thereby contributing to a diverse energy matrix that addresses both economic and environmental objectives [541, 542].

Each fuel type contributes distinct strengths to the energy grid—some offer stability and dispatchability, while others support sustainability and cost reductions. The future of energy grids lies in integrated systems that combine various fuels with digital technologies and storage to deliver reliable, low-carbon electricity in all conditions. The optimal mix depends on a country's resources, infrastructure, and policy goals.

Chapter 9

Industry and Manufacturing

Following the exploration of transportation and electricity generation, *Fuel, Energy and Net Zero* now turns to another cornerstone of modern civilisation: industry and manufacturing. These sectors are fundamental to economic development but are also among the most energy-intensive and carbon-emitting parts of the global system. Chapter 9 examines how fuel and energy are used in industrial processes—from steel and cement production to chemicals and heavy manufacturing. It highlights the reliance on fossil fuels for high-temperature processes, discusses current energy efficiency measures, and explores pathways to decarbonisation, including electrification, hydrogen use, and carbon capture technologies. As industries grapple with the dual challenge of maintaining productivity while reducing emissions, this chapter provides insight into how innovation, regulation, and fuel diversification are reshaping the industrial energy landscape.

Fuel Use in Steel, Cement, and Chemical Production

The production of cement, steel, and chemicals is a major contributor to global greenhouse gas (GHG) emissions, accounting for a significant portion of emissions across industrial sectors. These industries are labelled as "hard-to-abate sectors," largely due to their reliance on fossil fuels not only for energy but also as feedstocks or chemical inputs.

In the case of cement production, this sector alone contributes approximately 5-8% of global CO_2 emissions [543-545]. The primary source of these emissions arises from the calcination process of limestone to produce clinker, essential for cement manufacturing [546, 547]. Furthermore, the combustion of fossil fuels necessary to achieve the high temperatures required in kilns (around 1500 °C) accounts for about 40% of total emissions associated with cement production [543, 546]. Innovations such as the use of alternative fuels like municipal solid waste (MSW) have been suggested as viable methods to reduce net emissions, potentially positioning the cement industry to achieve "negative emissions" under certain conditions [548, 549].

In steel production, another significant contributor to GHG emissions, processes are similarly energy-intensive. Primary steelmaking predominantly utilizes coke, a fossil fuel derived from coal, which leads to substantial carbon emissions. The sector's contribution to global CO_2 emissions is estimated to account for about 7-9% of total emissions globally [545, 550]. Difficulties in transitioning from traditional methods to more sustainable practices are attributed to the high temperatures required and the economic feasibility of alternative methods like hydrogen-based steelmaking, which remains under exploration [551].

The chemical industry, also classified among the hard-to-abate sectors, is responsible for approximately 6% of global GHG emissions [551]. Here, processes often involve the use of fossil fuels as feedstock for producing a variety of chemicals, complicating decarbonization efforts. Research suggests that electrification of chemical processes may provide a pathway to reduce emissions, although the demand for conventional chemical products continues to challenge this transition [551].

Decarbonization efforts in these sectors face several challenges, including technological limitations, high operational costs, and infrastructural constraints [552]. However, emerging alternatives, such as the integration of carbon capture and storage (CCS) technologies, present viable pathways for mitigating emissions in cement and steel production [549]. The strategic use of biomass and hydrogen as energy and feedstock mitigates some reliance on fossil fuels, demonstrating promising potential in the sector's green transition [553].

Overall, addressing the GHG emissions from the cement, steel, and chemical industries necessitates a multifaceted approach involving innovative technologies, alternative materials, and significant policy support to invest in sustainable production processes.

Fuel, Energy and Net Zero

Steel Production: Conventional Fuel Use, Emissions, and Emerging Alternatives

Steel production remains one of the most energy-intensive and emissions-heavy industrial processes in the world. The predominant method for producing steel globally is the blast furnace–basic oxygen furnace (BF–BOF) process. This traditional route involves transforming iron ore into molten iron using coking coal—a refined form of coal known as coke. Coke serves two essential functions: it generates the intense heat required to melt the iron ore, and it acts as a chemical reductant, removing oxygen from the iron ore (iron oxide) through a high-temperature chemical reaction. The result is molten iron, which is then further refined into steel.

In addition to coking coal, natural gas is sometimes used as an energy source in a different process known as direct reduced iron (DRI). This technique is more common in regions where natural gas is abundant and relatively inexpensive. In the DRI method, natural gas reacts with the iron ore at lower temperatures to remove oxygen, yielding a solid form of iron that can later be melted in an electric furnace.

The emissions profile of conventional steel production is significant. The process accounts for approximately 7–9% of global CO_2 emissions, a consequence largely attributed to the combustion of coke in blast furnaces. Because coke is a carbon-rich fossil fuel, its use results in substantial carbon dioxide release. The scale of global steel production, combined with the chemical nature of the process, makes the industry one of the largest single contributors to industrial greenhouse gas emissions.

In response to the environmental impact of steelmaking, several emerging alternatives are under development and early-stage deployment. One of the most promising innovations is hydrogen-based direct reduced iron (H-DRI). This method replaces coke with hydrogen gas as the reducing agent. When hydrogen reacts with iron ore, it produces water vapor instead of CO_2, making it a near-zero-emissions process—provided that the hydrogen is produced from renewable sources (i.e., green hydrogen).

Another pathway is the use of electric arc furnaces (EAFs), which are already common in regions with established scrap metal recycling industries. EAFs melt down scrap steel using electricity, which can be sourced from renewable energy, thereby significantly reducing emissions compared to the BF–BOF route. This method, however, depends on the availability of high-quality scrap steel and a clean electricity grid.

Additionally, researchers and industry leaders are exploring the use of biomass-derived reductants and synthetic alternatives to coke. These alternatives aim to provide the necessary heat and chemical reduction capabilities while lowering the overall carbon footprint of the process. While promising, such approaches are still being optimized and face challenges in terms of cost, scalability, and consistency of supply.

Steel production is a vital but carbon-intensive process. While traditional methods remain dominant, a growing focus on decarbonisation is driving investment in hydrogen, electrification, and alternative reductants. These innovations, supported by policy, financing, and technological advancement, represent a crucial step toward aligning steel production with global climate goals.

The amount of fuel and energy used to produce steel varies significantly depending on the production method, the types of fuels involved, and the overall efficiency of the plant. Globally, the two dominant steelmaking methods are the Blast Furnace–Basic Oxygen Furnace (BF–BOF) route and the Electric Arc Furnace (EAF) route, each with distinct fuel and energy profiles.

The BF–BOF method is the traditional and most widely used steel production route, accounting for around 70% of global steel output. This method primarily relies on coking coal, which serves both as a fuel and a reductant to extract iron from ore. On average, producing one tonne of crude steel through this process requires approximately 300 to 400 kilograms of coke, 100 to 150 kilograms of pulverised coal (in operations using PCI to partially substitute coke), and 50 to 100 kilograms of limestone or other fluxes to remove impurities. Electricity is also used for auxiliary operations, amounting to about 400 to 600 kilowatt-hours (kWh) per tonne. In terms of total energy, this translates to 20 to 30 gigajoules (GJ) per tonne of crude steel, or 5,500 to 8,300 kWh. Most of this energy comes from coal (70–75%), with some contributions from natural gas and electricity. CO_2 emissions are high, ranging between 1.8 and 2.2 tonnes per tonne of steel, largely due to the combustion of coke and the chemical reduction of iron ore.

In contrast, the Electric Arc Furnace method, used for around 30% of global steel, is generally cleaner and more energy-efficient, particularly in regions with abundant scrap steel. This process typically consumes about 1,000 kilograms of scrap steel and between 350 to 700 kWh of electricity per tonne of steel. If Direct Reduced Iron (DRI) is used instead of or alongside scrap, an additional 1 to 3 GJ of natural gas or coal may be needed. Overall, EAF energy use ranges from 2 to 8 GJ per tonne, significantly lower than the BF–BOF method. CO_2 emissions are correspondingly lower as well, at 0.3 to 0.7 tonnes per tonne of steel, and can be near zero if powered entirely by renewable electricity.

An emerging low-carbon alternative involves combining hydrogen-based DRI with EAF technology. In this route, hydrogen replaces coke as the reductant, reacting with iron ore to produce iron and water vapor instead of CO_2. Producing one tonne of steel via this method requires around 50 to 60 kilograms of green hydrogen and 350 to 500 kWh of electricity for the electric arc furnace. However, the energy requirement is relatively high due to the energy-intensive electrolysis process used to produce hydrogen, amounting to roughly 30 to 35 GJ per tonne. If both the hydrogen and electricity come from renewable sources, this method can achieve near-zero CO_2 emissions.

To summarise, the BF–BOF process is energy-intensive and carbon-heavy, while EAF offers a more energy-efficient and cleaner pathway, especially when integrated with scrap recycling and renewable power. Hydrogen-based steelmaking is promising but demands substantial energy and infrastructure investment. As steel production is one of the most energy-consuming and emissions-intensive industrial processes, shifting toward cleaner methods like EAF and hydrogen-based DRI is essential for decarbonising the global steel sector.

Table 6: Summary Table (Approximate).

Method	Energy Use (GJ/tonne)	CO$_2$ Emissions (tonne/tonne)	Primary Fuels
BF–BOF	20–30 GJ	1.8–2.2	Coke, PCI coal, electricity
EAF (scrap-based)	2–8 GJ	0.3–0.7	Electricity, some gas
H$_2$-DRI + EAF	30–35 GJ	~0 (if green H$_2$/electricity)	Hydrogen, electricity

The production of steel utilizing renewable energy sources necessitates a transformative shift in production methodologies along with substantial infrastructure investment. The prevalent method of steel production—the Blast Furnace–Basic Oxygen Furnace (BF-BOF) route—currently accounts for approximately 70% of the global steel output and is highly carbon-intensive, emitting between 1.8 to 2.2 tonnes of CO$_2$ for each tonne of steel produced. This method primarily relies on coal, both as a heat source and as a chemical reductant, making it difficult to convert entirely to renewable energy sources without significant alterations to the existing production processes [554, 555]. Specifically, renewable energy cannot easily replace the coking coal used in blast furnaces, as coke is integral to the chemical reduction of iron ore, serving as both a fuel and reactant [554, 556].

In contrast, the Electric Arc Furnace (EAF) technology, which primarily leverages recycled scrap steel, presents a more sustainable alternative capable of operating entirely on renewable electricity. The EAF process is notably less carbon-intensive, resulting in emissions ranging from 0.3 to 0.7 tonnes of CO$_2$ per tonne of steel. With an appropriate renewable energy supply—such as wind, solar, or hydroelectric—EAFs can achieve close to zero emissions. However, a significant limitation remains; the availability of scrap steel is insufficient to satisfy global demand, particularly in growing industrial regions [555, 557].

An emerging and promising avenue for primary steel production involves the hydrogen-based Direct Reduced Iron (DRI) technique, in which green hydrogen—produced via electrolysis from renewable electricity—serves as a substitute for coke in the reduction process. This method only emits water vapor as a by-product and is considered effectively carbon-neutral. Nevertheless, transitioning to hydrogen-based DRI is highly energy-intensive, requiring around 30-35 GJ of energy per tonne of steel, predominantly due to hydrogen production needs [555, 558]. Thus, while power generation from renewable sources presents a viable pathway, there is an urgent requirement for significant advancements in renewable energy capacities, electrolysis efficiencies, and hydrogen storage technologies, all coupled with strong policy frameworks to drive necessary investments and innovations [537, 559].

In conclusion, while the transition to renewable energy-based steel production is not only feasible but partially underway through the EAF process, scaling this model globally—especially via hydrogen-based DRI—remains essential for decarbonizing the steel industry. This transition will necessitate extensive investments in renewable energy infrastructures and innovative technologies, which are critical for achieving global climate objectives and ensuring sustainable industrial growth [560, 561].

Cement Production

Cement production is one of the most carbon-intensive industrial processes in the world, contributing significantly to global greenhouse gas emissions. At the heart of cement manufacturing is the production of clinker, which involves heating a mixture of limestone (calcium carbonate) and other materials in a kiln to approximately 1,450°C. This high-temperature process, known as calcination, transforms limestone into calcium oxide (lime), which is then ground with other minerals to produce cement. The extreme temperatures required for this reaction mean that the process is highly dependent on thermal energy, traditionally supplied by burning fossil fuels.

The conventional fuels used in cement kilns vary by region, but coal and petroleum coke (petcoke) are the most widely used sources of energy due to their high energy density and affordability. In some regions, natural gas and fuel oil are also used, depending on availability and price competitiveness. These fuels provide the necessary heat to drive the calcination reaction but also contribute directly to carbon dioxide (CO_2) emissions through their combustion.

However, what makes cement production uniquely challenging from a climate perspective is that more than half of its CO_2 emissions arise not from burning fossil fuels, but from the calcination process itself. When limestone ($CaCO_3$) is heated, it breaks down into calcium oxide (CaO) and releases CO_2 as a by-product of the chemical reaction. This means that even if cement kilns were powered entirely by renewable energy, substantial emissions would still occur from the material transformation at the core of cement making.

Overall, cement manufacturing is responsible for about 8% of global CO_2 emissions, making it a major contributor to climate change. This includes emissions from both fuel combustion and process-related calcination. The dual sources of emissions—energy and chemical—make decarbonizing cement production particularly complex.

To address these environmental challenges, the industry is increasingly turning to emerging alternatives. One approach is the use of waste-derived fuels, such as shredded tires, plastic waste, and biomass, to replace coal and petcoke. These fuels can reduce net emissions, especially if they are derived from renewable or non-fossil sources. Another critical solution is carbon capture and storage (CCS), which can directly capture CO_2 released from the calcination process and prevent it from entering the atmosphere. Given that process emissions are unavoidable through combustion changes alone, CCS is considered essential for deep decarbonization of the sector.

In addition, there is a growing push toward alternative cements that reduce or eliminate the need for traditional clinker. Geopolymer cements and blended cements, which incorporate industrial by-products like fly ash or slag, can dramatically cut emissions by reducing the proportion of clinker in the final product. These innovations not only lower the carbon footprint of cement but also promote circular economy principles by making use of waste materials.

While cement production presents significant environmental challenges, a combination of fuel substitution, process innovation, and carbon capture technologies can help transform the industry. Achieving meaningful reductions in emissions will require a systemic approach involving new technologies, supportive regulations, and market incentives to accelerate the adoption of cleaner practices and alternative materials.

The amount of fuel and energy used to produce cement varies depending on the production technology, plant efficiency, and type of fuel employed. However, the cement industry remains one of the most energy-intensive industrial sectors globally. Producing cement involves heating raw materials—mainly limestone and clay—in a rotary kiln to around 1,450°C to form clinker, the key intermediate product. This process requires substantial thermal energy, ranging from 3.2 to 6.3 gigajoules (GJ) per tonne of clinker. The most energy-efficient systems, such as modern dry-process kilns with preheaters and precalciners, operate closer to 3.2–4.2 GJ/tonne, while older wet kilns often consume 5.5–6.3 GJ/tonne due to the extra energy required to evaporate water.

The fuels used to supply this thermal energy are primarily coal and petroleum coke (petcoke), though natural gas, oil, and alternative fuels such as biomass and waste-derived materials are also employed based on regional cost and environmental regulations. On average, producing one tonne of clinker requires approximately 100 to 200 kilograms of coal or petcoke, or an equivalent energy value from other fuel sources.

Electricity is another essential energy input in cement production, primarily used for crushing and grinding raw materials and clinker, operating fans, pumps, and conveyors, as well as packaging and dispatching the final product. Electricity consumption typically ranges from 90 to 150 kilowatt-hours (kWh) per tonne of cement produced.

When both thermal and electrical energy inputs are combined, the total energy consumption for producing one tonne of cement typically falls between 3.5 and 6.0 GJ. This is roughly equivalent to 1,000 to 1,700 kWh per tonne. A significant proportion of cement's carbon footprint is linked to energy use. Around 40% of CO_2 emissions result from fuel combustion, while 50–60% come from the calcination process itself, in which limestone ($CaCO_3$) decomposes into lime (CaO) and carbon dioxide (CO_2). As a result, cement manufacturing accounts for approximately 8% of global CO_2 emissions.

Improving energy efficiency and transitioning to cleaner fuels are essential for decarbonising the cement industry. Dry kilns with preheaters and precalciners offer greater efficiency, while innovations such as waste heat recovery systems, the use of alternative and waste-derived fuels, and the development of blended cements with reduced clinker content can significantly lower energy demand and emissions. In summary, cement production relies heavily on both thermal fuels and electricity, and meaningful environmental improvements will depend on adopting more efficient technologies and shifting to low-carbon energy sources.

Cement production is a significant contributor to global carbon emissions, particularly due to the high-temperature processes required, especially the calcination of limestone (calcium carbonate). This process occurs at approximately 1,450°C in rotary kilns and releases substantial amounts of CO_2. Research indicates that addressing both the energy demand and the inherent chemical emissions associated with cement production is crucial for transitioning to renewable energy sources in this sector [562, 563].

The current reliance on fossil fuels, such as coal, petcoke, and natural gas, is due to their efficiency in achieving the necessary high temperatures for clinker production. Alternatives such as biomass and waste-derived fuels are being investigated as renewable options. For instance, studies have highlighted the potential of biomass in conjunction with refuse-derived fuels (RDF) to partially replace traditional fossil fuels in cement kilns, offering pathways to both renewable energy solutions and reduced carbon footprints [563]. However, significant challenges persist due to the variable availability, consistency, and calorific value of these alternatives, which affect their scalability [563].

Electrifying kiln processes is another proposed strategy for utilizing renewable energy. Technologies such as electrified kilns and plasma torches, which can generate the required high temperatures using electricity from renewable sources, are still under development [564]. These methods, while promising, face considerable technical and economic challenges. The required energy inputs are projected to be between 3.5 to 6.0 gigajoules per tonne of cement, necessitating substantial increases in grid capacity and investments in renewable infrastructure [564].

Moreover, the calcination process contributes approximately 50–60% of CO_2 emissions associated with cement production. Even if thermal energy needs are satisfied with renewable sources, achieving significant overall emissions reductions will require the integration of carbon capture and storage (CCS) or carbon capture and utilization (CCU) technologies into cement manufacturing processes [563, 564]. Such innovations present additional complexities and need supportive policy frameworks for effective implementation [563].

While transitioning to renewable energy in cement production is technically feasible, it poses substantial challenges related to the thermal requirements for clinker production and the inherent emissions from calcination. A combination of renewable thermal solutions, electrification, and carbon capture technologies will be vital in achieving a low-carbon future for the cement industry. This multifaceted approach underscores the importance of strategic planning, investment, and innovative technologies to ensure both economic and environmental sustainability in cement manufacturing.

Chemical Production

The chemical industry is a major industrial energy consumer and greenhouse gas emitter, primarily because it relies heavily on fossil fuels not just for generating energy, but also as feedstocks—the raw materials used in synthesising a wide range of chemicals. This dual role of fossil fuels in the industry makes decarbonisation particularly challenging, yet critically important in the context of climate change.

A key fuel in the sector is natural gas, which serves both as a source of thermal energy and as a primary feedstock for producing hydrogen, particularly for ammonia synthesis via the Haber-Bosch process. Ammonia is essential for nitrogen-based fertilisers, making this process vital to global food production. Hydrogen derived from natural gas is also used extensively in the production of methanol, another fundamental industrial chemical used in plastics, paints, and fuels. In addition to natural gas, oil and naphtha are crucial feedstocks for petrochemical production, forming the building blocks for plastics, synthetic rubber, solvents, and numerous other consumer and industrial products. These hydrocarbon derivatives are processed in steam crackers to yield basic chemicals such as ethylene, propylene, and benzene, all central to modern materials and manufacturing.

Coal also plays a role in chemical manufacturing, especially in China, where coal-to-chemical processes are used to produce syngas (a mix of hydrogen and carbon monoxide) for synthetic fuels and chemicals. Although this route is economically attractive in coal-rich regions, it is extremely carbon-intensive, contributing significantly to both local air pollution and global CO_2 emissions.

The overall emissions profile of the chemical industry reflects this dependency. It is responsible for an estimated 3–5% of global CO_2 emissions, with particularly high emissions from the production of ammonia, methanol, and ethylene. These emissions stem from both the combustion of fossil fuels for energy and the chemical reactions inherent in transforming fossil-based feedstocks.

To mitigate these environmental impacts, a range of emerging alternatives are being explored. One of the most promising is the use of green hydrogen, produced from renewable electricity through electrolysis, to replace fossil-based hydrogen in ammonia and methanol production. This would eliminate a significant source of process emissions. Additionally, the electrification of heat processes—traditionally reliant on burning fossil fuels—is gaining attention. Techniques such as electric resistance heating, microwave heating, and plasma-based technologies can provide high-temperature heat using renewable electricity, potentially replacing fossil-fuel-based boilers and furnaces.

Other decarbonisation pathways include carbon capture and utilization (CCU), which involves capturing CO_2 emissions and converting them into value-added products like synthetic fuels, chemicals, and building materials. While still developing, CCU could help reduce the net emissions of unavoidable chemical processes. There is also growing interest in using bio-based feedstocks, such as plant-derived sugars, starches, and lignocellulosic biomass, as renewable alternatives to fossil-derived naphtha and natural gas. These can be processed into similar chemical products with a lower carbon footprint, provided their cultivation and harvesting are sustainable.

The chemical industry's reliance on fossil fuels for both energy and raw materials makes it one of the hardest-to-abate sectors. However, with strategic investments in green hydrogen, electrification, carbon capture, and bio-based feedstocks, it is possible to significantly reduce its environmental impact. A successful transition will require technological innovation, supportive policy frameworks, and a strong commitment from industry stakeholders to adopt low-carbon alternatives while maintaining the supply of critical chemical products for global economies.

The production of ammonia, primarily through the Haber-Bosch process, is one of the most energy-intensive industrial activities. This is largely because the process demands substantial quantities of hydrogen—typically derived from fossil fuels—and operates under extreme conditions of temperature (400–500°C) and pressure (150–300 atmospheres). The reaction at the heart of ammonia synthesis involves combining nitrogen from the air with hydrogen to form ammonia ($N_2 + 3H_2 \rightarrow 2NH_3$). The main technical challenge lies in generating the hydrogen, which is both energy-intensive and emission-heavy when derived from conventional sources.

The most common method for ammonia production globally is through steam methane reforming (SMR) of natural gas. Producing one tonne of ammonia via this method consumes approximately 28 to 35 gigajoules (GJ) of natural gas, along with 0.8 to 1.0 megawatt-hours (MWh) of electricity, primarily for compression and ancillary operations. Of the natural gas consumed, 60 to 70% is used as a feedstock to produce hydrogen, while the remaining 30 to 40% is combusted to generate process heat. This method typically results in carbon dioxide emissions ranging from 1.6 to 2.0 tonnes per tonne of ammonia produced.

In China, where coal is more abundant than natural gas, a coal gasification route is widely used. This approach requires between 40 and 50 GJ of coal-derived synthesis gas (syngas) per tonne of ammonia, making it more energy-intensive due to the additional steps needed to extract and purify the hydrogen. As a result, coal-based ammonia production has a significantly higher emissions profile, with carbon dioxide emissions ranging from 2.5 to 4.5 tonnes per tonne of ammonia, depending on the quality of the coal and plant efficiency.

An emerging and promising alternative is green ammonia, which uses renewable electricity to produce hydrogen via electrolysis, eliminating the need for fossil fuel-based feedstocks. Producing one tonne of ammonia through this method requires approximately 8.5 to 10 MWh (30 to 36 GJ) of electricity. While less efficient overall—due to energy losses in electrolysis and the need for high-pressure operations—this route offers near-zero carbon emissions if the electricity used comes entirely from renewable sources.

The method of ammonia production greatly influences both its energy consumption and environmental impact. Natural gas-based processes are more energy-efficient than coal-based ones but still result in significant emissions. Green ammonia provides a clean, low-emission alternative, but it is energy-intensive and requires major investment in renewable electricity generation and electrolyser infrastructure. As ammonia is a critical input for global fertiliser production and an emerging energy carrier, scaling up green ammonia will be essential for reducing industrial carbon emissions and transitioning to a low-carbon economy.

Ammonia production has been a focal point of contemporary research due to its substantial implications in agriculture and energy sectors. Traditional methods of ammonia production, specifically the Haber-Bosch process, are heavily reliant on hydrogen derived from fossil fuels such as natural gas and coal, making them carbon-intensive and contributing significantly to global CO_2 emissions. In contrast, green ammonia production offers a prospective pathway that integrates renewable energy sources into its production process, significantly reducing its carbon footprint.

Green ammonia is synthesized through the electrolysis of water using renewable electricity from sources like solar or wind. This process extracts hydrogen, which is then combined with nitrogen drawn from the atmosphere in the Haber-Bosch reaction to produce ammonia, similar to traditional methods while eliminating the reliance on fossil fuels [565]. The theoretical framework of green ammonia production highlights that one tonne of green ammonia typically necessitates approximately 8.5 to 10 megawatt-hours (MWh) of electricity, indicating a higher energy requirement compared to fossil-based methodologies [566]. However, this energy investment is offset by the elimination of direct CO_2 emissions, rendering green ammonia a viable candidate for decarbonizing the ammonia sector, which currently generates about 1.44% of global carbon emissions [567].

Despite the potential of green ammonia production, several challenges hinder its widespread adoption. The energy efficiency of this production method is lower than that of traditional processes, as losses occur during electrolysis and high-pressure synthesis [568, 569]. Additionally, the initial capital investments required for electrolyser technology are high, and the stability of renewable electricity supply poses a significant challenge due to its intermittent nature [569, 570]. The need for infrastructure advancements for storage and transport is also critical, particularly for applications where ammonia may serve as an energy carrier in maritime or power generation systems [571].

Recent research emphasizes the urgent need for supportive policy frameworks and investment to scale up green ammonia production efficiently. Countries with abundant renewable resources possess unique advantages in leading this transition towards sustainable ammonia production [572, 573]. Furthermore, the collaboration between energy and industrial policies will enhance the feasibility of green ammonia, ensuring that the technology not only develops but also integrates seamlessly into existing agricultural and energy infrastructures [574, 575].

While green ammonia is technically viable and offers substantial environmental benefits, achieving its full potential hinges on tackling current economic barriers, enhancing energy efficiency, and establishing the necessary infrastructure for deployment. The successful implementation of green ammonia production will facilitate significant reductions in global greenhouse gas emissions and contribute to the transition towards a sustainable hydrogen economy.

Outlook

Fuel use in steel, cement, and chemical production is deeply embedded in the industrial processes that underpin modern economies. These sectors are among the most energy-intensive and emission-heavy, relying heavily on fossil fuels not only for thermal energy but also as feedstocks in chemical transformations. Their complexity and high-temperature requirements make them particularly difficult to decarbonise, yet they are essential for construction, manufacturing, and agriculture.

Despite these challenges, emerging technologies offer promising pathways for reducing emissions. Advances in green hydrogen, process electrification, carbon capture and storage (CCS), and alternative materials are beginning to reshape what is possible in industrial production. These innovations could significantly reduce dependence on fossil fuels while maintaining industrial output.

Achieving effective decarbonisation in these sectors will require coordinated and sustained action. Policy support must align with industry goals to ensure a stable transition framework, while significant investment in research and development is needed to scale up and commercialise alternative processes. International mechanisms such as carbon pricing or market-based incentives can help level the playing field and drive innovation across borders. Furthermore, the strategic deployment of low-carbon infrastructure—such as renewable-powered hydrogen production or CCS networks—will be crucial in enabling industry-wide transformation.

If these measures are implemented effectively, the global economy can transition toward cleaner industrial production without compromising economic growth or material supply. This balance is vital for ensuring both climate responsibility and continued industrial resilience.

Conversely, transitioning a country's fuel and energy mix to support clean energy initiatives while potentially undermining foundational sectors such as steel, aluminium, cement, and chemicals can lead to significant socioeconomic repercussions. This transformation must be strategically planned to prevent declines in productivity and job security within these critical industries.

Industrially, the immediate impact of shifting away from fossil fuels—historically relied upon for their high-temperature thermal energy—can lead to declines in output and competitiveness. Edomah [576] discusses how poorly implemented energy policies can contribute to deindustrialization by failing to meet the operational demands of energy-intensive sectors. Studies indicate that moving these industries to regions with stable energy policies can offset domestic production capabilities, leading to facility closures and job losses in skilled labour markets [577]. Levinson's [578] research further highlights the correlation between energy policies, industrial declines, and economic resilience, confirming the nuanced balance needed between energy efficiency and industrial output.

Economically, deindustrialization in favour of clean energy initiatives can increase reliance on imported materials, exposing countries to global market volatility and trade imbalances. Gasim [579] points out that as industrialized nations offshore their energy-intensive production, it affects their competitiveness, particularly as higher domestic production costs inhibit market viability. Price volatility can ripple through supply chains, potentially impacting food security and manufacturing capabilities [580]. The loss of domestic industrial capacity undermines economic strength and increases vulnerability to external pressures and crises, as highlighted in the economic literature [581].

The social repercussions of energy transitions can be concerning. Regions reliant on heavy industry may face increased unemployment and economic distress, potentially provoking resistance against climate policies perceived as unjust [582]. Paterson and P-Laberge emphasize that the political economy plays a crucial role in shaping responses to climate change, highlighting tensions between economic growth and meeting decarbonization targets [583]. This sociopolitical instability can obstruct meaningful climate action, especially within the affected communities.

Infrastructure development, inherently linked to national growth and quality of life, may also suffer setbacks from energy transitions. Steel and cement are essential for constructing infrastructure, including renewable energy installations. Delays in local production can jeopardize progress toward sustainability goals, highlighting the complexities involved in these transitions [584].

From a national security perspective, there is a clear connection between a country's energy independence and its ability to respond to crises. Energy and industrial capacities are strategic assets vital for national resilience. A decline in domestic production of critical materials can compromise a nation's preparedness for emergencies, exacerbating vulnerabilities during crises [585].

Transitioning to a cleaner energy mix is vital for addressing climate change, but the approach must be carefully considered to avoid undermining economic, industrial, and social stability. Sustaining industrial viability alongside decarbonization requires significant investments in low-carbon technologies, diversification of energy sources, and carefully managed, region-specific transition plans that account for current and future socioeconomic landscapes [586]. These strategies are essential to achieve climate goals without threatening the foundational structures of national economies.

The transition away from fossil fuel dependence in the steel, cement, and chemical production industries is a multifaceted challenge that requires a nuanced understanding of various socio-economic factors, institutional capabilities, and governance structures. Some countries, including Germany, Sweden, Canada, Japan, and emerging economies like India and China, demonstrate varying degrees of readiness to effectively manage this transition. Successfully navigating this complex landscape necessitates robust frameworks in policy, technical innovation, and international cooperation, as well as addressing social equity concerns.

Germany has established itself as a leader in industrial decarbonization, leveraging strong governance and diversified economies to support ambitious climate policies. The implementation of frameworks like Carbon Contracts for Difference aligns with EU climate goals, enhancing policy stability amid concerns over energy costs [587]. However, Germany also faces challenges, including high energy prices and reliance on imported gas, underscoring the need for ongoing public support and innovation in green technologies such as hydrogen networks and fossil-free steel initiatives [587].

Sweden is another country demonstrating a strong commitment to low-carbon innovation, primarily due to its reliance on hydro and nuclear energy. Projects like HYBRIT show tangible steps toward achieving fossil-free steel production, further aided by social trust and cohesive policy frameworks that support environmental objectives [587]. However, Sweden must confront limitations regarding domestic mineral resources and the high upfront costs associated with transitioning its industrial base, necessitating careful long-term planning and resource allocation [587].

Canada benefits from its vast natural resources, including hydroelectric power and minerals critical for low-carbon technologies. Initiatives like CleanBC and the Net-Zero Accelerator illustrate Canada's commitment to decarbonization [587]. Nonetheless, political opposition from fossil-fuel-intensive provinces and logistical challenges inherent in remote areas may impede its transition efforts. However, Canada is well-positioned to emerge as a significant player in exporting clean hydrogen and developing integrated low-carbon supply chains across North America [587].

Australia's renewable energy potential offers considerable opportunities for leading the transition in green steel and other low-carbon products. Policy fragmentation and an aging energy infrastructure pose barriers, but state-level initiatives and private investment are starting to reshape the energy landscape [587]. Nonetheless, Australia's ongoing reliance on fossil fuels and public concerns about living costs present significant challenges that need to be addressed for a successful transition [588].

Japan's strategy focuses on leveraging technological innovation to address energy constraints and enhance energy independence through hydrogen and ammonia investments. Collaborative projects, such as importing hydrogen from Australia, reflect Japan's proactive stance. However, high costs of energy imports pose risks to its energy strategy, despite Japan's capacity for advanced manufacturing and clean fuel trade [589].

Emerging economies like India and China face unique challenges as they attempt to balance rapid industrial growth with climate commitments. India's commitment to green hydrogen and renewables illustrates its ambitions, yet widespread energy poverty and regional disparities complicate progress [587]. International finance and technology transfer are crucial to ensure India can undertake a just transition while promoting economic growth [587]. China plays a significant role in global industrial production, yet its dependence on coal complicates national transition strategies, highlighting the necessity of regional approaches to decarbonization [590].

The complexities of this transition underscore the importance of ensuring policy coherence across energy, industrial, and climate domains—using tools like tax incentives, carbon pricing, and social protection measures to foster job transformation in transitioning sectors [591]. Investment in renewable energy infrastructure is vital for facilitating the transition, while international cooperation is essential to ensure access to finance and technology [587]. Community engagement initiatives will be critical to building public trust and ensuring that transitions are equitable, especially for communities heavily reliant on fossil fuel industries [592].

Countries such as Germany, Sweden, Canada, Australia, and Japan are equipped to spearhead the industrial transition towards sustainable energy due to their advanced policy frameworks and technological capabilities. Emerging economies like India and China also possess notable potential, although they will require targeted support and strategies attuned to their contexts. A balanced approach is vital for achieving sustainable decarbonization without compromising energy security or economic resilience.

Heating and Process Fuels

Heating and process fuels are essential energy sources used to generate the thermal energy required in various industrial, manufacturing, commercial, and residential applications. These fuels support a wide range of high-temperature processes such as metal smelting, cement production, chemical synthesis, glass manufacturing, food processing, and space heating. The type of fuel used in these applications significantly affects energy efficiency, operational costs, greenhouse gas emissions, and the overall environmental sustainability of the system.

Fossil fuels remain the dominant source of heat energy in industrial settings. Natural gas is widely used due to its high energy content, clean combustion, and ease of control. It is particularly common in food processing, chemical plants, and commercial heating systems. Coal, although traditionally used in power generation and heavy industries such as steel and cement production, is being phased out in many regions due to its high carbon emissions. Oil and various grades of fuel oils are used in boilers and furnaces, especially where natural gas infrastructure is lacking. Petroleum coke (petcoke), a by-product of oil refining, is also used in high-temperature applications, particularly in the cement and metal industries.

Electricity is increasingly used for electric heating in industrial processes that require precise temperature control. Technologies such as induction heating, electric arc furnaces, and resistive heating are employed in manufacturing and laboratory environments. With the growing integration of renewable energy into power grids, electricity is becoming a low-carbon option for process heat, especially in low- to medium-temperature applications.

Biomass and biofuels offer renewable alternatives to fossil fuels. Solid biomass, such as wood chips and agricultural residues, is often used in boilers for medium heat applications. Biogas and bio-oils can substitute for natural gas and heating oil in suitably adapted systems. These fuels are particularly viable in regions with abundant agricultural resources and supportive renewable energy policies.

Hydrogen is an emerging option for decarbonising high-temperature industrial heating. It can be produced in several forms: grey hydrogen from natural gas, blue hydrogen with carbon capture and storage, and green hydrogen via electrolysis using renewable electricity. Hydrogen has the potential to replace fossil fuels in sectors like steelmaking, cement manufacturing, and chemical production. However, its widespread use is currently limited by infrastructure needs and high production costs.

In industrial applications, heating and process fuels are tailored to the specific thermal demands of each sector, with energy intensity and temperature requirements varying widely across industries.

Steel and metal processing is one of the most heat-intensive sectors, requiring extremely high temperatures for smelting, rolling, casting, and heat treatment of metals. Traditionally, these processes rely heavily on coal (especially in the form of coke in blast furnaces) and natural gas. However, the industry is increasingly exploring low-carbon alternatives such as hydrogen-based reduction and electric arc furnaces powered by renewable electricity to reduce carbon emissions.

Cement manufacturing also demands very high temperatures—around 1,450°C in rotary kilns—to produce clinker, the key ingredient in cement. This process has historically depended on coal and petroleum coke (petcoke) due to their high calorific value. In recent years, however, there has been a growing shift toward using alternative fuels such as waste-derived materials and biomass to reduce fossil fuel use and emissions.

Chemical production requires both feedstocks and process heat. Natural gas plays a dual role here, acting as both an energy source and a raw material, especially in the production of ammonia, methanol, and hydrogen. As the sector seeks to decarbonise, innovations in electrification of heat and the adoption of green hydrogen as a feedstock are being tested and deployed.

In the food and beverage industry, thermal energy is crucial for maintaining product quality, safety, and longevity. Various processes, including drying, baking, brewing, pasteurization, and sterilization, rely on the controlled application of heat to transform raw ingredients into safe, consumable products. Efficient thermal energy management not only ensures the quality of the final product but is also vital for preventing microbial growth and ensuring adherence to safety standards [593, 594].

Drying is one of the most energy-intensive methods for food preservation, effectively removing moisture from products such as grains, fruits, vegetables, and meats. The aim is to extend shelf life and prevent microbial growth while also reducing weight for transportation [595]. The efficiency of drying processes is highly dependent on the consistency of heat supply, airflow control, and ambient conditions. Studies have demonstrated that drying operations can account for a significant portion of the energy consumption in food processing—approximately 10% to 25% in industrial contexts [596]. Moreover, innovations such as intermittent drying and microwave-assisted processes are emerging to optimize energy use while preserving food quality [597, 598].

Baking is another critical thermal process, transforming dough into various baked goods through intense heat application. This process not only facilitates chemical transformations such as starch gelatinization and protein denaturation but also plays a significant role in moisture evaporation. Natural gas-fired ovens are prevalent in this sector due to their fast response times and uniform heat distribution, improving product consistency and quality while also emitting lower levels of pollutants compared to other fossil fuels [599]. The importance of energy reliability in baking cannot be overstated, as irregular heating can lead to inconsistent product quality, thus impacting safety and marketability.

Fuel, Energy and Net Zero

In beer production, thermal energy is essential during key steps like mashing and boiling, where mixtures of water and malted grains are heated to extract sugars [593]. Accurate temperature control is vital for both efficiency and quality, with many breweries opting for steam generated from natural gas due to its ability to deliver hygienic and precise heat applications. This reliance on fossil fuels, however, raises concerns about sustainability; hence, some breweries are beginning to explore biomass and other renewable energy options for their thermal needs [600].

Pasteurization involves heating liquids like milk and juices at controlled temperatures (typically between 60–90°C) to eliminate harmful microorganisms without compromising taste or nutritional value [601]. This thermal treatment is crucial for ensuring the safety of consumables and typically utilizes systems powered by natural gas or electric boilers. Sterilization, which applies higher temperatures along with pressure, is essential for producing shelf-stable foods like canned goods. The energy demands for these processes are considerable, necessitating efficient energy sources to ensure both food safety and quality [602].

Natural gas remains the dominant fuel source in the food and beverage industry due to its clean combustion and controllability, making it a reliable option that supports food safety and quality management. Additionally, the adoption of biomass for thermal energy is increasing, particularly in regions where agricultural waste can be utilized to meet sustainability goals and reduce carbon footprints [593, 603]. As these fuel sources evolve, their balance will play a crucial role in shaping the thermal management strategies employed across various food processing sectors [600].

The food and beverage industry is a significant consumer of energy globally, impacting fossil fuel utilization across agriculture, processing, transportation, and storage. Although precise global energy consumption data can be challenging to ascertain due to discrepancies in reporting methods, studies indicate that food systems account for approximately 15% of total global fossil fuel use annually [604, 605]. This substantial energy demand underscores the importance of adopting energy-efficient practices and transitioning towards sustainable energy sources within the sector.

Figure 34: JBT Stein TwinDrum PRoYIELD™ 600 Spiral Oven. Kearns jbt, CC BY-SA 4.0, via Wikimedia Commons.

In the United States, the food and beverage industry consumed around 1,935 trillion British thermal units (TBtu) of primary energy in 2018, with natural gas comprising approximately 83% of this consumption, primarily for process heating applications such as food dehydration [605]. This reliance on fossil fuels highlights critical operational processes. Similarly, in India, the food processing industry consumes about 5,300 kilotons of oil equivalent (ktoe) annually, where energy-intensive activities like washing, cooking, and cooling account for a significant part of energy consumption, with electrical energy constituting roughly 50% of total consumption [605]. This illustrates the diverse energy needs across different regions while emphasizing a common challenge in meeting energy demands sustainably.

In the European Union, the food processing sector accounts for approximately 28% of the total energy use within the food and beverage industry, highlighting the potential for energy efficiency improvements in this sector [606]. The differences in energy consumption patterns across regions indicate a need for integrated energy strategies to optimize consumption. Shifting towards renewable energy sources, alongside improved energy efficiency technologies, can significantly mitigate environmental impacts while meeting the growing food demand, projected to rise by about 60% by 2050, which will entail a corresponding increase in global energy consumption [607].

The high energy consumption in the food and beverage industry has implications beyond resource use; it includes significant environmental considerations. Implementing energy-efficient technologies and optimizing processes can reduce energy waste and greenhouse gas emissions, thus supporting global sustainability initiatives [605, 608]. Additionally, reducing food waste yields substantial energy savings since wasted food represents the energy invested in its production, processing, and distribution [604, 608]. Addressing these interconnected areas allows the food and beverage industry to establish more sustainable practices, contributing significantly to global energy conservation efforts.

The textiles and pulp and paper industries rely on process heat in operations such as drying, bleaching, washing, and dyeing. Steam is a common medium for heat delivery in these sectors, typically produced by burning natural gas or biomass. Direct heating methods are also employed depending on process requirements and fuel availability.

Each of these industries faces unique challenges and opportunities in the transition to cleaner heating and process fuels, but collectively they represent key areas for emissions reduction and energy innovation.

Heating and process fuel use in industry has significant environmental and economic implications, particularly as governments and industries work toward reducing greenhouse gas emissions.

One of the most pressing concerns is carbon emissions. The combustion of fossil fuels—such as coal, oil, and natural gas—to generate industrial heat is a major source of greenhouse gases. These emissions contribute substantially to climate change and make industrial sectors some of the highest-emitting globally. Reducing reliance on carbon-intensive fuels is therefore a central focus of decarbonisation efforts.

Improving energy efficiency is a key strategy for reducing both emissions and operational costs. Enhancements such as more efficient burner designs, better insulation of equipment, and the implementation of heat recovery systems can significantly reduce the amount of fuel needed to achieve required temperatures. These improvements can lead to substantial energy savings and lower emissions without fundamentally altering existing processes.

Fuel switching is another avenue being explored to lower the carbon footprint of industrial heating. Transitioning from fossil fuels to electricity (especially from renewable sources), hydrogen, or biofuels can dramatically reduce emissions. However, such shifts often require retrofitting or redesigning existing equipment and may lead to increased capital and operating costs, at least in the short term. The feasibility of switching also depends on the availability and affordability of low-carbon energy sources.

Regulatory pressure is playing a growing role in accelerating cleaner fuel adoption. Governments are introducing measures such as carbon pricing, emission caps, and mandatory energy efficiency standards, all of which incentivise industries to reduce emissions and invest in cleaner technologies. These regulatory frameworks are gradually reshaping the economics of industrial energy use, making low-carbon options more competitive over time.

Overall, environmental and economic considerations are driving significant change in how industrial heat is produced and managed. While challenges remain, the transition to more efficient and lower-emission fuel sources is both necessary and increasingly viable.

Emerging trends in industrial heating are focused on increasing efficiency and reducing emissions through innovative technologies and alternative energy sources. These developments are reshaping how industries generate and use heat across various processes.

One of the most significant trends is the electrification of heat, particularly in low- and medium-temperature applications below 400°C. Electrification offers precise temperature control, lower emissions when powered by renewable electricity, and reduced on-site air pollution. While high-temperature processes have traditionally been more challenging to electrify, new technologies such as plasma heating, microwave heating, and resistance heating are now being explored as viable solutions for high-heat industrial needs. These technologies hold promise for sectors like steelmaking and chemical processing, where temperatures often exceed 1,000°C.

Another important development is the recovery of waste heat from industrial processes. Waste heat recovery systems capture thermal energy from hot exhaust gases, steam, or liquids that would otherwise be lost. This recovered heat can be reused within the same process or redirected to other operations, significantly improving overall energy efficiency and lowering fuel consumption.

The use of combined heat and power (CHP) systems is also gaining traction. CHP, also known as cogeneration, simultaneously produces electricity and captures usable heat from the same fuel source. This approach greatly improves the total energy yield compared to separate heat and power generation, making it an attractive option for industries seeking to maximise energy use and reduce greenhouse gas emissions. CHP systems can be powered by a range of fuels, including natural gas, biomass, or even hydrogen, depending on availability and sustainability goals.

The energy demands of the food and beverage, textiles, and pulp and paper industries can be effectively met through the adoption of renewable energy solutions, particularly for low- and medium-temperature thermal processes and electricity requirements. These industries utilize thermal energy within a range that aligns well with renewable technologies. The food and beverage sector, for instance, generally operates within the moderate heat range of 60°C to 200°C, which supports processes such as baking, boiling, and sterilization. Similarly, the textiles industry consumes thermal energy for dyeing and drying, while the pulp and paper sector relies on steam for pulp digestion and paper drying, all of which are compatible with renewable heating systems like biomass boilers and heat pumps powered by renewable electricity [609].

Electrification also offers a pathway for decarbonization in these sectors. Numerous components, including motors and refrigeration systems, can be efficiently powered by renewable sources such as solar photovoltaic, wind, or hydroelectric systems. Industrial heat pumps can facilitate the delivery of low- to medium-temperature heat, especially when run on clean electricity, making them an appealing option for these industries [610]. Furthermore, agro-industrial operations often generate biomass waste, which can be harnessed to produce renewable energy for heating or biogas, thus promoting a circular economy and reducing reliance on fossil fuels [611].

Despite the potential benefits, the transition to renewable energy sources poses several challenges. Certain processes requiring high temperatures—exceeding 400°C—remain difficult to address with current renewable technologies. While advancements such as concentrated solar thermal systems and green hydrogen could offer long-term solutions, their deployment is not widespread or economically viable at present [609]. Additionally, infrastructure upgrades to support the shift from fossil fuels can be financially burdensome for smaller enterprises, resulting in hesitance to invest in necessary technologies and systems [612].

Supply reliability is also a significant concern. The intermittent nature of solar and wind power necessitates reliable energy storage solutions to ensure a consistent supply, a particularly crucial factor for industries operating continuously. Although technologies like thermal energy storage and battery systems are advancing, they introduce complexity and cost to the energy management systems [613]. Furthermore, regional disparities in resource availability, such as biomass, solar radiation, and grid infrastructure, may necessitate customized solutions for different facilities, especially those situated in rural areas [614].

Real-world case studies illustrate the effective application of renewable technologies in these sectors. Notable examples include breweries like Heineken and Carlsberg investing in biomass boilers and solar thermal systems as part of their energy strategy, while textile factories in developing countries like India and Bangladesh have started utilizing solar process heating and biogas to offset fossil fuel consumption. Similarly, European pulp and paper mills frequently adopt biomass cogeneration solutions, achieving substantial reductions in fossil fuel reliance [614].

Case Studies by Country

The industrial landscape is witnessing significant transformations as diverse sectors adopt innovative strategies to manage fuel use and transition towards more sustainable energy solutions. A notable example of this paradigm is the Kalundborg Eco-Industrial Park in Denmark, recognized globally for its industrial symbiosis model. This park features a network wherein businesses collaborate to optimize resource utilization and minimize waste. For instance, the surplus heat generated from the Asnæs Power Station is repurposed for heating local residences and supplying steam to neighbouring industries, including a pharmaceutical facility and an oil refinery. This synergistic approach has resulted in substantial savings on resources and reductions in waste, showcasing the ecological and economic benefits inherent to industrial symbiosis [615-617].

On a broader scale, major oil and gas corporations are also shifting towards renewable energy sources as part of their transition strategy. Companies are increasingly investing in renewable projects such as wind, solar, and green hydrogen initiatives. These efforts are designed not only to reduce carbon footprints but also to ensure long-term sustainability in response to evolving market demands and stringent regulatory frameworks [618]. This diversification of energy portfolios aligns with global sustainability goals, reflecting the industry's adaptive response to climate change challenges [618].

In the United States, industrial energy efficiency efforts have gained momentum, with numerous companies implementing comprehensive programs aimed at minimizing fuel consumption and lowering emissions. Strategies include upgrading technology, refining operations, and employing combined heat and power systems. These initiatives collectively contribute to substantial energy savings and operational cost reductions, indicating a proactive stance within various sectors to enhance efficiency and reduce environmental impacts [619, 620].

The shipping industry also illustrates an emerging trend as it explores green hydrogen as a viable fuel alternative, moving away from traditional heavy fuel oils. The production of green hydrogen using renewable energy sources represents a potential decarbonization pathway for maritime transport. Several pilot projects and collaborative efforts are currently underway to develop and test hydrogen-powered vessels, marking a significant shift towards greener maritime operations [621, 622].

Finally, Indonesia serves as a noteworthy case regarding the integration of renewable energy within its mining sector. Companies like Vale Indonesia are transitioning to using hydroelectric power to operate nickel smelters, demonstrating a commitment to renewable energy adoption amidst global pressure to reduce emissions [623]. This transition reflects a broader trend where industries, including mining, are increasingly recognizing the necessity of aligning with sustainability objectives as part of their operational frameworks [624].

Chapter 10

Residential and Agricultural Use

Having explored fuel's roles in major industrial and transport sectors, the focus now shifts to its critical presence in everyday life—namely, within homes and farms. Chapter 10, *Residential and Agricultural Use*, examines how energy is used to heat, cool, cook, irrigate, and power essential domestic and rural activities. This chapter highlights the intersection of energy consumption with basic human needs and food production, offering insights into the challenges and opportunities associated with decarbonising these more dispersed but significant energy use cases. As the energy transition accelerates, understanding fuel use in these settings is crucial for developing inclusive, practical, and sustainable energy solutions at both the household and agricultural scale.

Fuel for Cooking and Heating

The utilization of fuel for cooking and heating is a fundamental aspect of energy consumption globally, impacting both domestic and commercial entities and influencing crucial daily functions such as food preparation and indoor temperature regulation, particularly in colder regions. The choice of fuel types is heavily influenced by various factors, including geographic location, economic development, energy infrastructure access, and national environmental policies [625-627]. These choices have profound implications on household health, air quality, energy security, and global greenhouse gas emissions.

Among the prevalent cooking and heating fuel types is biomass—comprising firewood, charcoal, animal manure, and agricultural residue—which predominates in rural and impoverished regions, particularly across sub-Saharan Africa and parts of South and Southeast Asia [625, 628]. The combustion of biomass, often in inefficient stoves or open fires, is a major contributor to indoor air pollution, which is linked to a significant incidence of respiratory diseases that disproportionately affect women who traditionally partake in cooking activities [629, 630]. Furthermore, unsustainable biomass harvesting exacerbates deforestation and environmental degradation, highlighting the need for more sustainable alternatives [625, 631].

In contrast, Liquefied Petroleum Gas (LPG) represents a cleaner and more efficient alternative to biomass. Often utilized in urban areas for its portability and ease of use, LPG offers benefits over traditional fuels; however, its supply chain experiences vulnerability to price fluctuations and accessibility challenges in areas far from infrastructure [632, 633]. Similarly, natural gas, favoured in many developed nations for both cooking and heating, serves as a more consistent source of energy with lesser emissions compared to coal or oil, although it presents climate risks through possible methane leaks during its lifecycle [630, 633].

Electricity serves as another power source for household appliances, including stoves and heaters. Noted for its cleaner profile at the point of usage, electricity becomes increasingly low-carbon as renewable energy sources are integrated into power grids [626, 627]. Nonetheless, electricity access remains limited in numerous regions, and operational costs may exceed those of conventional fuels. On the other hand, kerosene persists in some low-income households due to the unavailability of alternatives; however, its combustion is associated with significant health risks from toxic emissions [626, 634]. Although coal and coke were once standard in urban heating, their use has diminished due to environmental concerns; they remain prevalent in certain locales, particularly in China and Eastern Europe [626, 627].

Heating systems also demonstrate considerable variability. Central heating, typical in colder climates, often relies on integrated energy sources such as natural gas, oil, or electricity, while supplementary options like space heaters are common in regions lacking central heating infrastructure [635]. The patterns of fuel use manifest distinct regional characteristics; traditional biomass remains dominant in sub-Saharan Africa and South Asia, while developed regions primarily leverage natural gas and electricity. In Latin America, urban areas show a transition from firewood to LPG and electricity [626, 627].

Fuel, Energy and Net Zero

The environmental and health repercussions of current fuel use are profound. The reliance on traditional biomass and coal leads to substantial indoor pollution, responsible for millions of premature deaths related to respiratory illnesses each year [627, 634]. From a climate perspective, all fossil fuels contribute significantly to greenhouse gas emissions. However, the integration of cleaner technologies, such as improved cookstoves, solar cookers, and heat pumps, represents an emerging trend toward sustainable energy practices [632, 634].

Support for this energy transition materializes through various initiatives and policies, such as the Clean Cooking Alliance, which advocates for the modernization of cooking fuels, alongside government subsidies aimed at promoting cleaner energy alternatives [634, 635]. The advancement of infrastructure, alongside educational campaigns, remains crucial. Collaboratively, these efforts seek to shift practices towards cleaner energies that ultimately foster public health and environmental sustainability [627, 634].

The most effective fuel sources for cooking and heating—including electricity generated from nuclear energy—are those that strike a balance between efficiency, reliability, affordability, environmental sustainability, and accessibility. The optimal choice often varies depending on regional infrastructure, household needs, and policy environments. Nevertheless, several key fuel types consistently stand out for their performance across these criteria.

Electricity generated from nuclear power is among the most effective options. Nuclear plants provide a stable and reliable source of baseload electricity, which is particularly well-suited for grid-powered cooking and heating systems. These plants produce virtually no greenhouse gas emissions during operation, helping to meet long-term climate goals. Furthermore, nuclear energy is highly scalable and capable of serving large populations. However, its expansion faces challenges such as the need for significant upfront investment, public concerns about nuclear waste and safety, and long regulatory approval processes. Additionally, nuclear-generated electricity is not practical for off-grid locations unless supported by energy storage or microgrid solutions.

Electricity from renewable sources—such as solar, wind, and hydro—is also highly effective. Electric cooking devices (like induction stoves) and heating systems (like electric heat pumps) produce no emissions at the point of use and are increasingly powered by clean energy grids. Renewables are particularly valuable in off-grid settings when paired with solar PV and battery systems. They also integrate well with smart appliances for increased efficiency. The main limitations include intermittency, which requires energy storage systems to maintain reliability, and limited grid infrastructure in some developing regions.

Liquefied Petroleum Gas (LPG) remains a widely used and practical option, especially in developing countries. Its high energy density and clean-burning flame make it highly efficient and well-suited for cooking. LPG is portable and relatively easy to distribute, making it accessible in both urban and some rural areas. However, as a fossil fuel, it contributes to CO_2 emissions and is susceptible to global price fluctuations and supply chain disruptions.

Natural gas is another effective fuel for both cooking and central heating, offering high efficiency and relatively lower emissions compared to coal or oil. It is widely used in urban areas with established pipeline networks. Despite these benefits, natural gas use involves concerns about methane leakage during extraction and transport, which can significantly impact its environmental footprint.

Modern biomass and biogas are renewable alternatives that can be particularly useful in agricultural or rural settings. These fuels are derived from organic waste and can provide both cooking and heating energy in a relatively sustainable manner if managed properly. However, they may still emit pollutants if burned inefficiently and typically require purpose-built systems like biogas digesters or efficient stoves to minimize health risks and maximize benefits.

Solar thermal energy is ideal for water and space heating, particularly in sunny climates. It produces zero emissions and has very low operating costs after installation. However, it is less effective for cooking and can be unreliable during periods of low sunlight or in colder regions.

Traditional biomass, including firewood, charcoal, and dung, is still widely used in many low-income and rural communities due to its availability and minimal infrastructure requirements. Despite this, it is the least effective option due to its low efficiency, significant indoor air pollution, and environmental degradation when harvested unsustainably.

In comparing these fuel sources, electricity from nuclear and renewables ranks highest in both efficiency and environmental performance. LPG and natural gas provide practical and reliable service, especially in areas without robust electric infrastructure. Modern biomass and solar thermal offer localized renewable options but are more limited in application. Traditional biomass, while accessible, poses significant health and environmental challenges.

Table 7: Fuel for Cooking and Heating Comparison Summary.

Fuel Source	Emissions	Efficiency	Accessibility	Suitability for Cooking	Suitability for Heating
Electricity (nuclear)	Very Low	High	High (where grid is available)	Excellent	Excellent
Electricity (renewables)	Very Low	High	Moderate to High	Excellent (with grid or battery)	Excellent (heat pumps)
LPG	Moderate	High	High	Excellent	Moderate
Natural Gas	Moderate	High	High (urban)	Excellent	Excellent
Modern Biomass/Biogas	Low–Moderate	Moderate	Moderate	Good	Good

Fuel, Energy and Net Zero

Fuel Source	Emissions	Efficiency	Accessibility	Suitability for Cooking	Suitability for Heating
Solar Thermal	None	High (for heat)	Moderate	Poor (cooking)	Excellent (water/space)
Traditional Biomass	High	Low	High (rural)	Moderate (basic cooking)	Poor (inefficient)

The most effective overall fuel sources for cooking and heating are electricity from nuclear and renewables, natural gas, and LPG, depending on the context. Nuclear electricity is especially important in countries with stable grid systems, offering both reliability and low emissions. To ensure long-term sustainability, countries should aim to electrify cooking and heating where infrastructure allows, promote LPG and biogas as transitional fuels, phase out traditional biomass, and continue investing in nuclear and renewable energy systems. These steps are essential for reducing emissions, improving public health, and ensuring equitable access to clean energy.

There is significant variation in consumer prices for fuel used in cooking and heating across the world, driven by a complex combination of economic, geographic, policy, and infrastructure-related factors. These disparities are shaped by how countries produce, source, distribute, and regulate their fuel supplies, as well as by their broader economic and political contexts.

One major factor is fuel source availability and dependence on imports. Countries with abundant domestic energy resources—such as natural gas in the United States, Russia, or Qatar—typically enjoy lower consumer prices due to reduced reliance on international markets. By contrast, nations that import fuels are exposed to global price volatility, shipping costs, and fluctuations in currency exchange rates. In rural areas of Africa and Asia, biomass like firewood or crop waste may be freely gathered or sold cheaply, but in urban areas where natural resources are scarce, these same fuels can become expensive commodities.

Government policy plays a key role through subsidies and pricing regulations. Many countries, including India and Indonesia, subsidise fuels like LPG and kerosene to improve affordability for low-income households. However, others have liberalised fuel pricing, allowing market forces to determine costs, which can result in higher price volatility and regional disparities. Some high-income countries implement carbon taxes or environmental levies, as seen in Sweden and Germany, to reflect the environmental cost of fuel use—this can raise prices but also encourages cleaner alternatives.

Infrastructure and distribution costs also significantly influence prices. Urban consumers often benefit from more reliable infrastructure such as gas pipelines or electricity grids, while rural consumers may face higher prices due to the cost of transporting bottled gas or fuel over long distances. In countries with unreliable electricity grids—such as parts of Sub-Saharan Africa and South Asia—households may have to rely on kerosene or charcoal, which are often more expensive and less efficient. Additionally, landlocked or geographically isolated countries with poor transport networks, like Niger or Nepal, face higher logistics costs that are passed on to end users.

Currency exchange rates and global commodity markets further contribute to pricing variation. Since oil, LPG, and coal are traded in U.S. dollars, countries with weak or unstable currencies face higher import costs. For example, Lebanon and Zimbabwe have seen dramatic increases in fuel prices due to economic crises and currency devaluation, even when global prices remain steady.

The structure of a country's energy mix and its market regulation also matter. Countries that rely on renewable energy or have a diverse energy portfolio—such as Norway with its hydroelectric power—are less vulnerable to fossil fuel price shocks. In contrast, deregulated or monopolistic markets can lead to sharp price swings depending on supply and demand conditions.

Demand patterns and seasonal factors influence pricing as well. In colder climates such as Canada or Russia, heating fuel demand surges in winter, pushing up prices. In low-income countries, where many people live in energy poverty and consume fuel in small quantities, the inability to benefit from bulk discounts or invest in efficient technologies keeps per-unit costs high.

Finally, conflict, sanctions, and political instability can dramatically affect fuel prices. War and unrest may disrupt supply chains and damage infrastructure, as seen in Yemen and Syria, leading to steep increases in cooking fuel prices. Economic sanctions, such as those imposed on Iran and Venezuela, have also led to domestic fuel shortages, despite those countries having substantial natural reserves.

The variation in global fuel prices for cooking and heating is shaped by a wide range of structural and contextual factors—not merely supply and demand. Addressing these disparities will require long-term investment in clean energy infrastructure, targeted subsidy reforms, improved governance, and tailored policies to ensure affordable and sustainable energy access for all.

Household fuel prices for cooking and heating vary widely across the globe due to a complex mix of factors, including domestic energy availability, government subsidy policies, infrastructure development, taxation levels, and market dynamics. These variables lead to significant differences in what consumers pay for natural gas, electricity, LPG, and biomass across different regions.

As of September 2024, the global average household price for natural gas stands at approximately USD 0.081 per kilowatt-hour (kWh). However, there are sharp contrasts between countries. Nations with abundant natural gas reserves and significant government subsidies, such as Iran, Algeria, and Russia, have some of the lowest prices, typically under USD 0.01 per kWh. In the United States, the price is around USD 0.04 per kWh, supported by strong domestic production and established infrastructure. On the other end of the spectrum, Sweden has the highest recorded household gas price at about USD 0.219 per kWh, followed closely by Singapore at USD 0.190 per kWh.

Electricity prices also show regional variation. Within the European Union, the average household electricity price in the second half of 2024 was €0.1233 per kWh. Sweden recorded the highest natural gas prices, while Hungary offered the lowest electricity prices at just €0.0320 per kWh. In the United Kingdom, household energy bills remain among the highest globally, despite recent declines. As of July 2025, the average UK household energy bill is projected to be £1,683 per year—still approximately 30% higher than pre-2022 levels.

In many developing countries, Liquefied Petroleum Gas (LPG) and biomass remain dominant cooking fuels. In Sub-Saharan Africa and South Asia, large portions of the population rely on biomass sources such as wood, charcoal, and animal dung, primarily due to limited access to cleaner alternatives. Countries like India and Indonesia have introduced subsidies to promote LPG use, making it more affordable in urban areas. However, significant urban-rural disparities remain, especially in terms of LPG distribution infrastructure.

Multiple structural factors drive the global variation in household fuel prices. Resource availability plays a major role; countries with ample domestic energy supplies often enjoy lower prices due to reduced reliance on imports. Government interventions, including subsidies, taxes, and price controls, can also heavily influence end-user prices. Additionally, infrastructure quality affects distribution costs, with remote or underserved areas often paying more for delivery. Lastly, market dynamics—including global commodity prices, currency exchange fluctuations, and geopolitical events—can cause rapid changes in fuel costs.

Understanding these disparities is essential for policymakers working to promote affordable and sustainable energy access. Coordinated investments in infrastructure, subsidy reforms, and energy diversification will be crucial in addressing price inequalities and supporting a just energy transition.

Fuel in Rural and Developing Areas

Fuel use in rural and developing areas is multifaceted, shaped significantly by necessity, resource availability, cultural traditions, and economic constraints. In numerous regions across sub-Saharan Africa, South Asia, Latin America, and Southeast Asia, households rely heavily on locally sourced and affordable energy for cooking and heating. This trend is visible in the predominant use of traditional biomass fuels such as firewood, charcoal, agricultural residues, and animal dung, which are often collected or purchased at minimal costs. These resources are favoured primarily due to their availability and low expense, despite their inefficiencies and the substantial indoor air pollution they produce, which contributes significantly to adverse health outcomes, particularly among women and children [628, 636-638].

Research indicates that the use of traditional biomass fuels can pose severe health risks due to the high levels of indoor air pollution generated from their combustion. Approximately four million premature deaths annually worldwide are attributable to household air pollution resulting from the burning of solid biomass fuels [638, 639]. Furthermore, women and children, who typically spend considerable time cooking and gathering these fuels, are disproportionately affected by the associated health risks [628, 640]. The environmental implications of biomass fuel use are equally troubling, as unsustainable harvesting practices can lead to widespread deforestation and biodiversity loss, while the inefficient combustion of these fuels contributes to the release of black carbon and other harmful pollutants [639, 641].

Kerosene represents another prevalent energy source in these regions. Although kerosene is considered more efficient and easier to transport than traditional biomass fuels, its affordability is frequently supported by government subsidies. However, kerosene use carries significant health and safety hazards, notably due to the toxic fumes it emits and its high flammability [639, 642]. Efforts to phase out kerosene usage are increasing among governments aiming to enhance household safety and reduce reliance on more harmful fuels [637, 642].

Liquefied Petroleum Gas (LPG) is gaining momentum as a cleaner alternative to biomass fuels. Many countries, including India and Indonesia, have initiated extensive LPG promotion programs, often accompanied by subsidies and infrastructure development aimed at boosting its accessibility [642, 643]. However, LPG adoption faces logistical challenges and upfront costs that limit its use in remote and economically disadvantaged areas, where access to clean energy is still minimal [643]. Approximately 2.4 billion people worldwide continue to depend on traditional biomass for cooking, underscoring the persistence of these energy inequalities [628, 637, 643].

Electricity is another crucial factor in energy access, though its availability is often limited and unreliable in many rural settings. Where it is accessible, electricity can support various household energy needs, including lighting and appliance use. However, in rural regions, the high costs and inadequate infrastructure can inhibit its widespread adoption for cooking [639, 644]. Nevertheless, progress is underway, with off-grid solar systems and renewable mini-grids increasingly providing reliable electricity options for rural households [642, 645].

Transitioning to cleaner energy solutions offers promising pathways for addressing health, environmental, and development challenges. Initiatives such as the promotion of clean cookstoves can significantly reduce fuel consumption and emissions associated with cooking [646, 647]. Additionally, biogas systems utilizing organic waste for methane production represent another viable sustainable cooking fuel option, contributing to both energy generation and waste management [648, 649]. Moreover, government and development organizations have recognized the importance of universal energy access goals, specifically those outlined in the UN Sustainable Development Goal 7, which emphasizes the need for equitable energy access [650]. Innovative financing and partnerships to support the deployment of clean energy solutions are critical for overcoming barriers to access and ensuring a just energy transition for these communities [650, 651].

Agricultural Fuels and Machinery

Agricultural fuels and machinery are essential components of modern agriculture, facilitating essential tasks such as ploughing, planting, harvesting, irrigation, fertilisation, and transporting produce. The mechanisation of these tasks leads to significant enhancements in productivity, directly contributing to improved food security. The transition from manual labour to mechanical operations, driven by agricultural machinery, enables farmers to achieve greater yields with less time and labour invested, highlighting the critical role of agricultural mechanisation in addressing global food demands [652].

Among the various types of machinery, tractors are foundational to mechanised agricultural operations, providing the power needed to tow and operate other equipment. Combine harvesters significantly contribute to the efficiency of crop processing for staple crops such as wheat and rice. Other essential machinery includes planters and seeders, which ensure proper seed distribution, and sprayers for the application of pesticides and fertilisers [653]. Moreover, modern irrigation systems, ranging from traditional diesel-powered pumps to innovative solar-powered solutions, are increasingly being integrated into farming practices, contributing to enhanced water management [654].

The primary fuel for agricultural machinery has traditionally been diesel, known for its high energy density and durability, making it suitable for the demanding environments of farming. However, the combustion of diesel is a significant contributor to greenhouse gas emissions and air pollution [655]. Alternative fuels, such as biodiesel and bioethanol, are gaining traction as more sustainable options. These biofuels can reduce lifecycle greenhouse gas emissions and provide rural communities with energy security [656]. Other alternatives, such as biogas from organic waste, are increasingly employed, particularly in small-scale farming operations [657]. Furthermore, electricity, increasingly sourced from renewable energy, is being adopted for powering more stationary machinery, thereby reducing reliance on fossil fuels [658].

Quantifying diesel consumption specifically for agriculture across countries is difficult due to limited availability of disaggregated data. Despite this challenge, existing information offers valuable insights into overall diesel usage and its importance within the agricultural sector.

In 2022, the global average diesel and heating oil consumption reached approximately 145.53 thousand barrels per day across 190 countries. The United States led with the highest consumption at 4,025.57 thousand barrels per day, followed by China at 3,228.44 thousand barrels per day. Other major diesel-consuming countries included India, Brazil, and Germany. These figures reflect the central role of diesel in powering transportation, industry, and agriculture worldwide.

Although exact figures for agricultural diesel use are scarce, it is widely acknowledged that diesel is the dominant fuel for farm machinery around the globe. In the United States, for example, diesel accounts for roughly 44% of direct energy consumption in agriculture. It powers essential equipment such as tractors, harvesters, and irrigation pumps. Similarly, in Australia, estimates suggest that about 85% of on-farm energy use comes from diesel. This heavy reliance reveals the sector's vulnerability to fuel price volatility and underlines the urgency of transitioning to more sustainable energy sources.

The absence of detailed, sector-specific data remains a significant obstacle in accurately measuring diesel use in agriculture. Without precise consumption figures, policymakers and industry leaders face challenges in planning and implementing energy transition strategies. This highlights the pressing need for improved data collection systems and reporting mechanisms that can support more informed decision-making.

Looking ahead, countries can better manage diesel consumption in agriculture by enhancing the granularity of energy use data, investing in research and development for alternative fuels, and creating supportive policies for the adoption of energy-efficient technologies. Such measures will be essential for reducing the sector's dependency on diesel, improving environmental outcomes, and increasing resilience in the face of fuel supply and price shocks.

The environmental implications of agricultural fuel usage are paramount in contemporary discourse. The extensive use of diesel engines and internal combustion machinery leads to substantial emissions of CO_2 and nitrogen oxides, exacerbating air quality concerns [655]. Advancements in technology and practices aimed at reducing environmental impacts, such as the adoption of biodiesel, biogas, and improvements in machinery efficiency, are becoming vital strategies to mitigate these effects [659, 660]. Additionally, conservation practices, including reduced tillage, hold promise for lessening the carbon footprint of agricultural operations [654].

Moving forward, the drive for decarbonisation in agriculture is gaining momentum. Government incentives and policies promoting cleaner energy alternatives are reshaping operational practices across the sector [661]. The growth of integrated energy-agriculture systems, where farms produce renewable energy locally, is vital for enhancing energy resilience and sustainability in agricultural operations [662]. Precision agriculture, enabled by modern technologies such as GPS and artificial intelligence, facilitates efficient fuel usage and resource management, laying the groundwork for a more sustainable agricultural future [663].

The abrupt cessation of diesel production would have significant repercussions for global agriculture, as diesel fuel is a critical component of mechanized farming, which enhances efficiency and scalability in food production systems. Agricultural machinery that primarily relies on diesel fuel—including tractors, harvesters, irrigation pumps, and other equipment—would soon become inoperable, effectively halting many agricultural operations. This shift would necessitate a reversion to manual labour or animal traction methods, both of which lack the efficiency of modern mechanized farming [655, 664]. Indeed, studies have reported that agricultural activities consume substantial quantities of diesel—over 101.48 litres per hectare is typical for operations like tilling and harvesting [665]. Consequently, without diesel, we could anticipate drastic reductions in crop yields, particularly among large-scale operations that rely heavily on machinery for productivity.

Furthermore, the ramifications would profoundly affect the logistics integral to the food supply chain. Diesel is indispensable for transporting agricultural inputs like seeds and fertilizers to farms, as well as moving perishable goods to market [664, 666]. This reliance creates a critical vulnerability; without diesel, logistical bottlenecks could emerge, causing food spoilage during transit and resulting in significant shortages, particularly in urban areas that depend on just-in-time delivery systems [664, 666]. Historical data suggests that such interruptions in supply chains can lead to sharp increases in food prices, contributing to substantial market volatility that disproportionately affects low-income populations and may result in widespread hunger and social unrest [664, 666].

Moreover, irrigation, which is crucial for agricultural success, would be severely compromised. Many regions, especially in areas like sub-Saharan Africa and India, depend on diesel-powered pumps to access water during dry seasons [655]. The sudden unavailability of diesel would drastically limit irrigation capabilities, potentially leading to decreased yields in areas already vulnerable to climate-induced water scarcity. Rain-fed agriculture would not fully compensate for this loss, further placing food production at risk [655, 667].

The livelihoods of farmers, particularly smallholders in developing nations, are also at risk. Diesel is instrumental not only for farming activities but also for transporting and marketing produce. A disruption in fuel supply could lead to crop losses and diminished income; many farmers might face increased debt, potentially forcing farm abandonment or migration to urban areas in search of alternative employment [668]. Additionally, livestock farming would be adversely affected due to deteriorations in feed supply and water access resulting from reduced transportation capabilities [666].

Further compounding the crisis would be the impact on the production and distribution of essential agricultural inputs like fertilizers and pesticides. Diesel is vital for both manufacturing and distribution processes of these products. Without diesel, there could be declines in crop health, exacerbating productivity losses [668].

While there are arguments that phasing out diesel could decrease greenhouse gas emissions, the immediate repercussions could be profound. Farmers might resort to burning biomass, contributing to environmental degradation through deforestation and increased indoor air pollution [655, 665]. Moreover, the lack of mechanization could lead to inefficient land use and encroachment into marginal lands, resulting in ecological consequences that could further stress agricultural ecosystems [658, 669].

Part 4

The Future of Fuel

Chapter 11

Environmental Challenges and Emissions

Having explored how fuel powers modern life—from electricity generation and transportation to industry, agriculture, and residential applications—this chapter shifts focus to the broader environmental consequences of our energy choices. While earlier chapters examined how fuels are sourced, processed, and consumed, Chapter 11 addresses the critical challenge that shadows all these activities: emissions and their environmental impacts.

As the global demand for energy continues to rise, so too does the urgency to understand and mitigate the harmful by-products of fuel use. Greenhouse gas emissions, air pollutants, and ecosystem disruptions are increasingly shaping international policy, national energy strategies, and public health outcomes. This chapter lays the groundwork for understanding these environmental challenges, exploring the scientific basis of emissions, the technologies for measurement and control, and the pressing need to align fuel use with sustainability goals.

By examining the emissions profiles of various fuel types and usage sectors, Chapter 11 connects the dots between energy systems and environmental health—an essential step in the journey toward a net-zero future.

Greenhouse Gases and Climate Change

Greenhouse gases (GHGs) are crucial components of the Earth's atmosphere, significantly influencing global temperatures through the greenhouse effect. This natural phenomenon allows GHGs to trap heat by absorbing and re-emitting infrared radiation emitted from the Earth's surface, enabling a habitable climate. The vital role of GHGs is underscored by their ability to regulate thermal energy; however, their concentrations have risen dramatically due to human activities since the Industrial Revolution. This escalation, primarily driven by fossil fuel combustion, deforestation, and various industrial operations, exacerbates the greenhouse effect and accelerates climate change [670, 671].

The primary GHGs include carbon dioxide (CO_2), methane (CH_4), nitrous oxide (N_2O), fluorinated gases, and water vapor. CO_2, the most prevalent GHG, is sustained mainly through human activities such as fossil fuel burning and deforestation [672]. Methane, while less abundant, is significantly more potent in trapping heat over short time frames and is released from agricultural practices, particularly livestock and rice paddies, and from waste management activities [673-675]. Nitrous oxide, linked predominantly to agricultural activities through synthetic fertilizers and livestock manure, poses considerable warming potential due to its ability to trap heat [670, 671]. Furthermore, synthetic fluorinated gases, although present in lower concentrations, are extremely effective at causing warming, while water vapor, the most abundant GHG, responds primarily to climate changes rather than direct anthropogenic inputs [671].

Understanding the linkage between human-induced GHG emissions and resultant climatic changes is essential. The intensification of the greenhouse effect through excess GHG emissions has resulted in global warming, defined as a long-term increase in the Earth's average surface temperature. This warming is characterized by disruptions to climatic patterns and ecosystems, illustrated by extreme weather events and rising sea levels [670, 671]. Specific anthropogenic activities such as electricity generation from fossil fuels, extensive use of transportation, and industrial processes significantly contribute to increased emissions [676-678].

The repercussions of rising GHG levels are alarming, leading to severe weather patterns, habitat loss, and health risks associated with heat-related illnesses and respiratory conditions [672, 679, 680]. The urgency of implementing strategies to curtail these emissions is recognized globally, with frameworks such as the Paris Agreement aiming to limit temperature rises to below 2°C above pre-industrial levels, advocating for significant transitions to renewable energy sources and improved energy efficiency [674, 678]. Such preventive measures must incorporate institutional reforms, technological advancements like carbon capture and storage (CCS), and investments in nature-based solutions to mitigate climate impacts effectively [673, 675].

Counter-arguments exist, suggesting that climate variability is natural and that historical cycles explain current changes. Proponents of this view highlight that climatic shifts have occurred throughout history due to solar cycles and geological activity [680, 681]. Nonetheless, the scientific consensus strongly affirms that despite natural variability, the unprecedented pace and scale of current warming correlate closely with anthropogenic emissions, particularly CO_2 from industrial activities [682, 683]. Furthermore, criticisms regarding the reliability of climate models, which often rely on numerous assumptions, fail to appreciate that these models have provided accurate long-term predictions and are regularly updated based on emerging data [684, 685].

While some argue for the benefits of CO_2 in agriculture through increased plant productivity, this perspective often neglects the broader adverse effects of climate change. A focus on economic arguments for continued reliance on fossil fuels, especially in developing nations, underscores the intricate balance between sustainable practices and developmental needs. However, as renewable technologies become more cost-effective, the argument for transitioning away from fossil fuels to mitigate climate change strengthens [674, 675]. Additionally, global cooperation via agreements like the Paris Accord is fundamental in addressing climate change as a shared global challenge, rather than a unilateral national issue [686, 687].

It is generally argued that while GHGs are essential for sustaining life, their excessive accumulation due to human activities has emerged as the primary driver of climate change. Addressing this multifaceted challenge necessitates a comprehensive response encompassing international collaboration, policy reforms, and technological innovations to mitigate the ongoing impacts of climate change on our planet's ecosystems and human health.

Addressing the conclusion that a comprehensive response is required to mitigate the impacts of climate change involves a combination of international cooperation, policy reform, and technological advancement. Several key components can shape this response, each contributing to a broader, coordinated global effort.

International collaboration plays a foundational role. Strengthening and enforcing global agreements, such as the Paris Agreement, is crucial to setting and achieving emissions reduction targets. Expanding financial mechanisms, including initiatives like the Green Climate Fund, enables developing countries to pursue climate adaptation and mitigation strategies. Facilitating the transfer of clean technologies and knowledge between nations helps bridge capability gaps, while carbon border adjustments in trade policies can prevent emissions outsourcing and encourage consistent global action.

Policy reforms are also essential to align economic incentives with climate goals. Implementing carbon pricing strategies, such as carbon taxes or cap-and-trade systems, helps internalize the environmental cost of emissions. Reallocating subsidies away from fossil fuels toward renewables and sustainable infrastructure can accelerate the clean energy transition. Furthermore, establishing robust regulations and efficiency standards for energy use, emissions, and building practices supports long-term decarbonization. Land use policies that prioritize reforestation, protect carbon sinks like wetlands and forests, and prevent land degradation are equally important.

Technological innovation underpins many climate solutions. Expanding renewable energy sources such as solar, wind, hydro, geothermal, and biomass is vital for decarbonizing power generation. The development of carbon capture and storage (CCS) technologies can reduce emissions from industrial processes and fossil fuel use. Investments in energy storage systems and smart grids enhance the reliability of renewable energy and improve demand management.

In agriculture and transport, advancing practices like precision farming, low-emission fertilisers, electric vehicles, and green hydrogen can significantly reduce emissions.

Responsibility for implementing these measures is shared across multiple sectors. Governments, at both national and local levels, must create and enforce climate policies, invest in infrastructure, and provide incentives for green innovation. Intergovernmental organizations such as the UNFCCC, IPCC, and World Bank play a critical role in guiding international cooperation and aligning funding mechanisms with climate objectives.

The private sector, particularly corporations and energy companies, must take the lead in reducing emissions across operations and supply chains, adopting sustainable technologies, and investing in research and development. Scientific and academic institutions are responsible for conducting climate research, monitoring progress, and developing new technologies and policy tools. Civil society and non-governmental organizations (NGOs) are essential in raising awareness, advocating for action, and ensuring accountability. Finally, individuals and communities must embrace sustainable lifestyles, reduce energy consumption, and support local and national climate initiatives.

Combating climate change requires a global, multi-stakeholder effort. Coordinated action—ranging from international agreements to individual behaviour changes—is essential to addressing the root causes of climate change and protecting ecosystems, public health, and future generations.

The relationship between climate action and economic strain presents a multifaceted challenge, particularly for countries burdened by high living costs, rising interest rates, and substantial public debt. These financial constraints often render large-scale climate investments politically challenging, as governments grapple with how to implement effective climate policies without exacerbating economic stress. Yet, despite these challenges, it is crucial to recognize that climate action need not be deferred. Strategic reframing of climate policies as integral components of economic reform, focused on smart fiscal planning rather than onerous spending, offers a productive pathway forward.

One effective approach is to prioritize high-return, low-cost interventions. Many climate actions—such as upgrading energy efficiency, retrofitting public buildings, and enhancing insulation—require minimal upfront investment and yield immediate savings on energy expenses. For instance, implementing energy-efficient technologies not only reduces emissions but also alleviates financial pressure on households and governments alike, contributing to overall fiscal health [116, 688]. This prioritization aligns with the broader goal of reducing overall sector costs without increasing debt levels or tax burdens, thereby creating a win-win scenario for citizens and the environment.

Reallocating existing public resources rather than seeking new revenue streams can facilitate climate initiatives without exacerbating fiscal constraints. Phasing out inefficient fossil fuel subsidies—which disproportionately benefit wealthier households—can free up substantial funds for green infrastructure investments [116]. This method serves not only as a means to promote environmental objectives but also as a strategic reform that redistributes resources in a way that supports broader economic and social goals, including cost-of-living reductions, rather than simply transferring fiscal burdens to taxpayers [116].

A critical aspect of successful climate policy is ensuring that it simultaneously addresses cost-of-living pressures for the populace. For example, investments in affordable public transport, incentives for rooftop solar installations, and support for local food systems can significantly reduce household expenses. Well-designed climate policies have the potential to function as anti-inflationary tools, easing financial burdens rather than complicating them [688]. This dual benefit—environmental improvement coupled with economic relief—is essential for garnering public support for climate policies, as evidenced by research indicating that comprehensive policy bundles that include social elements gain greater acceptance among populations [688, 689].

Moreover, maximizing private and international investment is pivotal for countries facing fiscal constraints. Governments can attract private capital by utilizing instruments such as green bonds, concessional loans, and risk guarantees to stimulate interest in green projects. Institutions like the World Bank and the Green Climate Fund can facilitate access to below-market rates, significantly reducing the financial load on national budgets. By viewing climate spending through the lens of preventative investment, governments can effectively classify these expenditures as necessary for avoiding future disaster costs, thereby recontextualizing them as fiscally prudent choices rather than burdensome outlays [116].

Protecting vulnerable groups is also paramount. Policies focused on equity, such as energy rebates and direct financial support, prevent the adverse impacts of climate action from disproportionately affecting low-income households [690]. Such targeted assistance is often more efficient and less costly than broad subsidies, thereby reinforcing social equity while supporting climate goals [688].

Integrating climate action into core economic planning frameworks solidifies its status as a priority in macroeconomic, industrial, and employment strategies [689]. By embedding climate initiatives within existing economic agendas—particularly through public procurement policies that emphasize low-emission materials—countries can foster green job creation while remaining within their current fiscal limits [688].

Finally, an incremental and predictable approach to implementation allows both markets and households the flexibility to adapt effectively. By establishing clear, long-term signals—such as phased fuel efficiency standards—governments can enable businesses and individuals to plan and invest, thus avoiding disruptive financial shocks [691].

Countries can pursue aggressive climate action strategies without succumbing to fiscal deterioration. Through innovative financial management, effective policy design, and prudent public investment frameworks, it is feasible to reduce emissions, bolster economic stability, and alleviate household pressures in a financially responsible manner, ultimately aligning climate policy with broader economic resilience goals.

Fuel Efficiency and Emission Standards

Fuel efficiency and emission standards are crucial regulatory measures aimed at reducing fuel consumption and the environmental impact of various sectors, notably transportation, industry, and energy production. These standards shape global fuel usage patterns by influencing vehicle design, manufacturing practices, and consumer choices across different regions [692].

In the United States, Corporate Average Fuel Economy (CAFE) standards are a key regulatory framework that mandates vehicle manufacturers to meet specified fuel efficiency targets for their fleets [693]. The Environmental Protection Agency (EPA) enforces stringent tailpipe emissions limits under Tier 3 regulations, targeting harmful pollutants such as nitrogen oxides (NOx), particulate matter, and carbon dioxide (CO_2) [693]. The establishment of these standards has led to significant improvements in vehicle efficiency, reduced gasoline consumption per vehicle, and considerable reductions in emissions related to smog since the 1970s [694]. Research indicates that these regulations have prompted technological advancements in the automotive industry, facilitating the development of more efficient engines and cleaner technologies such as catalytic converters and electric drivetrains [693].

The European Union (EU) has similarly adopted aggressive CO_2 emissions targets for new passenger vehicles, including a cap of 95 grams of CO_2 per kilometre [692]. The EU's Euro Emission Standards, which range from Euro 1 to the current Euro 6 (with Euro 7 pending), set stringent limits on harmful emissions, including NOx and particulate matter [692]. Initially, these policies resulted in increased diesel vehicle adoption; however, over time, they have significantly incentivized the transition toward electric vehicles, contributing to reductions in per-capita fuel consumption rates [692].

China's response to fuel efficiency standards is notable, having implemented a multi-phase fuel economy standard for both passenger and commercial vehicles, known as CN VI, which aligns closely with Euro 6 norms [692]. These stringent regulations have spurred advancements in vehicle technology, leading to improved air quality and accelerating the country's transition towards electric vehicles. The emphasis on stringent emissions standards is critical to mitigating the impact of urban pollution and enhancing public health [692].

India has made significant strides by following the Bharat Stage (BS) emission standards, akin to the EU framework, specifically leapfrogging from BS-IV directly to BS-VI in 2020 [695]. Additionally, the Corporate Average Fuel Consumption (CAFC) norms set efficiency benchmarks for automakers [695]. While these measures have led to cleaner vehicle fleets, they have raised production costs, presenting financial challenges for both consumers and manufacturers [695].

Japan's approach through the Top Runner Program promotes continuous improvement in fuel efficiency by setting benchmarks that reflect the best-performing products in the market [695]. This program has successfully cultivated a highly efficient national vehicle fleet and fostered the widespread adoption of hybrid technologies, thereby reducing overall fuel consumption [695].

Conversely, Australia presents a contrasting case, as it lacks mandatory fuel efficiency standards. Despite aligning its fuel quality with international benchmarks, the absence of binding regulations hinders progress in enhancing fuel efficiency and results in a continued reliance on high-emission vehicle imports [692].

Fuel efficiency and emission standards extend their influence beyond transportation, impacting industrial sectors and energy production. Emission limits in these areas encourage transitions from coal to cleaner energy sources such as natural gas and renewables, which are critical for meeting both domestic and international climate targets [692]. Furthermore, international organizations such as the International Maritime Organization (IMO) and the International Civil Aviation Organization (ICAO) set regulations for shipping and aviation, promoting the use of cleaner fuels and advanced engine technologies.

The measurable impacts of these standards are evident. Technological innovations born from regulatory requirements, including the employment of lighter materials and aerodynamic enhancements, have contributed to reduced fuel consumption [692]. Consumer demand for fuel-efficient and electric vehicles has notably surged, particularly in regions like the EU and China, where regulatory frameworks are robust and actively enforced.

However, despite successes, challenges remain. The cost of compliance is often passed on to consumers, creating potential political sensitivities [695]. Additionally, the variability of standards between countries complicates the operations of manufacturers in global markets [695]. High-profile scandals, such as the Volkswagen emissions scandal, highlight the need for stringent oversight and effective enforcement of these regulations to ensure adherence [692].

Global Initiatives to Reduce Fuel Impact

Governments, intergovernmental organizations, industries, and civil society groups worldwide are increasingly engaged in various initiatives to mitigate the environmental, economic, and health impacts associated with fuel use. These efforts are primarily focused on reducing greenhouse gas (GHG) emissions, improving air quality, enhancing energy security, and transitioning towards sustainable energy systems. Initiatives range from international agreements, such as the Paris Agreement, to regional collaborations, policy frameworks, and technological innovations aimed at fostering a global shift to cleaner energy sources.

One of the most significant global frameworks is the Paris Agreement, adopted in 2015 under the United Nations Framework Convention on Climate Change (UNFCCC). This agreement commits signatory countries to limit global temperature rise to well below 2°C, with aspirations of limiting it to 1.5°C, necessitating substantial reductions in greenhouse gas emissions. Countries' commitments, articulated in their Nationally Determined Contributions (NDCs), frequently involve enhancing fuel efficiency and transitioning from fossil fuels [696]. The success of the Paris Agreement is contingent upon countries' adherence to their commitments, which includes adhering to standards aimed at reducing emissions across various sectors [696, 697].

The International Energy Agency (IEA) plays a pivotal role globally in supporting initiatives designed to minimize the adverse effects of fuel use by providing data and policy analysis. It promotes energy efficiency measures, including fuel emissions standards and renewable energy adoption [698]. For instance, the Global Fuel Economy Initiative (GFEI) is an exemplary program aiming to double the average fuel economy of vehicles by 2050, reflecting the necessity for collaborative international efforts towards sustainable transport [698]. Complementarily, sector-specific organizations like the International Maritime Organization (IMO) and the International Civil Aviation Organization (ICAO) have established targeted strategies to curb emissions within the transportation sector, which is a significant contributor to global GHG emissions [697].

Within a regional context, the European Union's Green Deal is a landmark initiative, aimed at rendering Europe the first climate-neutral continent by 2050. It encompasses measures such as phasing out internal combustion engines by 2035 and promoting sustainable fuels [699, 700]. This transformative agenda is interlinked with national initiatives, as demonstrated by the U.S. rejoining the Paris Agreement and enacting comprehensive clean energy funding through the Inflation Reduction Act [701]. Similarly, China's investments in electric vehicles and renewable sources exemplify its commitment to peak emissions by 2030 and aim for carbon neutrality by 2060. These examples emphasize the collective responsibility and commitment required by nations to foster sustainable energy practices and policies [702, 703].

Fuel subsidy reform is another critical aspect of tackling fuel-related emissions. Institutions such as the International Monetary Fund (IMF) and the World Bank advocate for the gradual phase-out of fossil fuel subsidies, which often distort market practices and exacerbate overconsumption [697]. Redirecting these subsidies to bolster renewable energy initiatives can significantly alleviate fuel impact while fostering economic development, particularly in low- and middle-income countries [704, 705].

Furthermore, technology transfer and investment mechanisms like the Green Climate Fund (GCF) and Clean Technology Fund (CTF) are essential for aiding developing nations in overcoming their reliance on fossil fuels [705]. These frameworks support the development and proliferation of clean technologies, which include solar, wind, and biomass energy projects [704, 705]. Additionally, non-governmental organizations and corporate alliances actively engage in voluntary commitments to reduce emissions across supply chains. Programs such as the Science Based Targets initiative (SBTi) and the Race to Zero campaign encourage companies to enhance fuel efficiency and transition to renewable energy sources [705].

The wide array of initiatives underway globally underscores a growing consensus regarding the urgent need to transition from fossil fuel dependencies to cleaner, more sustainable energy systems. These multifaceted efforts combine regulation, financial incentives, innovations, and international collaboration to address pressing climate challenges, paving the way for a more energy-efficient and climate-resilient future [696, 701, 706].

The effectiveness of global initiatives to reduce the impact of fuel use has been mixed—marked by notable progress in some areas and ongoing challenges in others. The Paris Agreement has been pivotal in aligning international efforts around a common goal of limiting temperature rise, and nearly every country is now a signatory. However, current Nationally Determined Contributions (NDCs) remain insufficient to meet the 1.5°C target. According to the UN Emissions Gap Reports, the world is still on track for around 2.5–2.9°C of warming by the end of the century unless more ambitious action is taken.

Fuel, Energy and Net Zero

The International Energy Agency's work, particularly through the Global Fuel Economy Initiative (GFEI), has helped drive improvements in vehicle efficiency. For instance, several countries have implemented stricter fuel economy and emission standards, which have reduced per-vehicle fuel consumption and improved air quality. However, global fuel demand in transport continues to rise overall, especially in emerging economies, partially offsetting gains from efficiency improvements.

In international transport, the IMO and ICAO have made progress with frameworks like CORSIA and carbon intensity goals for shipping, but critics argue that these measures lack the enforcement strength and ambition needed to drive rapid decarbonisation. In aviation, the heavy reliance on carbon offsets rather than direct fuel-use reduction has also been a point of contention among climate experts.

Regional initiatives, such as the EU's Green Deal, have been more effective in driving systemic change. The EU has seen a notable decline in emissions from energy and transport and an acceleration in electric vehicle (EV) adoption. The United States has renewed its federal climate commitments through legislation like the Inflation Reduction Act, which includes billions in incentives for renewable energy and EVs. Meanwhile, China has emerged as a global leader in EV production and renewable energy investment, although it continues to be the largest global emitter due to its reliance on coal.

Fuel subsidy reform has proven difficult politically but is gaining traction. Countries like Indonesia and Iran have made partial progress, while others still struggle with the social and political risks of removing subsidies. Globally, however, fossil fuel subsidies increased in 2022 due to inflation and geopolitical tensions—especially after Russia's invasion of Ukraine—which undercut momentum for reform in many countries.

Technology transfer and climate finance mechanisms such as the Green Climate Fund have provided billions in support to developing countries, enabling renewable energy projects and adaptation strategies. Nevertheless, overall climate finance still falls short of the $100 billion annual target pledged by developed nations, which has created trust deficits in international negotiations and slowed progress.

Voluntary corporate initiatives like the Science Based Targets initiative (SBTi) have driven greater private sector engagement in emissions reduction. Thousands of companies have now set science-aligned targets, leading to investments in energy efficiency, sustainable fuels, and circular economy practices. However, these efforts remain uneven across industries and geographies, and accountability for progress varies.

While global initiatives to reduce the fuel impact have achieved significant successes—particularly in driving policy alignment, renewable energy growth, and technological innovation—major gaps remain. Ambition, enforcement, equity, and financing are the key areas where current efforts must be strengthened. Without faster and more coordinated action, the world risks falling short of the targets necessary to avoid the most dangerous impacts of climate change.

Net Zero

Net zero refers to the goal of balancing greenhouse gas (GHG) emissions produced with those removed from the atmosphere, effectively creating a neutral impact on the climate system [707]. This balance necessitates that any emissions released must be countered by equivalent absorption efforts, which can occur through natural carbon sinks, such as forests and oceans, or via technological innovations like carbon capture and storage (CCS) [708]. The ultimate aim of achieving net-zero emissions is to halt the rise of GHG concentrations in the atmosphere, thereby stabilizing global temperatures and mitigating climate change impacts [707, 709].

The significance of achieving net zero is underscored by findings from the Intergovernmental Panel on Climate Change (IPCC), which emphasizes that to limit global temperature increases to 1.5°C, emissions of carbon dioxide must reach net zero by approximately 2050. Other potent greenhouse gases, such as methane and nitrous oxide, necessitate even steeper reductions owing to their higher warming potential [709]. The transition to net zero global emissions involves a comprehensive approach that encompasses shifting energy sources to renewable alternatives, enhancing energy efficiency across various sectors, and implementing changes in agricultural practices [710].

To effectively progress toward net zero, multiple strategies are essential. Transitioning to renewable energy sources, such as solar, wind, and hydroelectric power, is critical, as is enhancing energy efficiency in buildings, transportation systems, and industrial operations [710]. Moreover, carbon removal strategies—particularly through the restoration of forests and enhancement of soil carbon reservoirs—are vital [708]. Some entities also engage in carbon offsetting, monetizing carbon credits to mitigate their emissions footprint, although this approach is often criticized for its potential lack of rigor in verification [711]. Promoting sustainable lifestyle choices, such as reduced food waste and increased public transportation usage, further supports the attainment of net zero targets [711].

Globally, more than 140 nations, including major economies like Australia, the U.S., and China, have committed to net zero emissions, typically by 2050, although timelines vary [712]. Numerous companies are also establishing net zero goals through frameworks such as the Science Based Targets Initiative (SBTi) [713]. Local governments are stepping up with independent climate action plans aimed at similar objectives.

Despite the increasing commitment to achieving net zero, several challenges persist. A significant issue is the prevalence of greenwashing, where some organizations proclaim net zero ambitions without substantiated action plans, over-relying on inadequate offsetting measures [714]. Furthermore, the verification of offset quality remains contentious, and many developing nations argue for equitable responsibilities, asserting that wealthier countries, which contributed more to historical emissions, should take the lead on global climate finance [715]. There is also concern regarding reliance on emerging or unproven technologies that may not be adequately scalable to meet ambitious climate targets [716].

Nevertheless, the potential benefits of achieving net zero are substantial. Attaining this goal could significantly reduce the risks associated with climate change, including severe weather events like heatwaves and flooding, while also enhancing public health through improved air quality and fostering energy independence [717]. Moreover, pursuing net zero goals can catalyse innovation, create employment opportunities, and promote sustainable development pathways [716].

Net zero is realistically achievable, but only under certain conditions. It requires unprecedented global cooperation, ambitious policy action, sustained investment, rapid technological innovation, and broad societal support. While science and technology offer many of the tools needed to reach this target, success depends just as much on political will, economic restructuring, and equitable implementation across nations.

From a scientific and technological standpoint, the path to net zero is clear but challenging. Mature renewable technologies like wind, solar, and hydro are increasingly cost-competitive and scalable. Energy efficiency measures in buildings, industry, and transportation have already proven effective, and electrification of transport—through electric vehicles and expanded public transit—is progressing rapidly. Additional solutions, including carbon capture and storage (CCS), direct air capture, reforestation, and soil carbon storage, are technically viable, though some are expensive or not yet widely deployed. The major challenge lies in scaling these solutions globally, especially in hard-to-abate sectors such as steel, cement, aviation, and agriculture.

Economically, net zero is viable and, in many cases, advantageous. Clean energy investments now tend to create more jobs and deliver greater long-term economic returns than fossil fuels. The cost of inaction—through disaster recovery, lost productivity, and health impacts—will likely surpass the costs of mitigation. Well-designed carbon pricing mechanisms can drive innovation without destabilizing economies. Nevertheless, the upfront costs are high, and support is essential for poorer nations or those reliant on fossil fuel exports to ensure fair transitions.

The political and social will to achieve net zero is growing. Over 140 countries, representing more than 90% of global GDP, have made net zero pledges. Cities, businesses, and local governments are also committing independently, and younger generations increasingly demand climate action. Still, many of these pledges lack legal backing or clear implementation pathways. There is also a risk of greenwashing and overreliance on carbon offsets. Political cycles and entrenched interests often delay or dilute meaningful action.

Equity and global cooperation are fundamental to the net zero journey. Developed countries, responsible for most historical emissions and better resourced, must lead by example. Developing nations require financial and technological support to grow sustainably. Just transitions are critical to protect workers, low-income communities, and vulnerable populations from bearing disproportionate costs. However, global cooperation is often fragile. Geopolitical tensions, economic instability, and mistrust between nations can undermine climate diplomacy and stall collective progress.

In summary, net zero is not a technical impossibility—it is a matter of speed, scale, and solidarity. Achieving it is theoretically possible and even cost-effective in the long run. But it will demand urgent, fair, and coordinated action across all sectors of society. Delays only make the path more difficult and more expensive in the future.

Net zero can also be achieved without nuclear power, but this path is considerably more complex and resource-intensive. Currently, nuclear power provides about 10% of global electricity and over 25% of low-carbon electricity. Countries such as France, Slovakia, and Hungary depend on it significantly, whereas others—like Germany and Australia—have phased it out or avoided it. To achieve net zero without nuclear, nations would need to replace its contribution with expanded renewables, enhanced grid infrastructure, energy storage systems, and more flexible demand-side solutions.

This nuclear-free route would require historically rapid deployment of renewables like wind and solar, as well as the build-out of extensive transmission networks—which often face planning and public opposition challenges. Intermittent renewables would also necessitate vast energy storage capacity, including both short-term battery systems and long-duration options like hydrogen or pumped hydro. Smart grids and demand management tools would be essential to match supply and demand without the steady baseload that nuclear provides. For sectors that remain hard to decarbonize, such as aviation and heavy industry, carbon removal technologies and nature-based solutions would need to play a bigger role.

Technical and economic models from the IEA and IPCC show that net zero without nuclear is possible—particularly in nations with rich renewable resources and supportive policy environments. However, these scenarios generally carry higher costs and depend on accelerated innovation in emerging technologies. Countries with limited land or less solar and wind potential may find it more difficult to follow a nuclear-free pathway.

Avoiding nuclear power also sidesteps valid concerns about radioactive waste, nuclear accidents, capital costs, and long construction timelines. In regions with strong public opposition to nuclear, a focus on renewables may yield greater political and social support. It can also drive broader innovation in energy efficiency and decentralized energy systems.

However, excluding nuclear entirely poses significant risks. Overdependence on intermittent renewables increases system vulnerability. In regions with weaker renewable capacity, backup fossil fuels may be used more, slowing emissions reductions. If renewable expansion or storage technologies falter, progress toward net zero could be significantly delayed.

Achieving net-zero emissions while incorporating fossil fuels into the energy mix is also technically feasible under strict conditions, reliant on the effective deployment of carbon management technologies, verified carbon offsets, and substantial emissions reductions across various sectors. The principle of net-zero is to balance greenhouse gas emissions with their removal from the atmosphere. Therefore, while a total cessation of emissions may not be immediately attainable, a limited and responsibly managed role for fossil fuels can persist, provided mechanisms are in place to adequately negate their climate impact [718].

One key strategy enabling fossil fuel use within a net-zero framework is Carbon Capture, Utilization, and Storage (CCUS). This technology allows fossil fuel power plants and industrial facilities to capture CO_2 emissions before they are released into the atmosphere. The captured carbon can then be permanently stored underground or utilized for various applications [719, 720]. For fossil fuels to maintain a viable long-term role, CCUS technologies must be implemented at significant scale and made economically viable [721]. Furthermore, innovations such as Direct Air Capture (DAC) demonstrate potential in complementing CCUS by extracting CO_2 directly from ambient air, thus offsetting emissions from sectors where decarbonization is particularly challenging, including aviation and cement production [722]. Nonetheless, the economic feasibility of DAC remains a substantial obstacle, as the technology is still developing and costly [723].

Another approach to achieving net-zero emissions involves carbon offsetting, which entails mitigating emissions through investments in projects like reforestation, soil carbon enhancement, or wetland restoration. However, the integrity of offsetting strategies can be questioned, due to concerns about permanence and verifiability, which can lead to accusations of greenwashing if offsets are not carefully monitored and supported by credible data [724]. Additionally, fossil fuels may primarily serve transitional roles in hard-to-abate sectors, where complete decarbonization is challenging. Industries such as aviation, maritime transportation, and steel production continue to depend on fossil fuels, creating a need for effective carbon management strategies, including offsets, as a bridge to achieving a net-zero future.

Moreover, negative emissions technologies, particularly Bioenergy with Carbon Capture and Storage (BECCS), hold promise for achieving net-negative emissions—removing more carbon from the atmosphere than they emit. This could potentially compensate for any remaining emissions from fossil fuels [722, 723]. However, the implementation of these technologies faces significant challenges, such as high costs, slow deployment rates, and substantial infrastructure requirements for the safe transport and storage of CO_2 [725].

Examples from various countries illustrate differing approaches to integrating fossil fuels within net-zero strategies. For instance, both the United States and Norway have incorporated natural gas use with CCUS into their decarbonization plans, while Saudi Arabia's 2060 net-zero roadmap places emphasis on carbon management technologies over reducing fossil fuel production [726, 727]. Similarly, Australia's Long-Term Emissions Reduction Plan allows for ongoing fossil fuel use, contingent upon the integration of CCUS and offsets as critical transitional tools [728].

The inclusion of fossil fuels in a net-zero framework is technically achievable, however it requires a rigorous approach to ensure emissions are thoroughly and credibly neutralized. The complexities of this transition highlight the importance of a temporary and tightly controlled role for fossil fuels, embedded within a comprehensive and well-regulated transition towards a low-carbon economy [729, 730].

Achieving net zero emissions by 2050 stands as a highly ambitious goal for many countries, contingent upon various critical factors such as economic capacity, political commitment, access to clean energy resources, technological readiness, and public support. High-income, technologically advanced nations are generally better positioned to meet this commitment, whereas developing countries might require additional time, financial assistance, and adaptable policies to progress toward this target [731, 732].

The United Kingdom exemplifies a nation well-prepared to achieve net zero by 2050, bolstered by its legally binding net zero legislation as outlined in the Climate Change Act. This robust framework establishes clear regulatory pathways and encourages substantial investment in renewable energy infrastructure [733]. The UK's success in decarbonizing its electricity supply, particularly through the phase-out of coal, showcases the effectiveness of structured carbon budgeting systems that pave the way for ongoing climate policies [733]. Such legislative measures serve as a model for other countries aiming to achieve similar targets [733].

The European Union (EU) provides another solid case, particularly with countries like Germany, Sweden, and Denmark, which are embracing the EU's commitment to net zero. These nations have introduced ambitious policy frameworks and possess high shares of renewable energy, supported by carbon pricing mechanisms that incentivize emissions reductions while enjoying considerable public backing [732, 734]. Sweden's ambition to reach net zero by 2045 reflects proactive governance and societal consensus around sustainability efforts [732].

Additionally, countries such as New Zealand and Canada have legislated their net zero goals, signifying political commitment at the national level. Both have made significant investments in clean energy innovation and are actively working to phase out coal, creating pathways for low-carbon transitions and emphasizing public engagement to bolster their credibility in climate action [732, 735]. In contrast, Japan and South Korea, with their advanced industrial bases, are fostering governmental support for clean technologies like hydrogen and electrification, positioning them well to transition their economies towards net zero [735].

Conversely, nations like Australia encounter more complex challenges, despite their vast renewable energy potential. Australia's heavy reliance on fossil fuel exports and inconsistent federal policies complicate its transition to net zero. Although technically capable of achieving this target, the variability in political commitment presents significant obstacles [732, 736]. Similarly, the United States possesses ample resources and technologies to achieve net zero; however, political polarization and policy volatility hinder sustained commitment to long-term climate goals. The Inflation Reduction Act marks progress, yet consistent actions across political administrations remain a critical barrier [732, 736].

In contrast, countries such as China and India are at higher risk of missing their emissions targets. China has committed to carbon neutrality by 2060, primarily influenced by its heavy coal dependence and substantial industrial energy demand. While its advancements in renewable energy deployment are commendable, reaching net zero by 2050 would require extraordinary efforts and restructuring [732]. India, targeting 2070 for its net zero goal, faces even more profound challenges due to widespread energy access issues and a heavy reliance on coal for economic development. Achieving net zero would necessitate significant international support in financing and technology transfers [732, 734, 735].

Fuel, Energy and Net Zero

Critical strategies to enable nations to achieve net zero by 2050 emerge as vital components of the discussion. A clean electricity grid underpins this transition, necessitating a focus on electrifying transport, heating, and industrial processes. Implementing comprehensive carbon pricing and regulatory frameworks, alongside scaling public and private investments in infrastructure and innovation, emerges as essential [736]. International cooperation will play a pivotal role, particularly concerning technology transfer and climate finance, highlighting the interconnected nature of global climate efforts [732].

Chapter 12

Innovation and the Energy Transition

Having examined the environmental challenges and emissions that stem from our reliance on traditional fuels in Chapter 11, the path forward becomes clear: innovation is not just desirable—it is essential. Chapter 12 shifts focus from the problems of the past and present to the solutions of the future. It explores how emerging technologies, policy frameworks, and collaborative global initiatives are driving a transformational shift in how energy is produced, distributed, and consumed.

This chapter introduces readers to cutting-edge innovations that are redefining the fuel and energy landscape, from advanced biofuels and green hydrogen to digital energy systems and carbon capture technologies. It also highlights the pivotal role of research, investment, and public policy in supporting a just and effective energy transition. As the world edges closer to net-zero goals, the innovations discussed here represent the bridge between current challenges and a more sustainable, resilient energy future.

Electric Vehicles and Battery Technology

Electric vehicles (EVs) and advanced battery technologies are critical components in the global energy transition. Countries are increasingly adopting EVs as an alternative to fossil-fuel-powered vehicles to mitigate greenhouse gas emissions and enhance energy security. EVs do not produce tailpipe emissions, making them significant players in decarbonizing transportation, one of the largest contributors to global carbon emissions. The ability of EVs to utilize electricity generated from renewable sources further enhances their environmental benefits in cities, reducing air pollution and thereby improving public health and quality of life [737, 738].

The linkage between EVs and battery technology is foundational. Innovations in battery chemistry, particularly the development of lithium-ion batteries, have incrementally improved the performance metrics of EVs, such as energy density, charging speed, and lifespan [738-740]. Advances in battery management systems are integral for enhancing safety, reliability, and overall vehicle performance, reinforcing the fundamental role of lithium-ion technology in the viability of electric vehicles [741, 742].

The integration of EVs into broader electricity systems exemplifies the concept of vehicle-to-grid (V2G) technology. By acting as flexible energy storage units, EV batteries can help balance the grid by storing excess renewable energy and providing power back to the grid when needed [743]. This bidirectional capability not only improves grid resilience but also promotes the use of variable renewable energy sources, thereby supporting national and international efforts toward achieving net-zero emissions [737, 738].

The growth and adoption of EV technology are bolstered by supportive government policies worldwide. Incentives for EV purchases, investments in charging infrastructure, and mandates for phasing out new petrol and diesel vehicles are increasingly prevalent [738]. China currently leads global EV production, while the European Union and the United States are rapidly expanding their domestic manufacturing capabilities through green industrial policies [738].

However, significant challenges persist. The sourcing of raw materials such as lithium, cobalt, and nickel raises both supply and ethical concerns, potentially jeopardizing the sustainable growth of battery manufacturing [738, 744]. Additionally, the charging infrastructure in many areas, particularly rural and developing regions, requires significant expansion to support widespread EV adoption [745]. Life cycle emissions from battery production also necessitate ongoing improvement efforts, focusing on sustainable practices in sourcing and recycling [739, 742].

Electric Vehicle Fuel Use for Production

Producing an electric vehicle (EV) involves a comprehensive and energy-intensive supply chain that relies significantly on various energy sources—primarily fossil fuels, unless offset by renewables. This process unfolds through several key stages, each contributing to the overall energy consumption and greenhouse gas emissions associated with EV manufacturing.

The foundation of EV production begins with the extraction and processing of raw materials such as lithium, cobalt, nickel, graphite, aluminium, copper, and steel. This stage is predominantly powered by diesel fuels for mining equipment, while electricity, often sourced from fossil fuels such as coal or natural gas, significantly contributes to energy use in these operations. The energy required for mining and processing minerals for a single EV battery (approximated at 60 kWh) has been estimated to range between 1,500 to 2,000 MJ of diesel-equivalent energy, which corresponds to approximately 40 to 50 litres of diesel, depending on the ore grade and mining methods employed [746]. The extraction processes can vary significantly in environmental impact based on the sources of electricity utilized [747].

Following raw material extraction, the battery manufacturing phase is a critical component of EV production, heavily reliant on electricity. The production of battery cells typically demands about 50 to 75 kWh of electricity for every kWh of battery capacity. Consequently, a 60 kWh battery could require as much as 3,000 to 4,500 kWh of electricity, translating to approximately 300 to 450 kg of CO_2 emissions if the electricity grid is fossil fuel-dominated [747]. This starkly highlights the carbon footprint associated with battery production and the pressing need for cleaner energy sources throughout this stage.

The assembly of EVs and the production of various components is another significant energy-consuming stage. The production process typically utilizes a mix of electricity, natural gas, and oil derivatives (plastics), with diesel fuels often required for transporting parts. Energy inputs for constructing an EV chassis, along with the electronics and drivetrain, amount to about 30 to 50 GJ, which is equivalent to approximately 800 to 1,400 litres of diesel equivalent energy [746].

Lastly, logistics and transportation add a further layer of energy consumption, predominantly through diesel and bunker fuels used for ships and trucks, as well as aviation fuels for air cargo. Transporting parts and finished vehicles generally accounts for around 10 to 15% of the total manufacturing energy usage [747]. This aspect underscores the broader impact of supply chain logistics in the EV production paradigm, further complicating efforts to decarbonize the sector and enhance sustainability.

While electric vehicles are celebrated for their potential to reduce operational emissions, the energy-intensive production processes largely depend on fossil fuels unless supported by renewable energy initiatives. As the demand for EVs continues to grow, optimizing the supply chain and sourcing materials sustainably will be key challenges to fully realize the climate benefits attributed to electric mobility.

Table 8: Summary: Energy and Fuel Input per EV (60 kWh battery example).

Stage	Main Fuels	Energy/Fuel Estimate
Mining & Material Processing	Diesel, Electricity	~40–50 L diesel equivalent

Fuel, Energy and Net Zero

Stage	Main Fuels	Energy/Fuel Estimate
Battery Production	Electricity (coal/gas)	~3,000–4,500 kWh (~1,000–1,500 kg CO_2 if fossil-based)
Vehicle Manufacturing	Electricity, Gas	~800–1,400 L diesel equivalent
Logistics & Transport	Diesel, Marine Fuel	~100–200 L diesel equivalent

For context and analysing Table 8, to estimate how many homes could be powered—and for how long—using the energy consumed in the production of one electric vehicle (EV), we must first convert the fuel and electricity usage involved in EV production into kilowatt-hours (kWh). This provides a common unit of comparison that can then be matched against average household electricity consumption.

In terms of energy inputs, various stages of EV production involve significant energy use. Mining and material processing typically consume around 45 litres of diesel, which converts to approximately 480 kWh using an average of 10.7 kWh per litre of diesel. Battery production is one of the most energy-intensive stages, consuming between 3,000 and 4,500 kWh of electricity—averaging around 3,750 kWh. Vehicle manufacturing itself uses approximately 1,100 litres of diesel, equating to about 11,770 kWh. Additionally, logistics and transport require roughly 150 litres of diesel, which amounts to around 1,605 kWh. Altogether, the total energy required to produce a single EV adds up to approximately 17,600 kWh.

To understand the significance of this energy figure, we can compare it to average household electricity use. In Australia, for example, a typical household consumes between 15 and 20 kWh per day. Using an average daily consumption rate of 18 kWh, we can calculate how long this 17,600 kWh of energy could support a household. Dividing the total energy by daily usage (17,600 ÷ 18) gives approximately 978 days of electricity use for a single home—equivalent to around 2.7 years.

Alternatively, if the energy were used all at once, it could power 978 homes for a single day. This illustrates the substantial embedded energy cost associated with EV production, particularly when fossil fuels are involved in the process. However, it's important to note that despite this high initial energy investment, electric vehicles generally result in significantly lower emissions over their operational lifetime compared to petrol or diesel vehicles—especially when powered by clean electricity sources. As such, the long-term environmental benefits of EVs tend to outweigh their production-related energy demands.

To put this in context, the U.S. Energy Information Administration (EIA) reports that the average U.S. household uses around 10,500 kWh per year, or about 29 kWh per day. Based on this average daily consumption, the energy used to manufacture one EV could power a single U.S. household for roughly 607 days, or approximately 1.66 years. Alternatively, the same amount of energy could supply electricity to 607 average homes for one full day.

The production of electric vehicles (EVs) typically involves higher upfront carbon dioxide (CO_2) emissions compared to traditional gasoline vehicles. A primary reason for this discrepancy is the energy-intensive manufacturing processes associated with EV batteries. Research indicates that the production phase for a typical petrol vehicle generates approximately 5 to 7 metric tonnes of CO_2, while electric vehicles equipped with a 60 kWh battery can emit between 8 and 14 metric tonnes of CO_2 [748-750]. It is particularly notable that battery manufacturing accounts for a significant portion (30% to 50%) of an electric vehicle's total production emissions, and this proportion tends to grow with larger battery sizes, potentially reaching up to 17 metric tonnes for longer-range models [751, 752].

The CO_2 emissions associated with EV production are influenced by several factors, including battery size and the energy mix used in manufacturing. Larger batteries require more raw materials and energy to produce, which elevates emissions significantly [753]. Moreover, the carbon footprint of EVs can vary notably based on the electricity sources utilized for their production. For instance, vehicles manufactured in areas reliant on coal, like China, exhibit a higher carbon footprint relative to those produced in regions that utilize renewable energy sources or nuclear power, such as Sweden or France [753, 754]. Furthermore, the demand for critical minerals such as lithium, cobalt, and nickel necessary for batteries underscores the complexity of sourcing materials sustainably, as mining and refining these materials can add to the overall carbon emissions unless sustainably managed [754, 755].

Figure 35: Vinfast's Electric Car Manufacturing in Vietnam in 2023. Taotop, CC BY-SA 4.0, via Wikimedia Commons.

Despite these higher initial emissions, electric vehicles achieve a lower overall carbon footprint over their lifetime, especially when powered by cleaner electricity sources. Studies suggest that while EVs might initially be more carbon-intensive to produce, they can compensate for this 'carbon debt' quickly, often within a timeframe of 6 to 24 months depending on local grid emissions [750, 756]. Once the breakeven point is achieved, EVs generally emit lower levels of CO_2 per kilometre since conventional gasoline vehicles produce approximately 180 to 250 grams of CO_2 per kilometre compared to less than 50 grams for EVs charged from a low-carbon grid [757, 758].

While the upfront CO_2 emissions from EV production can be significantly higher than those associated with gasoline vehicles, the long-term benefits in terms of reduced lifetime emissions strongly suggest that electric vehicles represent a more sustainable transportation option, particularly as battery technology continues to improve and as the global electricity grid transitions to lower carbon sources [751, 753]. As the industry advances, it is likely that these emissions disparities will continue to diminish, reinforcing the argument for the increased adoption of electric vehicles within a sustainable framework.

Battery Production

The production of lithium-ion batteries entirely without fossil fuels is technically feasible, but it is not yet common on a large scale. Current practices heavily rely on fossil fuels throughout multiple stages of production, from the extraction of raw materials to the manufacturing of battery cells and their transportation. Fossil fuels, particularly in the form of diesel and electricity derived from coal and natural gas, are significant in the mining and processing of essential materials like lithium, cobalt, nickel, and graphite [759, 760]. Mining operations, especially in regions like China, where coal remains a predominant energy source, frequently utilize diesel-powered machinery [761]. Moreover, manufacturing operations predominantly rely on electricity from fossil sources unless converted to renewable energy [762]. The logistical aspects of distributing these materials globally contribute significantly to their carbon footprint, as transportation mainly uses fossil-fuelled vehicles and vessels [759].

Despite these challenges, notable initiatives aim to decarbonize the battery supply chain. The electrification of mining operations and increased use of renewable energy sources such as solar, wind, and hydro for extraction and processing phases present viable pathways toward fossil-free production [759, 763]. Battery manufacturers like Tesla's Gigafactory in Nevada and Northvolt's facility in Sweden are pioneering efforts to operate entirely on renewable electricity [764]. Additionally, the transition to hydrogen-powered logistics and the use of recycled materials can significantly reduce energy consumption and emissions [765, 766]. Recycling effectively utilizes existing materials, diminishing reliance on newly mined resources while addressing waste generated from the battery lifecycle [759, 767].

Recent advancements illustrate progress in reducing the carbon footprint associated with lithium-ion battery production. For instance, Northvolt claims that its Swedish plant produces the "world's greenest battery," utilizing 100% hydropower [763]. Automotive giants like BMW and Volkswagen are actively sourcing batteries from suppliers committed to renewable energy, reflecting a significant industry shift [764]. Companies such as Redwood Materials are innovating recycling methods to produce batteries from reclaimed materials, further minimizing fossil fuel dependence [759, 765]. These developments suggest increasing momentum toward sustainable practices as the industry acknowledges the urgent need for cleaner manufacturing processes.

Effectiveness of the EV Transition

The global transition to electric vehicles (EVs) over the past decade represents significant progress, yet its effectiveness varies across different countries and regions due to different levels of policy support, infrastructure development, consumer incentives, and electricity grid decarbonization. As of 2024, global EV sales reached over 14 million annually, constituting approximately 18% of new car sales, with some nations like Norway seeing over 80% of new car sales being fully electric [768]. The dominant factors driving this increase include robust government policies, financial incentives, and comprehensive charging infrastructure, particularly evident in countries like Norway, China, and several nations in the European Union.

Norway serves as a prominent case study, where substantial fiscal incentives, such as tax exemptions and free access to charging stations, have led to extensive EV adoption. A study found that Norway's market penetration of battery electric vehicles (BEVs) reached approximately 43% by 2021, attributed to these policy measures [769-771]. In contrast, while China boasts the largest EV market by volume, with about 30% of new vehicle sales in major cities, it also faces challenges regarding sustainable practices, particularly concerning battery production emissions and sourcing of materials [772, 773]. Furthermore, the European Union implements strict emission regulations coupled with various incentives to promote EV uptake, leading to observable increases in countries like Germany and Sweden [774].

In the United States, the proliferation of EVs has been slower, resulting in approximately 8% of new vehicle sales by 2023. Adoption is higher in states such as California compared to those with less proactive policies [775]. Developing regions, including parts of Australia and India, often lag due to inadequate charging infrastructure, high upfront costs, and less aggressive governmental policies [776].

From an environmental standpoint, EVs effectively reduce local air pollution and lifecycle greenhouse gas emissions, especially when powered by renewable energy. However, the benefit-to-cost ratio heavily depends on local electricity grid conditions. In areas reliant on coal for electricity, while EVs reduce tailpipe emissions, they may not achieve a net reduction in emissions compared to conventional vehicles due to higher carbon emissions during electricity generation [777, 778]. Battery production also raises environmental concerns due to its reliance on rare minerals such as lithium and cobalt, prompting discussions about the sustainability of supply chains [779].

The limitations in the current EV transition are notable; high initial costs, range anxiety, and inadequate charging networks hinder widespread acceptance and equitable EV adoption. Additionally, there are growing concerns about the mining and processing impacts of essential minerals required for batteries, leading to calls for transparent and ethical sourcing practices [780, 781].

Finally, alternatives to electric vehicles, such as hydrogen fuel cells and advancements in renewable energy vehicles, are receiving attention amidst these challenges. Research is ongoing into these technologies as potentially viable solutions for sustainable mobility, demonstrating that the transition to zero-emission transportation might not solely rely on the exclusive adoption of EVs [781, 782].

While electric vehicles (EVs) are central to transport decarbonisation efforts, they are not a universal solution. Different contexts and transport needs require a variety of low-emission alternatives and complementary technologies.

One such alternative is hydrogen fuel cell vehicles (FCVs), which generate electricity onboard by combining hydrogen gas with oxygen. These vehicles offer faster refuelling times and longer driving ranges than battery EVs, making them more suitable for heavy-duty applications such as trucks, buses, and long-haul transport. However, hydrogen technology currently faces major barriers, including high costs, a limited network of refuelling stations, and a reliance on fossil fuel-derived hydrogen, which undermines its environmental benefits unless green hydrogen production is scaled up.

Biofuels such as ethanol, biodiesel, and synthetic fuels provide another avenue for reducing emissions, particularly in sectors where electrification remains challenging, like aviation, maritime shipping, and agricultural machinery. These fuels can often be used in existing internal combustion engines with minimal modification. However, concerns persist regarding the sustainability of large-scale biofuel production, including land use changes, the food-versus-fuel debate, and the overall scalability of supply.

Plug-in hybrid electric vehicles (PHEVs) offer a transitional solution by combining battery-electric propulsion with a petrol or diesel engine. They are especially practical in regions lacking robust charging infrastructure, allowing drivers to operate on electricity for short trips while maintaining a fuel-based backup for longer distances. However, studies have shown that some users rely heavily on the combustion engine, reducing the environmental advantage of the electric component.

Beyond individual vehicle options, public and active transport plays a crucial role in reducing emissions and promoting sustainable urban mobility. Investment in rail networks, bus systems, cycling infrastructure, and pedestrian-friendly design can significantly decrease the demand for personal car use. These options are often the most space-efficient, inclusive, and environmentally sustainable forms of transport, especially in densely populated urban areas.

Lastly, electric micromobility, such as e-bikes, electric scooters, and three-wheelers, is increasingly popular in cities—particularly across Asia and Africa. These modes offer low-emission, affordable, and efficient solutions for short commutes and last-mile connectivity. E-bikes, in particular, are emerging as a viable substitute for cars in urban settings, contributing to lower congestion and pollution levels.

Together, these technologies and strategies highlight the need for a diverse, integrated approach to sustainable mobility, recognising that no single solution will meet the full spectrum of transport demands.

Vehicle End of Life and Disposal

The lifespan of electric vehicles (EVs) and petrol vehicles is a critical area of inquiry, as both technologies continue to evolve and improve in terms of durability and environmental impact. Research indicates that EVs typically last between 10 to 20 years, highly contingent on battery health and usage patterns, which are significant factors in their longevity [783, 784]. Manufacturers of EVs frequently offer warranties that span from 8 to 10 years or up to 160,000 kilometres, although actual endurance can exceed these benchmarks due to improvements in battery technology and design [783]. Despite concerns over battery degradation, which can diminish the vehicle's driving range, the resultant effects do not typically render the vehicle inoperable, underscoring the resilience of modern EVs [785].

In contrast, petrol vehicles, or internal combustion engine (ICE) vehicles, have a comparable lifespan of about 12 to 20 years [786, 787]. Unlike EVs, petrol vehicles do not face battery degradation; however, they feature more mechanically complex systems that are subject to wear, necessitating more extensive maintenance [788]. This complexity increases the likelihood of component failures in systems such as gearboxes and exhaust mechanisms, which can significantly affect the overall lifespan and reliability of petrol vehicles.

Regarding end-of-life environmental impacts, EVs and petrol vehicles exhibit notable differences due to the nature of their components and materials. A primary concern associated with EVs is the disposal and recycling of batteries, which contain lithium, cobalt, and nickel, among other critical minerals. Improperly managed disposal could result in severe environmental repercussions, including toxic waste and groundwater contamination [789, 790]. Nevertheless, advancements in recycling technologies are promising; effective recycling and repurposing efforts can greatly mitigate these adverse effects, allowing for second-life applications such as stationary energy storage [783].

Conversely, while petrol vehicles incur simpler disposal processes, they contribute significantly higher greenhouse gas emissions throughout their operational lifecycle compared to EVs [791, 792]. The combustion of petrol fuels generates substantial quantities of air pollutants like carbon monoxide (CO), nitrogen oxides (NOx), and particulate matter, and the cumulative emissions from these vehicles over time dramatically eclipse the emissions associated with EV battery production and disposal [787, 793]. Thus, despite the resource-intensive production phase of EVs, their overall operational footprint can ultimately be lower, particularly when powered through low-carbon electricity [794].

The technological evolution of both EVs and petrol vehicles has led to a level of durability wherein their lifespans are increasingly comparable. Yet, the environmental ramifications of their end-of-life processes diverge markedly, highlighting the need for enhanced recycling processes for EVs and ongoing scrutiny of emissions from petrol vehicles. The shift toward improved battery technologies and recycling infrastructure is vital for safeguarding environmental integrity while favouring the long-term sustainability of electric transportation.

Hydrogen Economy

The Hydrogen Economy represents a transformative vision where hydrogen serves as a key energy carrier, enabling widespread decarbonization across various sectors, including transportation, industry, and power generation. As the most abundant element in the universe, hydrogen offers a clean combustion process—producing only water vapor when used as a fuel. This characteristic positions hydrogen as a critical component in efforts to mitigate greenhouse gas emissions associated with the utilization of fossil fuels [795, 796].

The concept of a hydrogen economy embodies the role of hydrogen as a key player in conjunction with low-carbon electricity to mitigate greenhouse gas emissions. By utilizing hydrogen, this economy aims to phase out fossil fuels, particularly in sectors where cheaper and more energy-efficient alternatives to clean energy are not feasible [797, 798]. The hydrogen economy encompasses both hydrogen production and its various applications, which collectively contribute to reducing dependency on fossil fuels and fostering long-term sustainability in energy usage.

Historically, the term "hydrogen economy" was first introduced by John Bockris in 1972, envisioning a system where hydrogen serves as a foundational energy carrier, replacing traditional fossil fuels amidst increasing energy demands and growing environmental concerns [799, 800]. Current production methods for hydrogen heavily rely on fossil fuels, with about 96% derived from non-renewable sources such as natural gas and coal [801, 802]. This dependence presents significant challenges, particularly in terms of carbon emissions associated with conventional hydrogen production methods. Nonetheless, advancements in hydrogen generation technologies, such as electrolysis driven by renewable energy sources, are gaining traction as promising alternatives [797, 803].

In terms of applications, hydrogen is gaining traction in heavily polluting sectors like transportation and industrial processes where direct electrification is challenging. Fuel cell vehicles, heavy-duty trucks, and aviation are pivotal areas that stand to benefit from hydrogen technology, enabling a shift away from traditional fuels [533, 804]. Additionally, the industrial domain, including steel production and chemical synthesis, is expected to transition towards hydrogen as a clean energy source, thus reducing reliance on carbon-intensive processes [540, 557]. Hydrogen also serves a vital role in energy storage, proving particularly advantageous in balancing the intermittency of renewable energy sources like wind and solar by storing excess generated energy for later use [805, 806].

Furthermore, the production methods of hydrogen are categorized primarily based on their carbon emissions profile. "Grey" hydrogen, produced from natural gas without carbon capture, remains prevalent but contributes significantly to CO_2 emissions. Conversely, "blue" hydrogen integrates carbon capture technology to mitigate emissions, while "green" hydrogen, derived from electrolysis powered by renewable energy, is increasingly sought after due to its zero-emission profile [533, 795, 796]. This transition towards green hydrogen aligns with global initiatives and governmental strategies aiming to enhance renewable hydrogen production capabilities and facilitate infrastructure development [796, 807].

The expansion of hydrogen use across global economies requires significant investment and incurs substantial costs across the entire value chain—from production to storage, distribution, and end use. Estimating the cost of hydrogen is inherently complex, as it depends on a wide range of variables, including the price of energy inputs (primarily electricity and natural gas), the production method used (such as steam methane reforming for grey hydrogen or electrolysis for green hydrogen), the type of electrolyser technology employed (e.g., alkaline or proton exchange membrane), as well as the storage and transport methods available. These costs are commonly consolidated into a single metric known as the Levelized Cost of Hydrogen (LCOH), expressed in US dollars per kilogram of hydrogen. The LCOH helps compare different hydrogen production pathways by factoring in assumptions about future costs and technology maturity.

Fuel, Energy and Net Zero

As of 2022, grey hydrogen—the most common and carbon-intensive form, produced from natural gas without carbon capture—is the cheapest to produce, particularly when no carbon tax is applied. For instance, according to the International Energy Agency (IEA), the cost of grey hydrogen ranged from $1.0 to $2.5/kg in 2021 but spiked to between $4.8 and $7.8/kg in 2022 due to the global energy crisis following Russia's invasion of Ukraine. Blue hydrogen, which includes carbon capture and storage (CCS), had a similar cost increase, rising from about $1.5–3.0/kg in 2021 to $5.3–8.6/kg in 2022. Although blue hydrogen reduces CO_2 emissions compared to grey hydrogen, its price remains tightly linked to volatile natural gas markets and may still face penalties for any uncaptured emissions. Long-term projections indicate only limited cost reductions for blue hydrogen, with forecasts for 2050 remaining relatively stable.

In contrast, green hydrogen—produced through the electrolysis of water using renewable electricity—currently has the highest production costs but is expected to see the steepest price reductions over time. Depending on the source and assumptions used, 2022 production costs for green hydrogen varied widely, from $2.2 to over $9.0 per kilogram. However, projections suggest that by 2030, green hydrogen could cost as little as $1.0–1.5/kg in regions with strong solar or wind resources, potentially dropping below $1.0/kg by 2050. For example, the International Renewable Energy Agency (IRENA) and Energy Transitions Commission foresee costs between $1.1 and $3.4/kg by mid-century. Much of this optimism is based on declining renewable energy costs and the anticipated scale-up of electrolyser manufacturing. It is particularly cost-effective to produce green hydrogen using otherwise curtailed renewable power, such as excess solar during midday or surplus wind during low-demand periods, which favours flexible electrolyser systems that can ramp production up or down.

Nevertheless, the path to cost-competitive hydrogen remains challenging. While the cost of electrolysers fell by 60% between 2010 and 2022, recent data show a reversal of that trend. Between 2021 and 2024, the cost of electrolyser units reportedly increased by around 50%, largely due to supply chain constraints and higher capital costs. Still, institutions such as the Oxford Institute for Energy Studies remain optimistic that green hydrogen costs will continue to decline over the next two decades, especially as technology matures and as large-scale projects become more common.

A 2022 report from Goldman Sachs projected that green hydrogen could reach cost parity with grey hydrogen globally by 2030, and potentially sooner if carbon pricing is widely adopted. However, it is important to note a fundamental cost reality: blue and grey hydrogen will always be more expensive than the fossil fuels they are derived from, while green hydrogen will always be more expensive than the renewable electricity used in its production. Therefore, hydrogen's viability depends not only on technological improvements and scale but also on regulatory frameworks, subsidies, and carbon pricing that support low-carbon alternatives over conventional fossil fuels.

Significant challenges remain on the path to a robust hydrogen economy, including high production costs, the development of adequate infrastructure for storage and transport, and safety concerns related to hydrogen's flammability [533, 808]. Additionally, while green hydrogen holds promise, today's production methods still incur higher costs compared to fossil fuels, necessitating further technological and economic advancements to achieve viability [804, 809]. The complexity of establishing an efficient hydrogen supply chain poses an additional barrier to widespread adoption, as it involves intricate logistics for production, transport, and end-use applications [810, 811].

The Hydrogen Economy is seen as pivotal for global decarbonization efforts, particularly in sectors where alternative energy sources face significant limitations. The successful integrated use of hydrogen will depend on overcoming current challenges through innovation and collaborative global strategies. As investment flows into research and development, the potential for hydrogen to fundamentally reshape energy systems and achieve net-zero emissions becomes increasingly tangible [540, 805, 812].

The global hydrogen production market was valued at over US$155 billion in 2022, with a projected annual growth rate of around 9% through 2030 [540, 813]. In 2021, the global production of molecular hydrogen (H2) reached approximately 94 million tonnes. Approximately one-sixth of this hydrogen was produced as a by-product within petrochemical industry processes, while the majority originated from dedicated production facilities [813, 814].

The predominant method for hydrogen generation remains fossil fuel consumption, with over 99% of dedicated hydrogen production derived from steam reforming of natural gas, accounting for about 70%, and coal gasification which contributes approximately 30%, particularly in China [813, 814]. As of now, low-carbon hydrogen options represent less than 1% of total production. Key low-carbon production methods include steam reforming combined with carbon capture and storage, green hydrogen from electrolysis, and biomass-derived hydrogen [813, 815].

In terms of environmental impact, CO_2 emissions from hydrogen production in 2021 were estimated at roughly 915 million tonnes, accounting for about 2.5% of energy-related CO_2 emissions and 1.8% of total global greenhouse gas emissions [816]. The main applications of hydrogen produced for the market focus on oil refining, which utilized around 40 million tonnes of hydrogen in 2021, alongside various industrial processes, which consumed an additional 54 million tonnes [817]. The refining industry primarily uses hydrogen in hydrocracking to convert heavier petroleum sources into lighter fractions suitable for fuel, while industrial applications mainly include ammonia production for fertilizers (34 million tonnes), methanol production (15 million tonnes), and the manufacturing of direct reduced iron (5 million tonnes) [814, 817].

Several countries are actively integrating hydrogen into their energy systems, each achieving varying levels of success. Hydrogen is increasingly viewed as a key enabler of clean energy transitions, especially in sectors that are hard to electrify. However, despite progress, the global hydrogen sector has experienced notable setbacks, including project cancellations and delays driven by high costs, infrastructure challenges, and inconsistent policy support.

Japan is one of the most advanced countries in hydrogen technology. It has developed a robust national hydrogen infrastructure, including the Fukushima Hydrogen Energy Research Field—one of the world's largest green hydrogen facilities powered by solar energy. Japan has also constructed a "hydrogen highway" of over 150 refuelling stations, supporting the growing adoption of fuel cell vehicles [818, 819].

South Korea is similarly investing heavily in hydrogen as part of its long-term energy strategy. The government provides generous subsidies for hydrogen vehicles and is rapidly expanding its hydrogen refuelling infrastructure. South Korea aims for hydrogen to supply 10% of its energy needs by 2030 and 30% by 2040, indicating a strong commitment to its hydrogen roadmap [820].

Germany is incorporating hydrogen into industrial decarbonisation efforts, especially in steelmaking. Projects like Thyssenkrupp's H2Stahl aim to replace coal with green hydrogen in high-temperature industrial processes [820]. Additionally, Germany is developing a national hydrogen network, such as the GET H2 initiative, to link hydrogen producers with industrial users [821].

India is making early progress in green hydrogen. The Indian Oil Corporation is developing a project at its Panipat refinery to produce 10,000 tonnes of green hydrogen annually, signalling a significant step toward clean energy goals and industrial transformation [818].

Saudi Arabia is investing on a massive scale, constructing what is expected to be the world's largest green hydrogen project in the NEOM city. The plant aims to produce up to 600 tonnes of hydrogen daily using 4 GW of solar and wind energy, showcasing the country's pivot toward clean energy leadership [818].

Despite these advancements, many hydrogen projects have faced significant challenges. In Europe, approximately 20% of hydrogen projects were cancelled in 2024, mainly due to prohibitive costs and a lack of viable infrastructure. Norway alone saw around 10 GW of potential capacity withdrawn [820]. In China, government permissions for six major green hydrogen projects in Inner Mongolia were revoked due to missed construction deadlines, reflecting broader execution difficulties.

Globally, a study of 137 green hydrogen projects over three years revealed that only 2% were completed on time. This highlights a serious gap between ambition and delivery. The same study estimated that implementing all announced projects would require around $1.6 trillion in subsidies, underscoring the immense financial commitment needed to scale hydrogen deployment [819, 822].

While nations such as Japan, South Korea, Germany, India, and Saudi Arabia are leading the charge in hydrogen adoption, the global hydrogen economy still faces significant hurdles. High costs, underdeveloped infrastructure, and policy gaps remain major barriers. For hydrogen to play a transformative role in decarbonising the global energy system, these challenges must be addressed through coordinated policy, innovation, and international collaboration.

Carbon Capture and Storage (CCS)

Carbon Capture and Storage (CCS) is recognized as a pivotal climate mitigation technology, aimed at curbing carbon dioxide (CO_2) emissions stemming from power generation and industrial processes. CCS operates by capturing CO_2 before it is released into the atmosphere and securely storing it underground. This technology is particularly crucial for achieving net-zero emissions targets in sectors that are notoriously difficult to decarbonize, such as cement, steel, and chemicals, as well as fossil fuel-based energy systems [823].

Numerous studies underscore the essential role of CCS in enabling decarbonization across various industries. For instance, Lau et al. emphasize that CCS supports not only the power sector but is also integral to decarbonizing the transport and industrial sectors, highlighting its versatility across hard-to-abate areas [824, 825]. The integration of CCS technologies has been shown to significantly enhance the potential for substantial emissions reductions in countries with high dependency on fossil fuels, such as those in Southeast Asia, where sufficient saline aquifers have been identified as ample storage sites for captured CO_2 [824, 826].

Specifically, the cement industry presents substantial decarbonization challenges due to its inherent carbon emissions from chemical processes. Research indicates that CCS is viewed as a viable long-term solution for mitigating emissions from cement production, offering a mature technology that can effectively sequester the carbon produced [827, 828]. Furthermore, the International Energy Agency outlines that various technological strategies, including CCS, are crucial for reducing emissions in cement manufacturing, positioning it as a key part of comprehensive decarbonization roadmaps [829].

The urgency of implementing CCS is accentuated by the global push for net-zero emissions. Significant investments in CCS are deemed essential for realizing aspirations of climate neutrality in both industrial and energy sectors. For instance, Lau and Tsai highlight that accelerating the deployment of decarbonization technologies, including CCS, is vital to meet imminent carbon reduction targets, as many industrial processes cannot transition away from fossil fuel reliance without these technologies [830, 831]. Moreover, systematic assessments suggest that geological formations offer substantial storage capacity that could accommodate many years of emissions from existing fossil fuel power plants and industrial operations [826, 832].

Lastly, it is critical to note that while CCS presents a viable solution for many emission-heavy sectors, it must be viewed as part of a broader strategy that includes improving energy efficiency, utilizing renewable energy, and innovating in energy technologies. This integrated approach will support increasingly ambitious global climate targets effectively, particularly given the limitations of existing low-carbon technologies in fully mitigating emissions from hard-to-abate sectors [833-835].

The process of Carbon Capture and Storage (CCS) consists of three key stages: capture, transport, and storage.

The first step, capture, involves separating carbon dioxide (CO_2) from the gases produced during electricity generation or industrial processes such as cement, steel, and chemical manufacturing. There are several methods for this. In post-combustion capture, CO_2 is removed from the exhaust gases after fossil fuels have been burned. In pre-combustion capture, the fuel is converted into a mixture of hydrogen and CO_2 before combustion, allowing CO_2 to be separated prior to energy generation. A third approach is oxy-fuel combustion, where fuel is burned in pure oxygen instead of air. This results in a flue gas that is primarily CO_2 and water vapor, making CO_2 easier to isolate.

Post-combustion capture is the most widely applicable method, especially for retrofitting existing power plants. In this approach, CO_2 is captured after the fuel—typically coal or natural gas—has been burned. The resulting flue gas, containing nitrogen, CO_2, water vapor, and other compounds, is treated using chemical solvents, most commonly amines such as monoethanolamine (MEA), which selectively absorb the CO_2. The CO_2-rich solvent is then heated to release the CO_2 for compression and transport, while the solvent is regenerated and reused in the system. This method is already in use in various pilot and full-scale facilities and is particularly common in coal- and gas-fired power stations, cement kilns, and waste incinerators. The main advantage is its compatibility with existing infrastructure, but it has drawbacks, including high energy consumption for solvent regeneration, which reduces the overall efficiency of the power plant.

Pre-combustion capture, by contrast, removes CO_2 before the fuel is burned and is typically applied in integrated gasification combined cycle (IGCC) plants or hydrogen production. The fuel—such as coal, natural gas, or biomass—is partially oxidized in a gasifier, producing synthesis gas (syngas) composed of hydrogen and carbon monoxide. This syngas undergoes a water-gas shift reaction, producing more hydrogen and CO_2. The CO_2 is separated using physical solvents such as Selexol or Rectisol and then compressed for storage, while the hydrogen is either combusted for electricity or used in industrial applications. This method is efficient and produces hydrogen, a valuable clean energy carrier. However, it is capital-intensive and not suitable for retrofitting conventional power plants, as it requires an entirely different technological setup.

Oxy-fuel combustion offers a third pathway, where the fuel is burned in pure oxygen instead of air, creating a flue gas composed mainly of CO_2 and water vapor. An air separation unit (ASU) isolates the oxygen required for combustion. After burning, the water vapor is condensed, leaving behind nearly pure CO_2, which is then compressed for storage. This approach simplifies CO_2 separation due to the absence of nitrogen in the flue gas. While promising in terms of CO_2 purity and potentially lower energy penalties than post-combustion methods, the technology is still largely in the research and demonstration phase. The energy and cost demands of the oxygen production process and the need to redesign combustion and heat recovery systems are significant challenges.

In practice, the selection of a CO_2 capture method depends on various factors including the type of facility, the kind of fuel used, the nature of the output (such as electricity or hydrogen), and broader economic and logistical considerations. As CCS technology evolves, there is growing interest in hybrid approaches and next-generation materials like solid sorbents and membranes, which may improve capture efficiency while reducing operational costs. Each method has unique strengths and limitations, and the most suitable option will depend on the specific industrial context and decarbonisation goals.

Table 9: Carbon Capture Methods Summary.

Method	When CO_2 Is Captured	Typical Use Cases	Key Pros	Key Cons
Post-Combustion	After combustion	Power plants, cement kilns	Retrofit-friendly, mature tech	High energy cost for solvent regeneration
Pre-Combustion	Before combustion	IGCC, hydrogen plants	High capture efficiency, produces H_2	Costly gasification, unsuitable for retrofits
Oxy-Fuel Combustion	During combustion	Cement, steel (R&D stage)	High CO_2 purity, simplified separation	Expensive oxygen production

The second step is transport. Once captured, the CO_2 is compressed into a high-pressure, liquid-like state to facilitate movement. It is then transported to a storage site, most commonly through pipelines. However, for projects located far from suitable storage formations, transportation by ship or truck is also possible.

The final step, storage, involves injecting the compressed CO_2 deep underground into geological formations designed to securely contain it for the long term. These formations include depleted oil and gas reservoirs and deep saline aquifers. To ensure safe and permanent storage, the rock structure must be porous enough to hold the CO_2, and capped by an impermeable layer of rock (known as a seal or cap rock) to prevent the gas from escaping back to the surface.

Together, these steps form a complete system designed to reduce CO_2 emissions from major sources and support efforts to mitigate climate change.

Figure 36 is a conceptual diagram that illustrates how offshore Carbon Capture and Storage (CCS) systems work, with a particular focus on methods of underwater CO_2 sequestration. It shows how carbon dioxide, captured from industrial sources on land, can be transported to offshore locations and stored securely beneath the ocean floor to prevent it from entering the atmosphere.

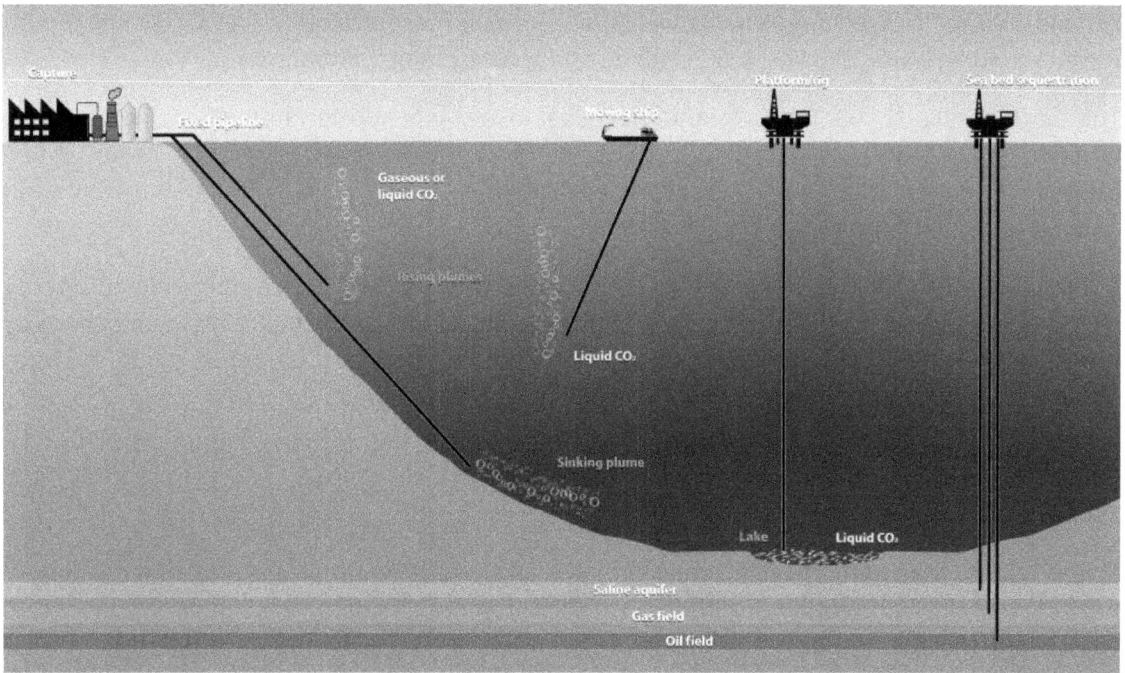

Figure 36: Carbon Storage at sea. The joy of all things, CC BY-SA 4.0, via Wikimedia Commons.

The process begins with CO_2 capture at industrial facilities such as factories and power plants. These operations produce CO_2 as a by-product of fossil fuel combustion or other industrial processes. Instead of releasing this CO_2 into the atmosphere, it is captured at the source using technologies that separate it from other gases.

Once captured, the CO_2 can be transported offshore in two main ways. One method uses fixed pipelines that carry the CO_2—either in gaseous or liquid form—from land to deep-sea injection points. During this descent, the CO_2 may form rising plumes if it is less dense than seawater, or sinking plumes if it is denser, depending on the pressure and temperature at various ocean depths. The second transport method involves moving ships that carry liquid CO_2 and inject it directly into the deep ocean. This method is often used in regions where fixed pipelines are impractical due to geographic constraints or high costs.

Upon reaching the injection point, CO_2 can be stored in several ways. Offshore platforms or rigs inject CO_2 through vertical wells into deep geological formations, such as depleted oil and gas fields or saline aquifers. These formations are well-suited for long-term CO_2 storage due to their porous rock structures and impermeable cap rocks that prevent leakage. Alternatively, sea bed sequestration systems use multiple wells to disperse CO_2 into various underground layers beneath the ocean floor.

In some scenarios, CO_2 can accumulate in specific forms depending on the oceanic conditions. At very deep and cold depths, CO_2 may settle into stable lakes of liquid CO_2 on the ocean floor. Rising plumes occur when the CO_2 is buoyant relative to seawater, potentially causing it to ascend unless managed properly. Conversely, sinking plumes develop when CO_2 is denser than the surrounding water, allowing it to descend and spread along the seabed.

The diagram also shows the different geological layers targeted for storage, including saline aquifers—rock layers saturated with saltwater, depleted gas fields, and oil fields. These formations provide stable, long-term storage options and, in the case of oil fields, may even support enhanced oil recovery.

Overall, this offshore CCS system is designed to store captured CO_2 permanently and safely beneath the seabed. It serves as a critical strategy in global efforts to mitigate climate change by reducing the concentration of greenhouse gases in the atmosphere, particularly from hard-to-decarbonise sectors such as heavy industry and fossil fuel energy.

Where CCS is Being Used

Carbon Capture and Storage (CCS) has emerged as a crucial technology for mitigating climate change. Various countries have implemented CCS projects with differing scales and successes, highlighting the importance of these initiatives in reducing carbon dioxide (CO_2) emissions. In Norway, both the Sleipner and Snøhvit projects have served as pioneering efforts since the 1990s, successfully storing CO_2 beneath the North Sea. These projects are frequently referenced as early successful examples of long-term geological CO_2 storage, demonstrating that engineered geological storage is a well-established technology with over fifty years of operational experience in this field [836, 837].

In the United States, the implementation of the 45Q tax credit scheme significantly supports CCS projects, offering financial incentives for capturing and storing CO_2 emissions. Among the notable initiatives is the Petra Nova coal plant in Texas, which, despite its recent shutdown, represented a significant stride in commercial-scale CCS application [838]. Additionally, the Illinois Industrial Carbon Capture and Storage Project continues to operate effectively, capturing emissions from an ethanol production facility and exemplifying the potential of CCS technologies in various sectors [839].

Australia's Gorgon Project stands as one of the largest CCS initiatives globally, integrated with a liquefied natural gas (LNG) facility and poised to store millions of tonnes of CO_2 in deep underground geological formations. This project underscores the scaling potential of CCS to address emissions from the fossil fuel industry [840]. Canada has also made notable contributions to the CCS landscape, particularly through the Boundary Dam project in Saskatchewan, which is recognized as the first commercial-scale CCS integrated into a fossil-fuel-based power plant. It captures a significant portion of the plant's CO_2 emissions while validating the feasibility of incorporating CCS technology in existing infrastructure [841].

These global efforts reflect a collective understanding of CCS as a pivotal component in achieving climate goals, especially in the context of the Paris Agreement. The successful implementation of CCS technologies can facilitate the transition towards a low-carbon economy while still relying on some fossil fuel-based energy sources [831, 842]. While there are challenges related to the economic viability and regulatory frameworks surrounding CCS, ongoing research and operational projects continue to refine the technology, thereby enhancing its role in future energy strategies [843].

The implementation of Carbon Capture and Storage (CCS) technologies represents a critical strategy for mitigating climate change. However, the economic implications of these systems are complex and influenced by a variety of factors including technology type, regional infrastructure, and regulatory frameworks. The costs associated with CCS can broadly be categorized into three segments: capture, transport, and storage.

A significant portion of CCS expenditures is incurred during the capture phase, where estimates suggest that it can constitute 60 to 80 percent of the overall cost of the system (Bui et al., 2018). Within this phase, post-combustion capture is notably prevalent for retrofitting existing facilities, with costs ranging from US$40 to US$100 per metric tonne of CO_2. The energy intensity of this method, attributed to the solvent regeneration process, results in what is often termed the "energy penalty" (Bui et al., 2018). Pre-combustion capture is commonly applied in hydrogen production plants and Integrated Gasification Combined Cycle (IGCC) systems, with per-tonne costs estimated between US$30 and US$70. However, this method requires new infrastructure, which can complicate initial capital expenditures [844]. The oxy-fuel combustion method, which generates a more concentrated flue gas, incurs costs typically between US$40 and US$90 per tonne due to the additional requirements for oxygen production [844].

Transporting captured CO_2 introduces further costs that are variable, hinged largely on distance, volume, and the mode of transportation. Pipeline transport remains the most cost-effective method, with estimates ranging between US$2 and US$14 per tonne, where costs generally decrease with larger volumes and shorter distances [845]. In cases where pipelines are not economically viable, alternative transport methods such as shipping or trucking are often significantly more expensive and context-dependent [845].

The final phase, storage, tends to be the least expensive of the three components yet still necessitates substantial investment for optimal site preparation, drilling, injection, and ongoing monitoring. Geological storage methods such as utilizing depleted oil and gas fields or deep saline aquifers carry costs approximating US$5 to US$20 per tonne [845]. Enhanced oil recovery (EOR) methods, which incorporate CO_2 for additional oil extraction, can partially offset storage costs but raise ethical considerations regarding fossil fuel dependence, potentially undermining climate goals [845].

Overall, the estimated total costs for implementing a CCS system vary widely, typically ranging from US$50 to US$150 per tonne of CO_2 avoided. For power plants implementing these technologies, this could increase electricity generation costs by 30 to 70 percent per megawatt-hour, depending on the capture method and the operational efficiency of the facility [845].

To mitigate these economic pressures, several factors could improve the viability of CCS technologies. Innovations in capture materials, like advanced solvents and membranes, could enhance the efficiency and reduce the energy requirements associated with CO_2 capture [844]. Economies of scale arising from scaling up CCS projects might also lead to cost reductions [846].

Furthermore, leveraging low-cost renewable energy sources to power capture processes has the potential to significantly reduce operational costs [844]. Strong policy measures, including carbon pricing and tax incentives, are also crucial in enhancing the financial attractiveness of CCS projects [846].

While CCS presents a financially demanding undertaking, understanding the nuances of its cost structure can lead to better planning and potentially more effective deployment of these essential technologies in the battle against climate change.

Countries that can afford to implement Carbon Capture and Storage (CCS) are typically those with strong economic resources, established industrial and energy sectors, supportive climate and energy policies, and access to suitable geological formations for CO_2 storage. These countries are either already operating CCS projects or investing significantly in their development.

Among high-income nations with strong climate commitments, the United States stands out as a global leader in CCS deployment. It benefits from extensive industrial capacity, advanced technology, and substantial federal incentives, such as the 45Q tax credit. Notable CCS projects include the Illinois Industrial CCS project and the now-decommissioned Petra Nova facility. The U.S. also possesses vast underground storage potential for CO_2. Canada is another major player, especially in oil-rich regions like Alberta and Saskatchewan. Its Boundary Dam and Quest projects are internationally recognized, and CCS plays a key role in Canada's strategy to reduce emissions from its oil sands industry.

Norway has operated CCS projects such as Sleipner and Snøhvit in the North Sea since the 1990s. Its early adoption of CCS is supported by strong climate policies and favourable geological conditions. The United Kingdom is also advancing CCS development through industrial decarbonisation hubs like HyNet and the East Coast Cluster. Government support and regulatory alignment are helping drive CCS in sectors that are difficult to electrify. Across the European Union, countries including Germany, the Netherlands, and Denmark are investing in CCS to meet climate targets. Offshore storage hubs and financial backing from the EU's Innovation Fund are enabling progress across the bloc.

Middle-income countries with strategic interests or significant oil revenues are also investing in CCS. Saudi Arabia, as part of its Vision 2030 plan, is using CCS to reduce emissions from oil and gas operations while supporting its transition to a hydrogen economy. The Uthmaniyah CCS project, operated by Aramco, demonstrates the country's technical capacity. The United Arab Emirates is similarly committed, operating the Al Reyadah facility through its national oil company ADNOC. These projects serve dual purposes of emission reduction and economic diversification. In China, large-scale CCS is still emerging, but significant investment is being directed toward pilot projects. CCS is viewed as essential to achieving China's 2060 carbon neutrality goal, especially in regions reliant on coal.

The affordability of CCS in these countries is driven by a combination of public funding, carbon pricing mechanisms, access to high-volume industrial emissions, suitable geological storage options, and existing infrastructure that can be adapted for CO_2 capture and transport. However, in low- and lower-middle-income countries, implementing CCS remains challenging. These nations often face high capital and operating costs, limited policy support, inadequate infrastructure, and more urgent development needs such as energy access and poverty reduction. While they may struggle to fund CCS independently, some may benefit from international cooperation, carbon credit markets, or financial assistance through mechanisms like the World Bank or the Green Climate Fund.

CCS is currently most feasible in wealthier and fossil-fuel-rich nations that possess the technological, economic, and geological capacity to support it. As CCS technology matures and costs decline, and as global funding for decarbonisation grows, more countries may begin to adopt CCS. Still, for the foreseeable future, the widespread implementation of CCS will remain concentrated in advanced economies.

Benefits, Challenges, Criticism and future outlook

Carbon Capture and Storage (CCS) offers several significant benefits that contribute to addressing climate change. One of the key advantages is its ability to reduce emissions from high-emitting sectors such as cement, steel, and chemical manufacturing—industries that are particularly difficult to electrify. CCS also plays a vital role in supporting hydrogen production, especially when used with natural gas to create blue hydrogen. Additionally, when combined with bioenergy in processes known as BECCS (Bioenergy with Carbon Capture and Storage), CCS can achieve negative emissions, effectively removing carbon dioxide from the atmosphere. Another important benefit is that it extends the life of existing fossil fuel infrastructure by reducing the emissions associated with its use, helping to ease the transition toward a renewable energy future.

However, CCS also faces a number of challenges and criticisms that hinder its widespread adoption. It is a capital-intensive technology with high operating costs, and in many regions, there are limited economic incentives or policy frameworks to support its implementation. The technology imposes an energy penalty, as capturing, compressing, and transporting CO_2 requires additional energy input, thereby reducing the overall efficiency of power plants or industrial facilities. Long-term CO_2 storage poses environmental risks as well, with concerns about the integrity of storage sites and the potential for leaks requiring ongoing monitoring and regulation. Deployment has also been slow, with only a limited number of large-scale projects in operation— far fewer than what is needed to meaningfully impact global emissions. Moreover, some critics argue that CCS may serve as a justification for prolonging the use of fossil fuels, diverting attention and investment away from renewable energy development.

The role of Carbon Capture and Storage (CCS) is pivotal in meeting global climate goals, with the International Energy Agency (IEA) estimating that it could contribute approximately 15% of the cumulative emissions reductions necessary by 2070 [831]. To effectively scale up CCS technologies, several key enablers must be established, including strong policy frameworks, carbon pricing mechanisms, public-private investment partnerships, and robust monitoring and verification systems.

Strong policy frameworks are essential for the effective deployment of CCS. This includes regulatory frameworks that facilitate investment and define clear guidelines for carbon pricing. Studies indicate that effective state policies can encourage CCS, particularly when combined with mechanisms like border adjustment measures, which protect the economic viability of CCS investments from international competition [831, 847]. Furthermore, the political landscape plays a crucial role in determining how these frameworks are implemented and sustained, underscoring the need for consistent governmental support [831].

Carbon pricing, whether through direct carbon taxes or carbon credits, serves as a crucial economic incentive for promoting CCS investments. Research highlights the importance of having a sufficiently high carbon price to trigger CCS investments across various sectors, including electricity generation and industrial processes [848-850]. In scenarios where carbon prices align closely with models of the social cost of carbon, CCS technologies become economically viable and help shift the energy landscape towards lower carbon emissions. Therefore, an integrated approach that combines market-driven pricing with incentive structures, such as tax credits, can significantly enhance the economic feasibility of CCS projects [851].

Public-private partnerships are important for mobilizing necessary capital for CCS deployment. These partnerships facilitate shared funding and risk management, enabling innovations in CCS technology and infrastructure development [831, 852]. The literature suggests that collaborative frameworks can improve the efficiency of project implementation and foster innovation through resource sharing between public entities and private investors [853]. Moreover, as CCS projects often require significant upfront capital, leveraging both public and private investments is critical to accelerate their commercial viability [854].

Lastly, robust monitoring and verification systems are integral to ensuring the effectiveness and accountability of CCS technologies. Monitoring systems facilitate the assessment of CO_2 emissions reductions and the overall performance of CCS projects, which is vital for regulatory compliance and public acceptance [853, 855]. These systems must be designed to provide reliable and transparent data regarding the carbon capture and storage processes. As climate policies evolve, continuous improvements in monitoring technologies will enhance the reliability of CCS contributions to climate goals [856].

Carbon Capture and Storage is a key but controversial climate solution. It holds promise for decarbonizing heavy industry and achieving net-zero emissions—especially when paired with clean energy and negative emission technologies. However, its success will depend on overcoming financial, technical, and societal barriers to ensure it plays a complementary role alongside renewables, energy efficiency, and systemic transformation.

Chapter 13

The Politics and Economics of Fuel

As we progress through the complex story of fuel, its extraction, use, and future potential, it becomes increasingly clear that technological innovation and environmental challenges are only part of the broader picture. Fuel is not just a physical resource—it is also a source of power, wealth, and geopolitical influence. In Chapter 13, *The Politics and Economics of Fuel*, we turn our attention to the critical forces that shape global and national energy agendas.

This chapter explores how economic policies, trade dynamics, market structures, and government interventions affect fuel availability, pricing, and strategic reserves. It also delves into the political tensions and alliances forged around energy supply, from oil embargoes to energy independence debates. By analysing both the macroeconomic and political dimensions of fuel, readers will gain insight into why decisions about fuel sourcing, subsidies, taxation, and international cooperation are deeply contested and consequential.

Whether discussing the impact of OPEC on global oil prices or how fuel subsidies influence consumer behaviour, this chapter positions fuel as a central player in national economies and international diplomacy. It lays the groundwork for understanding how energy decisions today are not merely technical but deeply political—and how they will continue to shape our shared future.

Oil Markets and OPEC

The global oil market is a complex system that plays a critical role in shaping international economics and geopolitics. At the heart of this market is OPEC—the Organization of the Petroleum Exporting Countries—which, together with its allies (collectively referred to as OPEC+), significantly influences the global supply, price, and politics of oil.

Founded in 1960, OPEC includes major oil-producing nations such as Saudi Arabia, Iran, Iraq, Venezuela, and the United Arab Emirates. OPEC's stated mission is to coordinate and unify petroleum policies among member countries to ensure stable oil markets, secure supply to consumers, steady income to producers, and fair returns on investments.

Since 2016, OPEC has worked closely with a group of non-OPEC oil producers—including Russia, Mexico, and Kazakhstan—forming OPEC+. This coalition enhances their collective ability to influence oil prices by adjusting production quotas in response to market conditions.

OPEC exercises political power by influencing oil prices through coordinated production cuts or increases. For instance, by reducing output, OPEC can raise global oil prices, increasing revenues for member states. However, this strategy is not without political consequences. Higher prices can hurt oil-importing countries, strain diplomatic relationships, and accelerate shifts toward alternative energy sources.

For OPEC members, oil revenues often fund state budgets, social programs, and infrastructure. Countries like Saudi Arabia, Iran, and Venezuela rely heavily on oil exports for economic stability. This reliance creates both leverage and vulnerability—leverage in controlling a key global commodity, and vulnerability to price fluctuations or shifts in global demand.

Oil prices are notoriously volatile. Political instability, war, pandemics (such as COVID-19), and technological change can send prices soaring or crashing. The 1973 oil embargo, the 1991 Gulf War, and the 2020 pandemic are examples of how geopolitical and economic shocks directly impact oil prices.

For oil-importing countries, price spikes can trigger inflation, slow economic growth, and increase political instability. For oil-exporting countries, price crashes can lead to budget deficits, currency devaluation, and social unrest.

OPEC's ability to influence oil markets has been challenged in recent decades. The United States became the world's largest oil producer due to the shale revolution. Countries like Brazil, Canada, and Norway also contribute significant non-OPEC oil supply.

Internal divisions among OPEC members can weaken compliance with quotas. Countries with urgent fiscal needs sometimes overproduce. Additionally, global climate policies, increasing electric vehicle adoption, and investment in renewables are gradually eroding oil's dominance.

OPEC nations are responding to these pressures in various ways. Some, like Saudi Arabia and the UAE, are diversifying their economies through long-term plans (e.g. Vision 2030). Others are investing in carbon capture, hydrogen, and renewable energy.

Nevertheless, oil will remain a vital global energy source for decades, especially for sectors like aviation, heavy transport, and petrochemicals. OPEC continues to adapt, seeking to maintain relevance in a decarbonizing world by managing output, shaping global supply expectations, and investing in clean energy technologies.

Oil as a Political Tool

Oil serves a significant role in global politics, functioning not merely as a commodity but as an instrumental tool of geopolitical power. This dynamic is particularly evident among OPEC nations, especially Saudi Arabia and Russia, who strategically manipulate their production capacities to achieve specific political objectives. This synthesis will explore three key mechanisms through which oil influences geopolitical landscapes: price control, strategic influence, and economic sanctions.

Firstly, OPEC+ countries exercise significant control over oil prices through their production decisions. By adjusting supply levels—whether increasing or decreasing—these nations can manipulate global oil prices, which have direct repercussions on inflation, energy costs, and overall economic growth in both oil-exporting and oil-importing countries. Research indicates that price volatility in oil markets is intricately linked to geopolitical risks, which can exacerbate fluctuations in oil prices and thus influence economic stability worldwide. For instance, it has been noted that geopolitical tensions, such as political unrest or conflict, can lead to significant changes in oil prices and may consequently impact stock markets [857-862].

In terms of strategic influence, oil-producing countries can wield substantial political leverage over nations that depend heavily on oil imports. Saudi Arabia and Russia have effectively utilized production decisions to sway Western foreign policy and international alignments. Evidence suggests that geopolitical risks shape oil prices and can alter relationships between producing and consuming nations, demonstrating how these countries position themselves as pivotal players in global affairs [863-866]. This influence is reflected in instances where oil diplomacy has been harnessed to achieve political ends, highlighting the power dynamics attributable to oil production.

Finally, the imposition of economic sanctions often revolves around oil, as seen in the examples of Iran and Venezuela. Such sanctions can disrupt oil flows and significantly alter the global market landscape. The conditional relationship between sanctions and oil prices illustrates how economic tools are projected in geopolitical strategies, particularly in exerting pressure on states perceived as threats or adversaries. Studies have shown that sanctions targeting oil exports can lead to drastic changes in oil supply chains and reshape the energy market [862, 867-869]. Furthermore, geopolitical events like war or severe political crises often catalyse sanctions and embargoes, illustrating how oil is frequently at the nexus of international power plays [861, 864, 866].

Through price control, strategic influence, and the leverage of economic sanctions, oil remains a fundamental component of geopolitical strategy. The interactions between oil markets and state behaviour underscore the profound implications of oil as more than a commodity, but as a pivotal instrument of political power.

Economics of the Oil Market

Oil prices are determined by a complex interplay of fundamental market forces and speculative behaviour. While the basic principle of supply and demand forms the foundation of oil pricing, broader factors such as geopolitical tensions, global economic trends, and the transition to alternative energy sources significantly influence market expectations and investor behaviour. This combination of tangible supply-demand metrics and future-oriented speculation makes the oil market uniquely dynamic.

One defining characteristic of oil markets is price volatility. Oil prices can fluctuate sharply over short periods, often triggered by external shocks. Geopolitical events—such as wars, sanctions, or political instability in oil-producing regions—can disrupt supply or create fears of disruption, which drives prices upward. Conversely, global crises that dampen demand, such as the COVID-19 pandemic, can lead to sudden and severe price collapses. For instance, global oil demand plummeted during the 2020 lockdowns, causing prices to crash, even turning negative for West Texas Intermediate (WTI) futures temporarily. By contrast, Russia's invasion of Ukraine in 2022 caused oil prices to spike as Western sanctions targeted one of the world's largest oil exporters.

Oil is traded globally through a system of benchmark pricing. The most prominent benchmarks— Brent Crude (based on North Sea production), West Texas Intermediate (WTI, based on U.S. production), and Dubai Crude (used for Middle Eastern and Asian trade)—serve as reference prices for the vast majority of oil contracts. Though each benchmark is tied to a specific region, they are increasingly influenced by global sentiment and interconnected market dynamics. For example, turmoil in the Middle East may impact Brent pricing even if the disruption does not directly affect North Sea oil.

Another key feature of oil markets is their cyclical nature. High oil prices tend to attract increased investment in exploration, production, and technological efficiency. This eventually leads to an oversupply, putting downward pressure on prices. Conversely, when prices are low, investment in new capacity declines, reducing future supply and creating the conditions for a price rebound. This boom-bust cycle is a persistent feature of the oil industry and poses significant challenges for producers, investors, and policymakers alike.

Overall, oil prices are shaped not just by barrels produced and consumed but also by complex feedback loops involving political decisions, market psychology, technological advances, and the shifting global energy landscape. As the world moves toward decarbonization, these dynamics are likely to evolve further, with new variables—such as carbon pricing, green investment, and electric vehicle adoption—playing a growing role in how oil is valued in global markets.

The Organization of the Petroleum Exporting Countries (OPEC) has long played a central role in the global oil market, aiming to coordinate production levels among member states to influence prices and maintain market stability. However, despite its historical significance, OPEC's ability to control the oil market is increasingly constrained by a range of challenges and structural limitations.

One major challenge to OPEC's influence is the growth of oil production from non-OPEC countries. In recent years, nations such as the United States, Brazil, and Canada have significantly increased their oil output, largely through the development of unconventional extraction technologies like hydraulic fracturing (fracking) and horizontal drilling. The United States, in particular, has emerged as a global energy superpower, at times outproducing even Saudi Arabia. This surge in non-OPEC supply has diluted the effectiveness of OPEC's production cuts, making it harder for the group to single-handedly manage global oil prices.

Internal divisions within OPEC also weaken its cohesion and credibility. Although the organization sets collective production targets, its members often have divergent economic needs and political priorities. For instance, some members, such as Saudi Arabia and the United Arab Emirates, possess significant financial reserves and can afford to reduce production to support prices. Others, like Venezuela, Nigeria, or Iran, are heavily reliant on oil revenues and tend to prioritize short-term income over long-term market strategies. These differences often lead to disagreements over production quotas, undermining compliance and limiting OPEC's capacity to present a unified front.

Moreover, the accelerating global transition toward renewable energy and decarbonization poses a long-term existential threat to OPEC. As governments and industries around the world push for carbon neutrality and cleaner energy sources, demand for fossil fuels—including oil—is expected to plateau or decline over the coming decades. Innovations in electric vehicles, battery storage, and alternative fuels are already chipping away at oil's dominance, especially in the transport sector. This shift puts pressure on OPEC members to diversify their economies and reduces the strategic power they have traditionally wielded in the global energy landscape.

While OPEC remains an important player in oil markets, its influence is no longer absolute. Rising non-OPEC production, internal fractures, and the ongoing energy transition all present significant obstacles to its ability to control oil prices and sustain its role as a central authority in global energy governance. OPEC's future relevance will depend not only on how it manages current production but also on how its members adapt to a world that is steadily moving away from fossil fuels.

In the contemporary global economy, oil retains its dominance in key sectors such as transportation, petrochemicals, and heavy industry. However, this supremacy is increasingly being challenged by various factors pertinent to climate change, technological advancements, and shifts in investor sentiment.

Climate Goals: Governments worldwide are progressively enacting policies aimed at reducing dependence on fossil fuels to meet climate commitments, such as the Paris Agreement. Strategies include carbon pricing mechanisms and legislative bans on internal combustion engine (ICE) vehicles, which are designed to limit fossil fuel consumption [870, 871]. The emergence of "anti-fossil fuel norms" among civil society actors signifies a changing socio-political landscape that is critically responding to climate imperatives [117, 872]. For instance, the International Monetary Fund (IMF) has emphasized the need for reform of fossil fuel subsidies, aligning economic activities with sustainable climate goals [873].

Technological Innovation: The rapid advancement of alternative technologies, notably electric vehicles (EVs), green hydrogen, and other renewable energy sources, is progressively reshaping energy demand. Current literature indicates that these technological innovations are gaining traction, compelling traditional fossil fuel markets to adapt or face decline [718, 874]. Significant investments in renewable infrastructure, coupled with breakthroughs in battery technology and energy efficiency, are fostering an environment conducive to the gradual phasing out of fossil fuel reliance [116].

Investor Sentiment: Investor behaviour is notably shifting as financial institutions and shareholders increasingly pressure oil companies to diversify portfolios and align with sustainability goals. There is a growing recognition of the financial risks associated with fossil fuel investments, including the potential for "stranded assets" due to shifts in regulatory environments and technological evolution [875, 876]. Studies show that institutional investors are actively engaging with fossil fuel companies to mitigate climate-related risks, advocating for increased accountability and the adoption of greener practices [870, 877]. This sentiment is reflective of a broader trend in which market participants are increasingly factoring climate considerations into their investment strategies, highlighting the financial implications of dependency on fossil fuels [878].

OPEC and the oil markets sit at the intersection of economics and politics. The organization's influence over oil supply and pricing has global ramifications—fuelling economic growth or recession, shaping foreign policy, and influencing energy transitions. As climate policies and alternative energy sources gain momentum, OPEC's future will depend on its ability to adapt economically and politically in a rapidly changing global energy landscape.

At its core, oil remains a crucial component of the global energy system, underpinning various sectors such as transportation, industry, and electricity. This reliance highlights oil's significant economic and political dimensions, particularly as the world grapples with the challenges of transitioning to a net-zero future. A comprehensive understanding of oil's role in energy security and international economic dynamics is vital for evaluating pathways toward sustainability and equitable development.

Globally, oil continues to be a dominant energy source, maintaining its position despite the rise of alternative energy sources. The Organization of the Petroleum Exporting Countries (OPEC) has historically wielded substantial influence over oil production and pricing, primarily through its coordinated production strategies. OPEC's ability to manipulate supply levels directly impacts global prices, inflation rates, and overall energy access [879-881]. This dynamic is crucial for sustainable development, as fluctuations in oil prices can inhibit or catalyse investments in renewable energy technologies [881, 882]. The interplay between OPEC's production decisions and global oil prices reflects a complex relationship between short-term economic interests and long-term decarbonization objectives.

Politically, oil serves as a potent tool for geopolitical manoeuvring among nations rich in natural resources. Countries like Saudi Arabia, Russia, and Iran leverage their production capabilities to influence international relations and domestic economies [883, 884]. These geopolitical strategies can significantly affect global energy transitions; when oil prices rise due to geopolitical tensions, the attractiveness of renewable alternatives may diminish, creating challenges in the race to lower carbon emissions. Moreover, OPEC's internal cohesion is often strained due to differing economic priorities among member nations, which affects their collective ability to adapt to changing global energy demands [885, 886].

From an economic perspective, oil prices exhibit a cyclical nature influenced by market dynamics and speculative trading practices [887, 888]. High prices can bolster revenues for oil-exporting nations, potentially delaying necessary economic reforms toward renewable sources. In contrast, low oil prices can deter investment in fossil fuel infrastructure, prompting a shift towards cleaner technologies. The influence of these cyclical patterns on investment decisions illustrates the complexity of achieving long-term climate goals while navigating short-term market realities.

However, the momentum towards achieving net-zero emissions is reshaping the oil market landscape. The Paris Agreement and national climate policies are pressuring oil-dependent countries to reconsider their futures. Notably, OPEC members, including the UAE and Saudi Arabia, are recognizing the importance of investing in diversification into renewable energy sources [889, 890]. This shift reflects a growing acknowledgment of the risks associated with prolonged reliance on oil, particularly amidst increasing global calls for decarbonization.

Despite these advancements, significant challenges remain in aligning oil market dynamics with carbon reduction targets. Non-OPEC producers have emerged as formidable players, complicating OPEC's attempts to manage global oil supply effectively [879, 887]. Additionally, transitioning away from oil dependence necessitates structural changes within the energy markets and bold policy reforms [891, 892]. As the world continues to navigate the tension between oil reliance and the urgent need for sustainable energy solutions, the decisions made by OPEC and global stakeholders will be pivotal in shaping the future trajectory of energy production and consumption.

Energy Security and Geopolitics

Energy security and geopolitics are deeply interconnected. Energy is a fundamental input to every economy and military system, and control over its supply, routes, and infrastructure has long been a driver of global power, conflict, and cooperation. As the global energy system transitions from fossil fuels toward renewables and decarbonisation, the nature of energy security is also evolving—but not without new risks and opportunities.

Traditionally, energy security has been conceptualized as the consistent availability of energy at predictable and affordable prices, particularly dominated by oil and natural gas as the primary fuels for industrial economies. This fundamental understanding has necessitated the establishment of stable supply chains, diversified sources of imports, and strategic reserves to counter temporary supply shocks. Countries that are highly sensitive to external disruptions—due to political instability, sanctions, or natural disasters—often develop comprehensive national strategies that include forming energy alliances and investing in infrastructure like pipelines and LNG terminals to safeguard their energy requirements [893, 894].

The landscape of energy security is shifting, particularly as the focus is increasingly moving toward cleaner and more sustainable energy sources. This transition signifies that energy security is no longer limited to fossil fuels; it now encompasses the affordability and resilience of electricity supply, which is becoming a predominant carrier of energy for essential services such as transportation and heating [895, 896]. The increasing reliance on renewable energy sources requires countries to ensure that their power systems remain stable and affordable for consumers while being resilient to fluctuations in energy markets [897].

Furthermore, diversifying energy sources has emerged as a critical pillar of modern energy security strategies. This diversification entails not just importing from various countries, but also integrating a wider array of energy technologies, including renewables such as solar, wind, and hydroelectric power. By decreasing reliance on a single form of energy or a single supplier, nations can significantly enhance their energy resilience [898, 899]. Recent geopolitical realities, including concerns about gas dependency from Russia, have intensified discussions regarding energy diversification [900, 901].

Moreover, the digitization and decentralization of energy systems make ensuring the reliability of electricity grids and the integrity of digital infrastructures paramount. Modern energy networks must navigate challenges posed by variable renewable energy generation, real-time demand management, and cyber threats that can lead to widespread energy disruptions [902, 903]. Enhancing grid security and robustness is essential not just for sustaining energy supply but also aligns with the increasing integration of smart technologies [904].

Additionally, the contemporary understanding of energy security includes the safeguarding of critical raw materials necessary for clean technologies, such as those used in batteries and solar panels. The supply chain for these materials, which is often concentrated in politically unstable or monopolized countries, poses significant risks to energy security. Nations are actively seeking to secure access to these vital resources through domestic sourcing, recycling initiatives, and international agreements, recognizing their importance in a future focused on clean energy solutions [905, 906].

In conclusion, the definition of energy security has evolved from ensuring the availability of conventional fuels like oil and gas to encompassing a multifaceted framework that includes renewable energy sources, robust digital infrastructures, and the strategic management of essential minerals. This broadening perspective reflects a global shift toward decarbonization and sustainable growth while addressing emerging challenges in the energy sector.

Fuel, Energy and Net Zero

Geopolitical Implications of Energy Dependence

Energy has long been a cornerstone of geopolitical power, shaping the strategic decisions of both exporting and importing nations. Access to reliable energy sources underpins economic stability, industrial capacity, and military readiness, making energy control and influence a central feature of international relations. Throughout the 20th and 21st centuries, the global reliance on oil and gas has enabled resource-rich nations to wield significant foreign policy and economic leverage. Countries such as Saudi Arabia, Russia, Iran, and Venezuela have repeatedly used their fossil fuel wealth to strengthen their international standing, influence regional politics, or exert pressure on rivals. These nations, through organizations such as OPEC and OPEC+, have coordinated production quotas to manipulate global oil prices, generating both profit and strategic influence. In some cases, the control or disruption of energy exports has served as a political tool, shaping everything from diplomatic negotiations to military interventions.

Geopolitical tensions are frequently heightened by energy transit routes, which are essential for the delivery of oil and gas from producing to consuming nations. Pipelines, maritime shipping lanes, and narrow chokepoints such as the Strait of Hormuz, Suez Canal, and Strait of Malacca play a critical role in global energy logistics. These routes are often subject to geopolitical risk. For example, the Strait of Hormuz—through which a large percentage of the world's oil supply travels—has been the site of repeated military tensions between Iran and Western powers. Similarly, the war in Ukraine has underscored the strategic importance of natural gas pipelines running through Eastern Europe, with Russia's actions and Europe's response illustrating how energy transit vulnerabilities can escalate into full-scale geopolitical crises. Disruptions in these corridors due to conflict, terrorism, piracy, or trade embargoes can trigger energy price shocks and global economic volatility.

In response to such risks, many energy-dependent countries have pursued strategic alliances and foreign policy initiatives aimed at securing long-term access to energy. For example, the European Union, Japan, and South Korea have all engaged in extensive energy diplomacy, forging economic partnerships and investing in infrastructure to diversify their energy sources. The United States, historically a major energy importer, has maintained a significant military and political presence in the Middle East, driven in part by a desire to protect energy flows from this geopolitically sensitive region. Even as the U.S. has become more energy independent in recent years due to the shale oil revolution, its strategic posture in the Persian Gulf remains linked to global energy market stability.

Overall, energy remains deeply embedded in the fabric of international power dynamics. Whether through resource control, transit infrastructure, or political alliances, nations continue to treat energy access and influence as vital components of their geopolitical strategies. As the global energy system shifts toward renewables and decarbonisation, these dynamics may evolve, but the intersection of energy and geopolitics will continue to shape international relations for decades to come.

The global energy transition—driven by climate imperatives, technological innovation, and shifting economic priorities—is rapidly reshaping the geopolitical landscape. While the movement away from fossil fuels reduces dependence on traditional oil and gas producers, it simultaneously introduces new geopolitical vulnerabilities and power centres. This evolving context marks a shift from a world dominated by oil-rich states to one in which control over critical minerals, clean energy technologies, and digital infrastructure becomes central to national and global power.

One of the most pressing issues in this transition is the rising importance of critical minerals, which are essential for renewable energy systems. Technologies like electric vehicle batteries, solar panels, and wind turbines rely heavily on minerals such as lithium, cobalt, nickel, and rare earth elements. However, the production and processing of many of these minerals are concentrated in a limited number of countries. For example, the Democratic Republic of Congo supplies a significant portion of the world's cobalt, China dominates rare earth refining and battery supply chains, and Chile and Indonesia are major lithium and nickel producers, respectively. This concentration creates new geopolitical chokepoints, replacing oil transit routes with mineral supply chains vulnerable to trade restrictions, political instability, or resource nationalism. Additionally, there are growing concerns about environmental degradation, labour conditions, and human rights associated with mining practices in some of these regions, which may spark international tensions and ethical dilemmas.

As the world races to decarbonise, countries that lead in green technology development are becoming new centres of geopolitical influence. Nations such as China, Germany, and South Korea have made massive investments in battery production, solar photovoltaic manufacturing, wind turbine engineering, and hydrogen technology infrastructure. Their dominance in these sectors gives them a strategic edge in the emerging global energy economy. For example, China's leadership in solar panel manufacturing not only makes it a key supplier to many countries but also gives it significant leverage in global climate diplomacy and trade negotiations. This shift implies a realignment of global power from oil-rich states to tech-savvy, innovation-driven economies.

Another major transformation lies in the potential for energy independence through renewable sources. Solar, wind, hydro, and geothermal energy can often be produced domestically, which reduces reliance on imported fossil fuels and enhances national security. This shift allows countries—especially those previously energy-import dependent—to regain control over their energy systems. However, transitioning to renewables is not without its own challenges. It demands modernized electricity grids, large-scale energy storage systems, and heightened cybersecurity measures, especially as energy infrastructures become increasingly digitized and decentralized. Cyber threats, data vulnerabilities, and supply chain disruptions could become as significant to future energy security as physical disruptions are today.

Finally, emerging sectors such as hydrogen exports and international carbon trading are likely to reshape alliances and global competition. Countries investing in green hydrogen production (using renewable electricity to split water into hydrogen and oxygen) are positioning themselves as future exporters of clean energy. Australia, Saudi Arabia, and Germany, for example, are all pursuing hydrogen strategies with global ambitions. Similarly, carbon markets—which involve the trading of carbon credits between countries or corporations—are evolving into significant mechanisms for financing climate action and influencing global trade patterns. As rules governing carbon accounting and trading mature, they may serve as another axis around which international energy diplomacy turns.

The energy transition is not only an environmental imperative but also a source of new geopolitical fault lines. While it reduces reliance on fossil fuels, it also concentrates power in the hands of those who control clean energy technologies, critical minerals, and digital infrastructure. The countries that manage to balance sustainable development, secure supply chains, and technological leadership will likely emerge as the dominant players in the next phase of global energy geopolitics.

In the contemporary landscape of energy security, the focus has shifted significantly from a narrow concentration on the steady supply of fossil fuels to a broader, multi-dimensional framework that integrates various risks, including cyber threats, climate-induced disruptions, and fluctuations in fossil fuel markets. This evolution highlights the intricate interplay of energy systems that are increasingly interconnected and diversified, necessitating an adaptive approach to safeguarding energy reliability at both national and global levels.

Cybersecurity has emerged as one of the most pressing risks to modern energy systems. The integration of smart grids and digital controls into the energy sector has introduced new vulnerabilities, making critical infrastructure susceptible to cyberattacks. Incidents such as the cyber intrusions targeting Ukraine's power grid in 2015 and 2016, as well as the ransomware attack on the Colonial Pipeline in 2021, demonstrate the capacity of cyber threats to disrupt energy supply chains and create significant economic and political fallout [907]. As energy systems increasingly rely on automation and cloud-based architectures, prioritizing cybersecurity measures becomes imperative to fortify against potential attacks that could compromise grid stability and functionality [908].

Another dimension of modern energy security is the risk stemming from climate-related disruptions. Climate change has led to an increase in the frequency and severity of extreme weather events that threaten energy infrastructure. For instance, the 2021 winter storm in Texas revealed the frailties of an isolated power grid when millions were left without power due to system overloads caused by freezing temperatures [909]. Additionally, natural disasters such as hurricanes and floods have resulted in long-term outages for numerous regions worldwide, underscoring the necessity for investment in resilient infrastructures and rapid response strategies to mitigate these impacts [910]. Addressing these vulnerabilities requires innovative solutions, including the development of weather-adaptive technologies and robust emergency preparedness frameworks [911].

Despite significant strides in renewable energy adoption, fossil fuel markets continue to exert considerable influence on global energy security dynamics. The volatility of fossil fuel prices remains a critical issue; events such as the COVID-19 pandemic, which precipitated a drastic drop in oil prices, and the ongoing geopolitical tensions associated with the Russia–Ukraine war, which have driven prices up, exemplify the fragility of dependency on fossil energy sources. These fluctuations can result in widespread economic instability, particularly for countries with heavy reliance on fossil fuel exports or imports, necessitating a dual strategy of managing price volatility while transitioning to a more sustainable energy portfolio [912]. This transition involves a balanced approach that diminishes reliance on volatile fossil fuels and promotes the development of renewables as a pathway toward long-term energy security [913].

Energy security and geopolitics are in flux. While the traditional dynamics of fossil fuel control and geopolitical competition persist, the energy transition is creating a new map of global influence. Nations that manage this shift effectively—securing resources, investing in resilience, and building diplomatic ties around clean energy—will shape the future of both energy and power. At the same time, international cooperation will be essential to avoid replacing one set of energy insecurities with another.

Fuel Subsidies and Pricing

Fuel subsidies and pricing are complex issues that play a significant role in shaping the politics and economics of energy on a global scale. Governments often implement fuel subsidies with objectives such as ensuring affordability, protecting vulnerable populations, and supporting economic development. However, these subsidies bring about substantial fiscal burdens, market distortions, and detrimental environmental consequences [914-916].

The discussion surrounding fuel subsidies encompasses the idea that they often disproportionately benefit higher-income groups rather than the intended low-income households. Coady et al. [915] assert that subsidies are poorly targeted, primarily benefiting higher-income groups, which contributes to inequitable income distribution and exacerbates fiscal pressures, particularly in developing nations where subsidy costs may account for significant portions of national budgets. For instance, in Nigeria, fuel subsidies represented approximately 20% of the total public budget at one point, highlighting the substantial financial implications of these policies on a government's fiscal strategy [917].

Fuel subsidies are financial policies or mechanisms that governments use to lower the cost of fuel below its true market value. These subsidies can take several forms. One common method is direct price control, where governments set fuel prices artificially low regardless of fluctuations in global oil markets. Another form involves tax reductions or exemptions, such as removing fuel taxes or offering rebates to reduce the overall price for consumers. Additionally, some governments provide producer support by offering financial assistance to oil companies to offset exploration, production, or refining costs.

Subsidies can be either explicit or implicit. Explicit subsidies appear directly in government budgets as public expenditures on energy. Implicit subsidies, on the other hand, are not formally documented but arise from failures to fully account for the social and environmental costs of fuel use—such as air pollution, greenhouse gas emissions, and damage to infrastructure caused by increased vehicle traffic and fossil fuel consumption.

Governments—particularly in low- and middle-income countries—often justify fuel subsidies for a range of socio-economic reasons. A primary motivation is to shield low-income households from energy price shocks and volatility, ensuring that energy remains affordable for the poor. Subsidies also aim to stimulate economic growth by lowering production and transport costs for businesses, thereby improving competitiveness and economic stability. In many cases, subsidies are introduced or maintained to prevent inflation, as fuel price hikes can increase the cost of goods and services across an economy.

Fuel subsidies are also used as a tool for maintaining social stability, especially in regions where rising energy prices have historically led to civil unrest or protests. In oil-exporting countries, subsidized fuel is often regarded as a means of sharing the nation's resource wealth with the population. While this approach may generate short-term political goodwill, it often fosters long-term economic inefficiencies, encourages wasteful fuel consumption, and places heavy financial burdens on national budgets.

Economically, the removal of fuel subsidies could lead to a general equilibrium effect where an initial increase in fuel prices subsequently raises transportation and production costs across various goods and services [918, 919]. These cascading effects can inflate overall living costs, disproportionately impacting low-income households [919, 920]. Furthermore, studies indicate that while subsidies are intended to stimulate economic growth, their long-term consequences include inefficiencies and potential environmental degradation due to encouraged excessive energy consumption [914, 915, 921]. This contradiction presents a critical challenge for policymakers seeking to balance immediate economic advantages against sustainability goals [116, 922].

Moreover, the environmental impacts of fuel subsidies cannot be overlooked. By effectively lowering the market price of fossil fuels, subsidies incentivize increased use, which directly contributes to higher carbon emissions and exacerbates climate change [116, 921]. The literature suggests that substituting fossil fuel subsidies with targeted social programs could potentially mitigate negative outcomes while still supporting vulnerable populations [922].

The dynamic nature of fuel pricing and subsidies demands a nuanced understanding, as evidenced by various studies that reflect on the socio-political complexities of reformation efforts. For example, local corruption and vested interests can significantly impede subsidy reforms [914, 917], while broader political contexts influence public support for such reforms [38, 116]. Policymakers are thus confronted with crucial decisions as they navigate a landscape where fiscal sustainability, social equity, and environmental integrity must converge [116, 923].

Consequences of Fuel Subsidies

While often politically appealing, fuel subsidies come with substantial economic, environmental, and social costs. One of the most notable issues is economic distortion. Subsidies reduce the incentive for consumers and industries to invest in energy efficiency or transition to cleaner energy sources. They also divert valuable government funds that could otherwise support essential services such as healthcare, education, or infrastructure development.

From an environmental perspective, fuel subsidies contribute significantly to climate and pollution challenges. By artificially lowering the cost of fossil fuels, they encourage overconsumption, leading to increased greenhouse gas emissions, air pollution, and traffic congestion. The International Energy Agency (IEA) has consistently cited fossil fuel subsidies as a major barrier to effective climate action, as they undermine global efforts to reduce emissions and shift to low-carbon economies.

Fuel subsidies are also often regressive in nature, meaning they disproportionately benefit higher-income households. Wealthier individuals typically consume more fuel through private vehicle use and electricity consumption, whereas lower-income populations rely more on public transport and consume less energy overall. As a result, subsidies that aim to support the poor can inadvertently reinforce inequality, benefiting those who need it least.

On the fiscal side, maintaining subsidies can place significant pressure on national budgets, especially in countries that are heavily reliant on imported oil or have volatile revenues from domestic oil production. When oil prices rise, the cost of sustaining subsidies can balloon, leading governments to cut essential public services or increase borrowing, which can weaken long-term economic stability.

In response, many governments have begun reforming fuel pricing systems to address these issues. Reforms may include phasing out blanket subsidies and replacing them with targeted support, such as direct cash transfers to low-income households. Others have introduced automatic pricing mechanisms that adjust domestic fuel prices in line with global oil markets, increasing transparency and reducing fiscal risk.

Governments are also investing in public transport and clean energy infrastructure to provide affordable, sustainable alternatives to fossil fuel use. However, such reforms are politically sensitive and often provoke public backlash. Notable examples include widespread protests in Iran, Nigeria, France (the Yellow Vests movement), and Indonesia, where sudden fuel price hikes triggered unrest.

To succeed, fuel subsidy reforms must be carefully planned, accompanied by strong public communication strategies and social safety nets that protect the most vulnerable. Gradual implementation, transparency, and inclusive policymaking are key to navigating the trade-offs between economic efficiency, environmental responsibility, and political stability.

Fuel Pricing and Global Markets

Fuel prices result from a complex interplay of international market dynamics and domestic economic policies, reflecting both external and internal factors. A primary external influence on fuel prices is the global price of crude oil, which is determined by varying elements such as international supply and demand, geopolitical events, production levels of OPEC and non-OPEC nations, and wider economic developments. For instance, Berument et al. [924] illustrate that oil price shocks can have significant ramifications on the growth of economies, particularly in oil-importing countries that face higher fuel costs when global oil prices surge due to conflicts or production cuts). Additionally, Moshiri and Kheirandish [925] note that historically, significant oil price increases have been accompanied by economic slowdowns, especially when accompanied by a weakening national currency, making imports even more costly. This aligns with findings that a depreciated currency, particularly against the U.S. dollar in which oil is predominantly traded, exacerbates the burden of high oil prices on domestic consumers [926].

Governments of oil-importing nations are often placed in a difficult position regarding fuel pricing strategies in response to these external pressures. They must decide between absorbing the rising costs through subsidies, which may protect households from immediate financial strain but can create long-term fiscal sustainability issues, or allowing market-driven price increases that might enhance energy conservation but could incite public discontent [927]. For example, Aliyu [928] discusses the various economic policies adopted in response to oil price volatility and the subsequent impact on consumer prices and household incomes. The policy decisions taken by governments not only affect economic stability but also have substantial social implications, particularly in countries where low-income households rely heavily on affordable fuel for daily activities [929].

In contrast, the domestic fuel pricing strategies adopted by oil-exporting countries often reflect a form of energy nationalism. Countries such as Venezuela and Saudi Arabia maintain low domestic fuel prices, viewing them as entitlements derived from natural resource wealth, with the goal of fostering public support and social stability [930]. Nonetheless, this approach carries its own set of challenges. Keeping fuel prices artificially low can signal a short-term gain for consumers but might lead to long-term economic inefficiencies, discourage investments in sustainable energy sources, and perpetuate excessive consumption practices [931]. Furthermore, Moshiri highlights that these practices can inhibit necessary structural reforms aimed at diversifying economies beyond oil reliance, which is crucial for sustainable long-term growth [925].

Ultimately, in both importing and exporting contexts, fuel pricing is an intricate matter that intersects with political considerations, social welfare, and long-term energy strategies. The diverse influences of exchange rate fluctuations, changes in taxation, and global oil supply and demand necessitate a careful balancing act by governments, as they strive to reconcile the need for affordable energy with sustainability imperatives and economic resilience.

Fuel prices are influenced by a complex interplay of international market dynamics and domestic economic policies, shaped by factors including geopolitical tensions and economic strategies. A significant external driver is the global price of crude oil, which fluctuates largely due to supply-demand imbalances and political events. For instance, the escalation of hostilities between Russia and Ukraine in 2022 resulted in a rise in global oil prices, driven by Western sanctions on Russian oil exports. This situation imposed substantial inflationary pressure on oil-importing nations such as India, which affected the government's decisions regarding fuel taxation and subsidies [932, 933].

The COVID-19 pandemic serves as another notable illustration of the volatility in oil prices. In 2020, the global demand for oil dropped significantly, resulting in unprecedented price declines, including instances of U.S. West Texas Intermediate (WTI) futures trading at negative values. This decline resulted in lower import costs for oil-dependent countries, although the long-term implications included reduced investment in oil production, leading to potential supply constraints as demand rebounded [934, 935].

Currency fluctuations also play a critical role in determining domestic fuel costs. Countries experiencing currency depreciation relative to the U.S. dollar see increased import costs for crude oil, regardless of whether they are oil producers. Nigeria exemplifies this scenario; the devaluation of the naira inflated fuel import costs, leading the government to either enhance fuel subsidies—a measure that strains national finances—or risk social unrest due to rising prices at the pump [936, 937].

Domestic policies are vital in managing fuel pricing mechanisms. In Indonesia, for example, significant reforms in fuel subsidies were implemented in 2014 after they began consuming a considerable portion of the national budget. The government transitioned to a floating price system linked to international oil prices while also introducing direct cash transfers to lower-income households to mitigate the transition's impact. Although this reform aimed to stabilize fiscal health, it initially incited public protests due to rising prices [938, 939].

Conversely, oil-exporting nations often employ policies that keep domestic fuel prices low for political and social stability, which can have adverse effects. For instance, in Venezuela, heavily subsidized petrol became virtually free, leading to inefficient consumption and extensive smuggling. Saudi Arabia has historically kept fuel prices subsidized but began reforming these policies under its Vision 2030 initiative to enhance fiscal sustainability while managing public expectations [940, 941].

These dynamics indicate that fuel pricing is not merely an economic issue but a strategic policy tool with far-reaching implications. Governments are tasked with balancing these pricing strategies against broader economic factors such as inflation, public welfare, fiscal health, and international relations, navigating a complex landscape that affects both domestic and global markets [933, 942].

In summary, fuel subsidies and pricing are powerful tools that influence national economies, household welfare, and global climate efforts. While subsidies can provide short-term relief and economic support, they often undermine long-term sustainability and fiscal health. Reforming fuel subsidies is a key step toward aligning energy policy with environmental and economic resilience in a net-zero world.

As described above, fuel prices vary significantly across the globe due to a combination of international market forces and domestic economic policies. Key influences include crude oil benchmarks, currency exchange rates, taxation systems, government subsidies, and each country's fuel pricing strategy. These factors interact to create dramatic differences between the cheapest and most expensive places to buy fuel.

As at mid-2025, among the countries with the lowest fuel prices are Iran, Venezuela, Libya, Algeria, and Egypt. In Iran, fuel costs as little as approximately US $0.03 per litre due to heavy government subsidies, which in some years have accounted for nearly 27% of the national GDP. Venezuela's fuel is priced even lower, around US $0.02 per litre, a result of long-standing political strategies to provide subsidised energy to the public. Similarly, Libya and Algeria offer fuel at roughly US $0.03–0.34 per litre. These extremely low prices are maintained through state intervention in fuel pricing, often intended to share oil wealth with citizens or ensure social stability. However, such subsidies can lead to excessive consumption, energy inefficiencies, and fiscal strain on national budgets.

On the opposite end of the spectrum are countries with the world's most expensive fuel prices. Hong Kong tops the list, with prices around US $3.40 per litre, followed by Singapore, Iceland, Norway, Denmark, Switzerland, France, Germany, and the United Kingdom, where fuel prices range from US $2.10 to US $2.83 per litre. These high costs are primarily driven by high fuel taxes, such as excise duties, VAT, and carbon taxes. Additionally, many of these nations are heavily dependent on fuel imports, which raises the base cost of fuel. Expensive refining infrastructure, strong environmental regulations, and advanced distribution networks also contribute to higher prices at the pump.

The stark contrast in global fuel prices can be explained by several factors. Firstly, subsidies and price controls in oil-rich nations allow for cheaper fuel domestically, though they often create budgetary burdens and undermine energy efficiency efforts. Secondly, in developed nations, fuel is taxed heavily to discourage fossil fuel use, raise public revenue, and fund infrastructure. Thirdly, the value of a country's currency relative to the US dollar plays a major role, especially in oil-importing countries—when a local currency weakens, the cost of importing fuel rises significantly. Fourthly, the distinction between oil-producing and oil-importing countries affects pricing. Producers like Saudi Arabia and Kuwait can afford to subsidise fuel, while importers must contend with global price fluctuations and transport costs. Lastly, local refining and distribution capabilities affect the final price paid by consumers, with regions like California experiencing particularly high prices due to stringent standards and limited refining capacity.

As of mid-2025, fuel prices reflect these global dynamics. Countries like Iran, Venezuela, Libya, and Algeria continue to offer the cheapest fuel. Nations such as Kuwait and Saudi Arabia provide relatively low prices due to state-supported subsidies. Countries with moderate prices include the United States, Canada, and China, which benefit from domestic production and relatively lower taxes. Meanwhile, the most expensive fuel is found in places like Hong Kong and several European countries with high taxes and import dependence.

In Australia, the average petrol price was around AUD 1.62 per litre (approximately USD 1.05) as of mid-June 2025. However, prices fluctuated significantly by region, with some areas like Kojonup in Western Australia experiencing prices as low as AUD 1.26/L, while more remote locations such as Alice Springs in the Northern Territory reached AUD 2.57/L. The key factors that influenced prices included global oil benchmarks (then approximately USD 74 per barrel), domestic fuel taxes, and transportation costs. Geopolitical tensions, particularly in the Middle East, were expected to add up to 12 cents per litre if the instability persisted.

In the United States, fuel remained relatively affordable, with an average price of USD 3.17 per gallon (approximately USD 0.84 per litre) for regular gasoline. Diesel prices were slightly higher, averaging USD 3.57 per gallon or USD 0.94 per litre. Regional differences were pronounced— West Coast prices exceeded USD 4.22 per gallon, while the Gulf Coast remained closer to USD 3.10. Prices were driven by crude oil costs, seasonal fluctuations in demand, refining capacity, and taxes ranging between 16% and 19%. With growing global instability, prices were anticipated to rise above USD 4 per gallon.

In the United Kingdom, petrol prices hovered around 132 pence per litre (approximately GBP 1.32 or USD 1.64), with diesel slightly higher at 138 pence per litre. These high prices were primarily the result of heavy taxation, reliance on imports, and refining costs. Recent geopolitical tensions led analysts to predict further increases of around 5 pence per litre. Future pricing was expected to be influenced by policy changes, especially in fuel duty.

Canada's national average fuel price stood at approximately CAD 1.38 per litre (around USD 1.02), though prices varied by province. For example, in Vancouver, regular petrol reached about CAD 1.73/L, and diesel was approximately CAD 1.52/L. Prices fluctuated weekly in response to international crude oil prices, provincial tax regimes, and transport costs. Wholesale and retail price gaps were also impacted by distinctive gas surcharges and distribution frameworks.

In India, fuel pricing was dynamic and tied to global crude oil rates, currency exchange fluctuations, and domestic tax policies. By mid-2025, petrol in Delhi was priced around ₹94.8 per litre (USD 1.14), with prices across states ranging from ₹92 to ₹109. Diesel prices ranged from ₹87 to ₹95 per litre (USD 1.05–1.15). Since 2017, India had operated under a daily market-linked pricing mechanism. During crises, the government often chose not to pass global price hikes onto consumers, although this approach added fiscal pressure.

In China, fuel prices were regulated through a monthly pricing formula that accounted for global oil benchmarks and the USD/CNY exchange rate. Petrol averaged around CNY 7.49 per litre (approximately USD 1.03–1.04), with diesel priced slightly higher—roughly 113% of the petrol price. Despite being the world's largest oil importer, China's state-controlled pricing model provided some buffer from immediate international price fluctuations.

Fuel price variations across these countries were shaped by a combination of global oil market dynamics, domestic currency values, tax and subsidy frameworks, and infrastructure conditions. Countries such as India and China adjusted prices through regulatory mechanisms to protect consumers and stabilize their economies. In contrast, the UK maintained high prices through fuel taxation to support environmental goals. Australia, the US, and Canada pursued hybrid models, balancing market-based exposure with regional pricing and policy interventions. Altogether, these diverse approaches reflected national priorities in economic management, social welfare, and environmental responsibility.

Table 10: Fuel pricing dynamics across Australia, the USA, the UK, Canada, India, and China Mid-2025.

Country	Petrol Price	Diesel Price	Key Drivers & Notes
Australia	AUD 1.62/L (~USD 1.05)	Slightly higher	Global benchmarks, taxes, regional variance
USA	USD 3.17/gal (USD 0.84/L)	USD 3.57/gal	Crude costs, seasonal demand, regional taxes
UK	132 p/L (~USD 1.64/L)	138 p/L	High taxes, import reliance
Canada	CAD 1.38/L (~USD 1.02/L)	~CAD 1.52/L	Provincial taxes, weekly fluctuations
India	~₹95/L (~USD 1.14/L)	~₹90/L	Daily revisions, local taxes, stable policy
China	CNY 7.49/L (~USD 1.04/L)	~CNY 8.45/L	Monthly formula-based adjustments

In summary, the cheapest fuel tends to be found in oil-exporting nations that subsidise domestic consumption, although this comes at a cost to long-term fiscal health and environmental sustainability. The most expensive fuel is found in wealthy, import-reliant nations with high taxation and strong regulatory frameworks. Countries in the middle manage to balance local production, modest taxes, and strategic pricing. Ultimately, global fuel price differences reveal the complex interplay of national policies, energy strategies, and international economic conditions that determine what consumers pay at the pump.

Chapter 14

Global Trends and Future Outlook

After exploring the technical, economic, and societal dimensions of fuel production and consumption, the focus now turns to the broader picture—how global energy systems are evolving in response to mounting environmental challenges, policy pressures, and technological advancements. Chapter 14, *Global Trends and Future Outlook*, builds on the foundation laid in previous chapters to examine the dynamic shifts that are reshaping the world's approach to fuel and energy. From decarbonisation strategies and renewable energy adoption to green innovation and geopolitical realignments, this chapter considers where we are headed and what forces will define the post-fossil fuel era. It provides insight into the emerging global consensus around net-zero targets, the potential and limitations of various energy technologies, and the roles different regions are likely to play in the decades to come. As the energy transition gains momentum, understanding these global trends is essential to anticipating future challenges, guiding investment, and formulating informed policy decisions.

Renewable Fuel Growth

Renewable fuels such as bioethanol, biodiesel, renewable diesel, green hydrogen, and synthetic fuels have gained substantial importance in the context of the global energy transition. As nations strive to decarbonize their transport sectors, diminish their reliance on fossil fuels, and meet net-zero emissions targets, the production and utilization of renewable fuels have expanded consistently across various regions and industries. These fuels are seen as critical solutions to both climate change and energy security [254].

In recent years, global production of renewable fuels has witnessed significant growth. The United States, Brazil, the European Union, China, and Indonesia are leading producers of biofuels, particularly bioethanol derived from sugarcane and corn, and biodiesel from vegetable oils and animal fats. This expansion also includes advanced biofuels such as renewable diesel created through hydrotreatment and cellulosic ethanol. However, their development faces challenges, including higher production costs and technical hurdles [943, 944]. Thus, the progress is coupled with a critical examination of technological and economic factors affecting broader adoption [254].

Policy frameworks have emerged as essential catalysts in this sector. National blending mandates and low-carbon fuel standards shape the dynamics of renewable fuel markets. In the United States, the Renewable Fuel Standard (RFS) sets blending requirements for ethanol and biodiesel. Similarly, the European Union's Renewable Energy Directive II (RED II) mandates member states to meet minimum shares of renewable energy in transportation. Countries like Indonesia and Brazil have enforced biodiesel blending mandates, such as the B35 and B20 standards, respectively, leading to increased consumption and investment in necessary infrastructure and supply chains [945, 946].

Diversification of feedstocks is another important trend in renewable fuel production. Global efforts are underway to broaden the spectrum of biomass sources used for fuel production to enhance sustainability and reduce competition with food crops. By utilizing waste oils, agricultural residues, municipal solid waste, algae, and forestry by-products, countries are supporting circular economies, thereby enhancing the environmental profiles of renewable fuels [947, 948].

The emergence of green hydrogen and synthetic e-fuels is also gaining momentum, particularly for sectors that are difficult to electrify, such as heavy transportation, aviation, and maritime industries. Green hydrogen, produced via electrolysis powered by renewable energy sources, shows promise in facilitating transitions towards lower emissions in these hard-to-reach areas. E-fuels, produced from captured CO_2 and green hydrogen, represent alternative fuels for applications where battery technology might not be feasible [949, 950]. Decarbonizing these sectors significantly relies on these technologies, demonstrating their critical role in addressing climate challenges [951].

Substantial investments and innovations are unfolding across the landscape of renewable fuels. Both government and private sectors are increasing financial commitments towards renewable fuel refineries, green hydrogen production facilities, and carbon capture systems. Companies like Neste and Shell are leading these investments, particularly in renewable diesel and sustainable aviation fuel (SAF), which are anticipated to be instrumental in mitigating transport-related emissions [952, 953]. This surge in investment reflects a broader movement towards achieving cleaner fuel systems underpinned by supportive policy frameworks, technological advancements, and emerging economic opportunities.

The future outlook for renewable fuels from 2025 to 2050 points to substantial growth, innovation, and global transformation in energy systems. As nations double down on climate commitments and shift away from fossil fuels, renewable fuels are expected to expand well beyond their current market share, particularly in sectors that are challenging to decarbonize.

One of the most significant areas of growth is anticipated in aviation and maritime shipping. The aviation industry is forecasted to experience the fastest rise in renewable fuel use, especially with Sustainable Aviation Fuel (SAF) playing a central role in decarbonizing long-haul flights. The International Air Transport Association (IATA) projects that SAF could constitute up to 65% of aviation fuel consumption by 2050 under global net-zero scenarios. Similarly, the shipping sector is expected to integrate a mix of low-carbon fuels—such as advanced biofuels, green ammonia, and methanol—to meet the International Maritime Organization's (IMO) greenhouse gas reduction targets. These fuels offer feasible alternatives to heavy fuel oil, especially for long-distance shipping.

Another critical development is the exponential scaling of green hydrogen. Green hydrogen, produced through electrolysis powered by renewable electricity, is expected to grow rapidly, with over 100 gigawatts of electrolyser capacity already in global development pipelines. This increase in capacity will support the creation of hydrogen-derived fuels, including ammonia, synthetic methane, and synthetic kerosene, which could reshape energy consumption in sectors such as heavy industry, freight transport, and chemicals.

Renewable fuels are also set to become closely integrated with global carbon markets and net-zero strategies. As more countries implement carbon pricing and emissions trading systems, fuels with lower lifecycle greenhouse gas emissions will enjoy economic and regulatory advantages. Lifecycle analysis (LCA) and sustainability certification schemes—such as the International Sustainability and Carbon Certification (ISCC), California's Low Carbon Fuel Standard (LCFS), and the Carbon Offsetting and Reduction Scheme for International Aviation (CORSIA)—will become vital tools for verifying emissions reductions and ensuring market access.

Technological innovation and economies of scale are expected to significantly reduce the cost of renewable fuel production. The cost of electrolysers, for example, has dropped by more than 60% in the past ten years, and further declines are projected by 2030. Similarly, improvements in feedstock conversion, biorefinery efficiency, and carbon capture integration will reduce production costs for advanced biofuels and synthetic fuels. This will help make renewable fuels more competitive with traditional fossil fuels, especially as carbon costs increase.

Geopolitically, regions with abundant renewable resources, such as Brazil, Australia, Chile, and parts of sub-Saharan Africa, are well-positioned to become global leaders in renewable fuel production and exports. These countries can leverage their access to solar, wind, biomass, or water resources to produce and ship renewable hydrogen and other fuels to energy-importing countries. International green energy trade routes—referred to as "green corridors"—are likely to develop, supported by bilateral partnerships and multilateral frameworks that standardize transport, storage, and certification.

Despite the promising trajectory, the sector will face several challenges in the years ahead. One concern is the availability of sustainable feedstocks, particularly where there is competition with food production or biodiversity conservation. Infrastructure adaptation for new fuels—such as retrofitting fuelling stations, upgrading shipping terminals, and ensuring safety standards—will also be necessary. Another critical issue is the harmonization of cross-border sustainability certification and emissions accounting, which remains fragmented. In some regions, policy uncertainty and fluctuating support mechanisms could slow investment and deployment.

The renewable fuels industry is transitioning from a niche market to a central pillar of the global low-carbon energy system. Its expansion is being driven by net-zero goals, falling technology costs, supportive government policies, and private sector investment. However, the speed and equity of this transition will depend on how well the world navigates issues of scalability, affordability, and sustainability. The next quarter-century will be decisive in shaping the renewable fuel landscape and determining its role in mitigating climate change on a global scale.

The anticipated implementation of renewable fuels on a global scale from 2025 to 2050 is expected to exhibit significant variations among countries, shaped by their economic, political, and geographical contexts. Leading economies with robust climate commitments, such as the European Union, the United States, Japan, South Korea, Australia, and Canada, are poised to accelerate the transition to renewable fuels. This acceleration is largely attributed to their ambitious net-zero targets, supportive regulatory frameworks, and substantial financial resources, which facilitate rapid advancements in technology and infrastructure for renewable fuel adoption [262, 954, 955].

Conversely, nations endowed with rich renewable energy potential—such as Brazil, Chile, Namibia, and various regions within Southeast Asia and Africa—are anticipated to emerge as key exporters of renewable fuels. These countries are well-positioned to capitalize on the growing global demand for alternatives like green hydrogen, sustainable aviation fuel (SAF), and biofuels, especially if they can forge international partnerships and secure necessary investments in infrastructure [956-958]. For instance, as indicated by IATA forecasts, the adoption of SAF in aviation could escalate significantly, potentially reaching over 60% of overall aviation fuel usage by 2050 under favourable scenarios, supported by appropriate policy mandates [959, 960].

In the context of transportation, sectors like aviation, maritime shipping, and heavy freight are projected to be primary drivers of renewable fuel demand. Renewable fuels are crucial in decarbonizing these traditionally hard-to-electrify sectors. For instance, while the current contribution of SAF to aviation fuel is less than 1%, significant scaling is necessary to meet future needs, as outlined by industry forecasts [958-960]. Additionally, the International Maritime Organization has identified the necessity for fuels such as green ammonia and methanol to eventually replace fossil fuels in maritime shipping, aligning with strategies for reducing greenhouse gas emissions by 40% by 2030 and achieving net-zero by 2050 [376, 958].

However, the transition to renewable fuels may not be universally smooth. Nations heavily reliant on fossil fuel exports, such as Russia, Iran, and Iraq, are likely to view the transition as economically threatening and may exhibit reluctance in diversifying their energy portfolios. Similarly, developing countries with limited budgets might struggle to invest in the required infrastructure or prioritize renewable fuel deployment, especially where energy poverty prevails [961, 962]. The lack of supportive policies—such as subsidies for renewable options or stringent mandates for blending—could hinder substantial market growth in these regions [955, 962].

Several challenges could impede the expected trajectory for renewable fuels. If anticipated reductions in the costs of technologies like green hydrogen and advanced biofuels do not materialize as quickly as projected, these fuels may remain financially uncompetitive against fossil fuels [262, 955, 963]. Additionally, fluctuations in global energy markets, including potential declines in oil prices, could undermine the economic viability of renewable transitions [262, 964]. Political factors, such as public opposition to fuel pricing reforms and land use changes associated with biomass production, may further complicate the landscape of renewable fuel adoption [962, 964]. Thus, decisive political leadership is essential to create a stable policy environment that can steer industries toward investments in renewable technologies [964, 965].

While the spatial distribution of renewable fuel adoption from 2025 to 2050 is likely to be uneven, the driving momentum remains robust. If multilateral cooperation strengthens and investments in infrastructure and innovation continue unabated, alongside equitable access to technology and financing, the global rollout of renewable fuels may reach the necessary scale to meet climate goals by mid-century. However, the success of this transition will hinge critically on international collaborations, national ambitions, and the ability to navigate complex trade-offs between energy access, economic development, and environmental sustainability [962, 964, 965].

Decarbonisation Strategies

The decarbonization strategies that are expected to dominate the global climate agenda from 2025 to 2050 serve as critical elements in the pursuit of sustainability, economic competitiveness, and geopolitical stability. As governments and industries worldwide aim to meet their commitments under the Paris Agreement, the urgency and scale of these strategies are being amplified. A pivotal finding is that key economic actors, including the European Union, China, the United States, and India, are increasingly aligning their policy frameworks towards achieving net-zero emissions by mid-century, fundamentally integrating decarbonization across energy, transportation, industry, and agriculture sectors [966, 967].

The electrification of various sectors emerges as a cornerstone of decarbonization, promoting a shift from fossil fuels to renewable electricity in areas such as transportation and residential heating. This transition hinges on the concurrent decarbonization of electricity grids through renewable sources like wind and solar, alongside implementing innovative technologies such as battery storage and smart grids [968, 969]. By 2040, electrification is anticipated to become predominant in passenger transport and residential energy use in developed nations, showcasing the essential role that green technologies will play in the energy transition [966, 967].

For sectors resistant to electrification, including aviation, maritime transport, and heavy industries, the development and scaling of renewable fuels and clean hydrogen solutions are integral to future decarbonization. Sustainable Aviation Fuel (SAF), advanced biofuels, and carbon-neutral synthetic fuels derived from captured CO_2 offer potential pathways for reducing emissions in a more challenging context [970, 971]. As carbon pricing mechanisms expand worldwide, encompassing taxes and emissions trading systems, they are likely to incentivize the adoption of these cleaner alternatives while creating more equitable trade frameworks through measures like border carbon adjustments [972, 973].

Additionally, Carbon Capture, Utilization, and Storage (CCUS) technologies are crucial for significantly cutting emissions in hard-to-abate sectors. Investments in large-scale CCUS infrastructure reflect a growing acknowledgment of its necessity, particularly as nations establish supportive regulatory environments [974-976]. The innovation landscape also includes the emergence of green hydrogen and biogas as viable resources for energy needs, designed to support the reduction of greenhouse gas emissions effectively [977].

Decarbonization is also fostering advancements in green industrial policy. Governments are implementing policies to nurture domestic production of low-carbon technologies, which is reshaping global trade dynamics. Initiatives like the U.S. Inflation Reduction Act and the EU's Green Deal Industrial Plan exemplify strategic efforts to position nations as leaders in the clean energy economy [978]. The interplay between urban development and decarbonization underscores the importance of cities in driving climate initiatives through energy-efficient buildings, sustainable transport, and smart urban technologies [972, 979].

Integrating nature-based solutions, such as reforestation and regenerative agriculture, incorporates ecological stewardship into decarbonization strategies, broadening their appeal and enhancing efficacy [972, 980]. Nevertheless, achieving these ambitious targets is not without challenges. Developing countries often lack the necessary resources for implementing comprehensive decarbonization strategies, necessitating targeted climate finance aimed at fostering equity in the transition [981, 982]. Barriers such as technological uncertainties surrounding CCUS and hydrogen deployment signify that coordinated international efforts are critical to overcoming these obstacles [976].

The global transition to decarbonization from 2025 to 2050 will be characterized by interconnected, cross-sector strategies that not only embrace innovative technologies but also prioritize inclusive economic growth. Nations that prioritize clear policies, strategic investments in clean technologies, and a focus on equitable transition mechanisms will secure not just environmental sustainability but also economic and social benefits in this evolving landscape.

The trajectory toward global decarbonization between 2025 and 2050 encapsulates a blend of optimism and complexity. Major economies such as the European Union, China, the United

States, and India have firmly positioned decarbonization within their policy frameworks, reflecting a robust international commitment to achieving net-zero targets. This commitment translates into actionable strategies that integrate decarbonization into economic, industrial, and infrastructural planning across various sectors [983-985]. Progress is notably evident in the transportation and energy sectors; electrification efforts, the expansion of renewable energy resources, and the emergence of Sustainable Aviation Fuel (SAF) are transitioning from pilot projects to commercial applications [984, 986, 987]. For example, the adaptation of green energy practices and the increase in investments in clean hydrogen technologies signify a paradigm shift in energy consumption patterns, poised for significant impact if supported by conducive policies and financial backing [988, 989].

However, the implementation of these strategies is heterogeneous globally and faces challenges that stem from differing national capacities and priorities. Wealthier nations with stable governance structures, such as those in Western Europe and parts of North America, typically have better access to the necessary capital, advanced technologies, and regulatory frameworks that promote a smooth transition to low-carbon economies [825, 990]. For instance, the positive industrial transformations noted in European countries highlight the effectiveness of policy frameworks and financial incentives in fostering technological maturation [726, 825]. Furthermore, regions in Asia are also emerging as critical green manufacturing hubs; China's aggressive clean hydrogen initiatives and urban climate strategies in European cities exemplify this trend [726, 825].

In stark contrast, developing countries often grapple with significant obstacles that hinder their decarbonization efforts. Countries in sub-Saharan Africa and smaller island nations frequently lack the foundational infrastructure, technological readiness, and financial resources essential for deploying extensive decarbonization projects [989, 991]. For many of these nations, immediate priorities may still align more closely with economic growth and electrification, rather than deep decarbonization efforts [992, 993]. The need for enhanced international support through climate finance, technology transfers, and capacity-building initiatives is paramount. Without these, developing nations risk lagging further behind in the global transition towards a sustainable low-carbon future [991, 992].

Moreover, geopolitical dynamics and market instabilities threaten to disrupt the forward momentum of decarbonization processes even in more advanced economies. Political uncertainties, weak enforcement of environmental policies, and public scepticism regarding the costs associated with energy transitions serve as critical barriers to achieving widespread adoption of decarbonization strategies [994]. Additionally, countries heavily reliant on fossil fuel exports may face unique challenges and resistance in embracing decarbonization due to perceived threats to their economic foundations, necessitating clear pathways for economic diversification to facilitate their transition [989, 994].

Although a promising foundation for global decarbonization efforts from 2025 to 2050 exists, achieving these ambitious targets necessitates a coordinated global approach. Countries that proactively invest in technological advancements and supportive policy frameworks are better positioned to lead this transition [995]. However, addressing the disparities that exist, particularly for developing nations, is crucial to ensure that all regions can contribute effectively to global climate goals. Thus, fostering an inclusive, equitable, and adaptable decarbonization strategy will be vital in steering the world towards a sustainable future.

What Comes After Fossil Fuels?

As the global energy landscape transitions toward a low-carbon future, the post-fossil-fuel era is emerging with a distinctive reconfiguration of energy production, distribution, and consumption dynamics. This transition does not entail a sudden eradication of hydrocarbons but is rather characterized by the systematic phasing out of fossil fuels—coal, oil, and natural gas—while promoting cleaner and more sustainable alternatives across various sectors, including power generation, transportation, industry, buildings, and agriculture [228, 996, 997].

Central to this transformation is the ascendance of renewable energy sources, particularly solar and wind power, as the backbone of electricity generation. Projections indicate that renewables could supply around 80–90% of global electricity by 2050, particularly through the integration of wind and solar technologies [998]. Moreover, hydropower and geothermal energy will retain their importance in specific geographical regions. To mitigate the inherent intermittent nature of renewable energy, innovations such as battery energy storage systems (BESS), smart grids, and demand-response technologies will become essential [999]. The synergy between solar and wind resources enables effective energy generation and increases reliability in energy supply [1000].

For sectors that are notably harder to electrify, such as heavy industries and aviation, green hydrogen and synthetic fuels are anticipated to play pivotal roles. Green hydrogen, produced from renewable electricity via electrolysis, presents an opportunity to replace fossil fuels like diesel and jet fuel in transportation, while also serving as an energy carrier for high-temperature industrial processes [996, 1001]. Additionally, synthetic fuels, derived from captured CO_2 and hydrogen, can offer carbon-neutral alternatives to conventional fossil fuels [949, 1002]. The adoption of hydrogen is fuelled by its versatility and potential to integrate into existing infrastructure, thereby helping drive the transition across various economic sectors [1003].

In conjunction with hydrogen, bioenergy and biomass will continue to hold significance, particularly in regions rich in agricultural and organic waste. Second-generation biofuels generated from non-food sources such as crop residues are a vital solution to the food-versus-fuel conflict [1004, 1005]. This transition reflects an evolving energy ecosystem where decentralized applications enable the utilization of biomass for diverse purposes, including electricity generation, transportation fuel, and industrial heating [1006].

The electrification of previously fossil-fuel-dependent sectors presents yet another key trend. Electric vehicles (EVs), electric heat pumps, and electric arc furnaces are gradually replacing traditional technologies reliant on fossil fuels [996, 1007]. However, this shift necessitates a resilient and low-carbon electricity grid capable of accommodating the growing energy demands driven by renewable sources [998, 1008]. The need for a more decentralized and digitally managed energy landscape is exemplified by the emergence of microgrids and peer-to-peer energy trading, which promise enhanced community empowerment and reduced energy losses [1007, 1009].

Despite the declining reliance on fossil fuels, some emissions will inevitably persist, particularly from sectors such as aviation and agriculture. Addressing these emissions through carbon removal technologies—such as carbon capture and storage (CCS) and direct air capture (DAC)—is imperative for achieving net-zero and ultimately net-negative carbon targets [228, 996]. Moreover, nature-based solutions, including reforestation and regenerative agriculture, will complement technological solutions to mitigate residual emissions [949].

The comprehensive approach required in the post-fossil world extends to rethinking societal production and consumption models through circular economy principles, where reuse, recycling, and remanufacturing practices will become vital [996, 997]. The development of low-carbon materials and the promotion of sustainable urban design will also be essential in fostering low-energy lifestyles, thereby catalysing both behavioural shifts and infrastructural changes necessary for a sustainable future [1010].

Geographically, the strategies to transition beyond fossil fuels will vary. Regions like Europe, Japan, and parts of North America aim for a rapid electrification and green hydrogen deployment, while countries such as China combine renewable energy with nuclear power, focusing on global clean energy supply chains [1002, 1003]. In contrast, countries in Africa and South Asia, with insufficient grid infrastructures, may leapfrog directly to renewables, assuming adequate international support [949, 997].

References

1.   Markowski, J., et al., *The Concept of Measurement of Calorific Value of Gaseous Fuels.* E3s Web of Conferences, 2020. **207**: p. 01025.
2.   Dryer, F.L., *Chemical Kinetic and Combustion Characteristics of Transportation Fuels.* Proceedings of the Combustion Institute, 2015. **35**(1): p. 117-144.
3.   Bock, T., H. Möhwald, and R. Mülhaupt, *Arylphosphonic Acid-Functionalized Polyelectrolytes as Fuel Cell Membrane Material.* Macromolecular Chemistry and Physics, 2007. **208**(13): p. 1324-1340.
4.   Lee, K.Y., H. Hwang, and W. Choi, *Advanced Thermopower Wave in Novel ZnO Nanostructures/Fuel Composite.* Acs Applied Materials & Interfaces, 2014. **6**(17): p. 15575-15582.
5.   Fries, S., *Introduction to Transforming Energy Systems.* 2021.
6.   Komarnicka, A. and A. Murawska, *Comparison of Consumption and Renewable Sources of Energy in European Union Countries—Sectoral Indicators, Economic Conditions and Environmental Impacts.* Energies, 2021. **14**(12): p. 3714.
7.   Гусаров, B.A., D. Pisarev, and A. Sporov, *Characteristics of Diesel and Gas Turbine Power Plants for Vehicles.* Elektrotekhnologii I Elektrooborudovanie v Apk, 2022. **1**(46).
8.   Sartika, Y. and S. Amar, *Pengaruh Perekonomian Dan Jumlah Penduduk Terhadap Permintaan Bahan Bakar Minyak Di Indonesia.* Jurnal Kajian Ekonomi Dan Pembangunan, 2020. **2**(3).
9.   Scott, N., M. Barnard-Tallier, and S. Batchelor, *Losing the Energy to Cook: An Exploration of Modern Food Systems and Energy Consumption in Domestic Kitchens.* Energies, 2021. **14**(13): p. 4004.
10.   Izmaylov, A.Y., *Decarbonization of Mobile Energy Vehicles Used in Agricultural Production.* 2024(1): p. 4-10.
11.   Shcheklein, S.E., A.M. Dubinin, and N.T. Alwan, *Obtaining fresh water from natural and synthetic fuels in the energy sector.* International Journal of Energy Production and Management, 2021. **6**(2): p. 193-201.
12.   Ilhak, M., et al., *Alternative fuels for internal combustion engines.* The future of internal combustion engines. London: Intechopen book series, 2019.
13.   Wang, Y., S. Zheng, and Z. Wang, *Energy efficiency and emissions analysis of ammonia, hydrogen, and hydrocarbon fuels.* J. Energy Nat. Resour, 2018. **7**: p. 47.
14.   Nguyen, A.T., *Examining the Relationships Between Energy Use, Fossil Fuel Consumption, Carbon Dioxide Emissions, and Economic Growth in Southeast Asia.* Journal of Asian Energy Studies, 2019. **3**(1): p. 45-59.
15.   Özden, R., et al., *Untitled.* Energy Environment & Storage, 2021. **1**(2).
16.   Hou, S.-S. and J.-C. Lin, *Performance Analysis of a Diesel Cycle Under the Restriction of Maximum Cycle Temperature With Considerations of Heat Loss, Friction, and Variable Specific Heats.* Acta Physica Polonica A, 2011. **120**(6): p. 979-986.
17.   Davis, F. and S.P.J. Higson, *Biofuel Cells—Recent Advances and Applications.* Biosensors and Bioelectronics, 2007. **22**(7): p. 1224-1235.
18.   Yuan, Z., et al., *Thermal Layout Analysis and Design of Direct Methanol Fuel Cells on PCB Based on Novel Particle Swarm Optimization.* Micromachines, 2019. **10**(10): p. 641.

19. Calvo, R., et al., *Territorial Energy Vulnerability Assessment to Enhance Just Energy Transition of Cities.* Frontiers in Sustainable Cities, 2021. **3**.

20. Kabel, T.S. and M. Bassim, *Reasons for Shifting and Barriers to Renewable Energy; A Literature Review.* 2020.

21. Müller, C. and H. Yan, *Household Fuel Use in Developing Countries: Review of Theory and Evidence.* Energy Economics, 2018. **70**: p. 429-439.

22. Masera, O., B.D. Saatkamp, and D.M. Kammen, *From Linear Fuel Switching to Multiple Cooking Strategies: A Critique and Alternative to the Energy Ladder Model.* World Development, 2000. **28**(12): p. 2083-2103.

23. González, P.H., J.M. Rojo, and V. Scapini, *Relationship Between Pollution Levels and Poverty: Regions of Antofagasta, Valparaiso and Biobio, Chile.* International Journal of Energy Production and Management, 2022. **7**(2): p. 176-184.

24. Munir, S. and A. Khan, *Impact of Fossil Fuel Energy Consumption on CO2 Emissions: Evidence From Pakistan (1980-2010).* The Pakistan Development Review, 2014. **53**(4II): p. 327-346.

25. Omar, N., et al., *Peukert Revisited—Critical Appraisal and Need for Modification for Lithium-Ion Batteries.* Energies, 2013. **6**(11): p. 5625-5641.

26. Cugnet, M., M. Dubarry, and B.Y. Liaw, *Peukert's Law of a Lead-Acid Battery Simulated by a Mathematical Model.* Ecs Transactions, 2010. **25**(35): p. 223-233.

27. Shamim, N., et al., *Valve Regulated Lead Acid Battery Evaluation Under Peak Shaving and Frequency Regulation Duty Cycles.* Energies, 2022. **15**(9): p. 3389.

28. Vlachogianni, T. and A. Valavanidis, *Energy and Environmental Impact on the Biosphere Energy Flow, Storage and Conversion in Human Civilization.* American Journal of Educational Research, 2013. **1**(3): p. 68-78.

29. Schmidt, M.U., et al., *Charcoal Kiln Sites, Associated Landscape Attributes and Historic Forest Conditions: DTM-based Investigations in Hesse (Germany).* Forest Ecosystems, 2016. **3**(1).

30. Oliveira, C., et al., *Woodland Management as Major Energy Supply During the Early Industrialization: A Multiproxy Analysis in the Northwest European Lowlands.* Land, 2022. **11**(4): p. 555.

31. Jonell, T.N., et al., *Limited Waterpower Contributed to Rise of Steam Power in British "Cottonopolis".* Pnas Nexus, 2024. **3**(7).

32. Nuvolari, A., *The Making of Steam Power Technology: A Study of Technical Change During the British Industrial Revolution.* The Journal of Economic History, 2006. **66**(2): p. 472-476.

33. Kennedy, C., *The Energy Embodied in the First and Second Industrial Revolutions.* Journal of Industrial Ecology, 2020. **24**(4): p. 887-898.

34. Wang, Z. and G.F. Naterer, *Integrated Fossil Fuel and Solar Thermal Systems for Hydrogen Production and CO2 Mitigation.* International Journal of Hydrogen Energy, 2014. **39**(26): p. 14227-14233.

35. Armaroli, N. and V. Balzani, *The Legacy of Fossil Fuels.* Chemistry - An Asian Journal, 2011. **6**(3): p. 768-784.

36. Kreps, B.H., *The Rising Costs of Fossil-Fuel Extraction: An Energy Crisis That Will Not Go Away.* American Journal of Economics and Sociology, 2020. **79**(3): p. 695-717.

37. Kouassi, K.E., et al., *Study of the Viscosity and Specific Gravity of the Ternary Used Frying Oil (UFO)-Bioethanol-Diesel System.* Journal of Materials Science and Chemical Engineering, 2024. **12**(04): p. 53-66.

38. Asselt, H.v. and K. Kulovesi, *Seizing the Opportunity: Tackling Fossil Fuel Subsidies Under the UNFCCC.* International Environmental Agreements Politics Law and Economics, 2017. **17**(3): p. 357-370.

39. Schuster, P.F., et al., *Atmospheric Mercury Deposition During the Last 270 Years: a Glacial Ice Core Record of Natural and Anthropogenic Sources.* Environmental Science & Technology, 2002. **36**(11): p. 2303-2310.

40. Crafts, N., S.J. Leybourne, and T.C. Mills, *Trends and Cycles in British Industrial Production, 1700-1913.* Journal of the Royal Statistical Society Series a (Statistics in Society), 1989. **152**(1): p. 43.

41. Wrigley, E.A., *Energy and the English Industrial Revolution.* 2010.

42. Fernihough, A. and K. O'Rourke, *Coal and the European Industrial Revolution.* 2014.

43. Јовановски, Г., B. Boev, and P. Makreski, *Chemistry and Geology of Coal: Nature, Composition, Coking, Gasification, Liquefaction, Production of Chemicals, Formation, Peatification, Coalification, Coal Types, and Ranks.* Chemtexts, 2023. **9**(1).

44. Anastassiadis, S., *Historical Developments in Carbon Sources, Biomass, Fossils, Biofuels and Biotechnology Review Article.* World Journal of Biology and Biotechnology, 2016. **1**(2): p. 71.

45. Rose, N.L., *Spheroidal Carbonaceous Fly Ash Particles Provide a Globally Synchronous Stratigraphic Marker for the Anthropocene.* Environmental Science & Technology, 2015. **49**(7): p. 4155-4162.

46. Day, J.W., et al., *The Coming Perfect Storm: Diminishing Sustainability of Coastal Human–natural Systems in the Anthropocene.* Cambridge Prisms Coastal Futures, 2023. **1**.

47. Novakov, T. and J.E. Hansen, *Black Carbon Emissions in the United Kingdom During the Past Four Decades: An Empirical Analysis.* Atmospheric Environment, 2004. **38**(25): p. 4155-4163.

48. Du, W., L. Chen, and Y. Chen, *Solid Fuel Combustion and Air Pollution: Filling the Data Gap and Future Priorities.* International Journal of Environmental Research and Public Health, 2022. **19**(22): p. 15024.

49. Ogbonna, C., *An Overview of the Effect of Biomass in-Door-Air Pollution on Household Members.* Journal of Epidemiological Society of Nigeria, 2017. **1**: p. 66-69.

50. Lichkovakha, D.V., *The Role of Russian Coal in Modern World Energy Sector.* Общество Политика Экономика Право, 2024(8).

51. Makarov, V.M., *Forecasting the Output of Coalproducts in Thepost-Warperiod in Ukraine.* System Research in Energy, 2024. **2024**(1): p. 35-44.

52. Evans, W.D., et al., *The Shamba Chef Educational Entertainment Program to Promote Modern Cookstoves in Kenya: Outcomes and Dose–Response Analysis.* International Journal of Environmental Research and Public Health, 2019. **17**(1): p. 162.

53. Amegah, A.K. and J.J.K. Jaakkola, *Household Air Pollution and the Sustainable Development Goals.* Bulletin of the World Health Organization, 2016. **94**(3): p. 215-221.

54. Boudewijns, E.A., et al., *Factors Critical to Implementation Success of Cleaner Cooking Interventions in Low-Income and Middle-Income Countries: Protocol for an Umbrella Review.* BMJ Open, 2020. **10**(12): p. e041821.

55. Sood, A.K., et al., *ERS/ATS Workshop Report on Respiratory Health Effects of Household Air Pollution.* European Respiratory Journal, 2018. **51**(1): p. 1700698.

56. Azanaw, J. and G.S. Chanie, *Spatial Variation and Determinants of Solid Fuel Use in Ethiopia; Mixed Effect and Spatial Analysis Using 2019 Ethiopia Mini Demographic and Health Survey Dataset.* Plos One, 2023. **18**(11): p. e0294841.

57. Chafe, Z., et al., *Household Cooking With Solid Fuels Contributes to Ambient PM$_{2.5}$Air Pollution and the Burden of Disease.* Environmental Health Perspectives, 2014. **122**(12): p. 1314-1320.

58. Wernecke, B., et al., *Indoor and Outdoor Particulate Matter Concentrations on the Mpumalanga Highveld – A Case Study.* Clean Air Journal, 2015. **25**(2): p. 12.

59. Hussain, A.M. and E.D. Wachsman, *Liquids-to-Power Using Low-Temperature Solid Oxide Fuel Cells.* Energy Technology, 2018. **7**(1): p. 20-32.

60. Farla, J., F. Alkemade, and R.A.A. Suurs, *Analysis of Barriers in the Transition Toward Sustainable Mobility in the Netherlands.* Technological Forecasting and Social Change, 2010. **77**(8): p. 1260-1269.

61. Jabbar, M.H.A., et al., *Performance and Durability of Metal SOFCs in Alternate Fuels.* Ecs Transactions, 2023. **111**(6): p. 2311-2320.

62. Fukuzumi, S., *Production of Liquid Solar Fuels and Their Use in Fuel Cells.* Joule, 2017. **1**(4): p. 689-738.

63. Zheng, Y., *Analysis of the Current Situation and Prospective Study of Hydrogen Preparation and Storage.* Highlights in Science Engineering and Technology, 2024. **96**: p. 40-46.

64. Steen, E.J., et al., *Microbial Production of Fatty-Acid-Derived Fuels and Chemicals From Plant Biomass.* Nature, 2010. **463**(7280): p. 559-562.

65. Shih, C.F., et al., *Powering the Future With Liquid Sunshine.* Joule, 2018. **2**(10): p. 1925-1949.

66. Walter, M.G., et al., *Solar Water Splitting Cells.* Chemical Reviews, 2010. **110**(11): p. 6446-6473.

67. Metzger, M., et al., *Liquid Fuels From Alternative Carbon Sources Minimizing Carbon Dioxide Emissions.* Aiche Journal, 2013. **59**(6): p. 2062-2078.

68. Ikäheimo, J., et al., *Role of Power to Liquids and Biomass to Liquids in a Nearly Renewable Energy System.* Iet Renewable Power Generation, 2019. **13**(7): p. 1179-1189.

69. Azad, A.K., et al., *Energy and Waste Management for Petroleum Refining Effluents: A Case Study in Bangladesh.* International Journal of Automotive and Mechanical Engineering, 2015. **11**: p. 2170-2187.

70. Boyd, M., et al., *Mitigating Imported Fuel Dependency in Agricultural Production: Case Study of an Island Nation's Vulnerability to Global Catastrophic Risks.* Risk Analysis, 2024. **44**(10): p. 2360-2376.

71. Datta, A., A. Hossain, and S. Roy, *An Overview on Biofuels and Their Advantages and Disadvantages.* Asian Journal of Chemistry, 2019. **31**(8): p. 1851-1858.

72. Jiang, Y. and D. Bhattacharyya, *Process Modeling of Direct Coal-Biomass to Liquids (CBTL) Plants With Shale Gas Utilization and CO2 Capture and Storage (CCS).* Applied Energy, 2016. **183**: p. 1616-1632.

73. Zaimes, G.G., et al., *Biofuels via Fast Pyrolysis of Perennial Grasses: A Life Cycle Evaluation of Energy Consumption and Greenhouse Gas Emissions.* Environmental Science & Technology, 2015. **49**(16): p. 10007-10018.

74. Osorio-Tejada, J.L., E. Llera-Sastresa, and S. Scarpellini, *A Multi-Criteria Sustainability Assessment for Biodiesel and Liquefied Natural Gas as Alternative Fuels in Transport Systems.* Journal of Natural Gas Science and Engineering, 2017. **42**: p. 169-186.

75. Lamb, B., et al., *Direct Measurements Show Decreasing Methane Emissions From Natural Gas Local Distribution Systems in the United States.* Environmental Science & Technology, 2015. **49**(8): p. 5161-5169.

76. Dai, Q. and C.M. Lastoskie, *Life Cycle Assessment of Natural Gas-Powered Personal Mobility Options.* Energy & Fuels, 2014. **28**(9): p. 5988-5997.

77. Meyer, P., et al., *Total Fuel-Cycle Analysis of Heavy-Duty Vehicles Using Biofuels and Natural Gas-Based Alternative Fuels.* Journal of the Air & Waste Management Association, 2011. **61**(3): p. 285-294.

78. Castillo, J.C., et al., *Natural Gas, a Mean to Reduce Emissions and Energy Consumption of HDV? A Case Study of Colombia Based on Vehicle Technology Criteria.* Energies, 2022. **15**(3): p. 998.

79. Greyson, K.A., *Vehicles Power Consumption: Case Study of Dar Rapid Transit Agency (DART) in Tanzania.* 2021.

80. Ogunlowo, O.O., A. Bristow, and M. Sohail, *A Stakeholder Analysis of the Automotive Industry's Use of Compressed Natural Gas in Nigeria.* Transport Policy, 2017. **53**: p. 58-69.

81. Zakirova, G., et al., *Storage of Compressed Natural Gases.* Energies, 2023. **16**(20): p. 7208.

82. Fedorov, M.P., et al., *Production of Biohydrogen From Organ-Containing Waste for Use in Fuel Cells.* Energies, 2022. **15**(21): p. 8019.

83. Schiro, F., A. Stoppato, and A. Benato, *Potentialities of Hydrogen Enriched Natural Gas for Residential Heating Decarbonization and Impact Analysis on Premixed Boilers.* E3s Web of Conferences, 2019. **116**: p. 00072.

84. Калініченко, А., V. Havrysh, and V. Perebyynis, *Evaluation of Biogas Production and Usage Potential.* Ecological Chemistry and Engineering S, 2016. **23**(3): p. 387-400.

85. Adánez, J. and A. Abad, *Chemical-Looping Combustion: Status and Research Needs.* Proceedings of the Combustion Institute, 2019. **37**(4): p. 4303-4317.

86. Wierzbicki, S., M. Śmieja, and M. Mikulski, *Effect of Fuel Pilot Dose Parameters on Efficiency of Dual-Fuel Compression Ignition Engines Fuelled With Biogas.* Applied Mechanics and Materials, 2016. **817**: p. 19-26.

87. Zhou, H., et al., *Effects of Injection Timing on Combustion and Emission Performance of Dual-Fuel Diesel Engine Under Low to Medium Load Conditions.* Energies, 2019. **12**(12): p. 2349.

88. Gracz, W., et al., *Multifaceted Comparison Efficiency and Emission Characteristics of Multi-Fuel Power Generator Fueled by Different Fuels and Biofuels.* Energies, 2021. **14**(12): p. 3388.

89. Llera-Sastresa, E., et al., *Exploring the Integration of the Power to Gas Technologies and the Sustainable Transport.* International Journal of Energy Production and Management, 2018. **3**(1): p. 1-9.

90. Puškár, M., et al., *Analysis of Various Factors' Influence and Optimization of Low-Temperature Combustion Technology.* Acs Omega, 2024.

91. Sahu, G.K., *Performance Analysis of Four Stroke Diesel Engine Working With Acetylene and Diesel.* International Journal for Research in Applied Science and Engineering Technology, 2017. **V**(VIII): p. 2038-2043.

92. Owczuk, M., et al., *Evaluation of Using Biogas to Supply the Dual Fuel Diesel Engine of an Agricultural Tractor.* Energies, 2019. **12**(6): p. 1071.

93. Xue, Y., W.Q. Liu, and X.C. Lu, *Experimental Study on LPG Composite Fuel Used on Spark Ignition Engine.* Advanced Materials Research, 2012. **516-517**: p. 607-613.

94. Mikulski, M., S. Wierzbicki, and A. Piętak, *Zero-Dimensional 2-Phase Combustion Model in a Dual-Fuel Compression Ignition Engine Fed With Gaseous Fuel and a Divided Diesel Fuel Charge.* Eksploatacja I Niezawodnosc - Maintenance and Reliability, 2015. **17**(1): p. 42-48.

95. Huijbregts, M.A.J., et al., *Is Cumulative Fossil Energy Demand a Useful Indicator for the Environmental Performance of Products?* Environmental Science & Technology, 2005. **40**(3): p. 641-648.

96. Epping, R., S. Kerkering, and J.T. Andersson, *Influence of Different Compound Classes on the Formation of Sediments in Fossil Fuels During Aging.* Energy & Fuels, 2014. **28**(9): p. 5649-5656.

97. Simoneit, B.R.T., et al., *Aerosol Particles Collected on Aircraft Flights Over the Northwestern Pacific Region During the ACE-Asia Campaign: Composition and Major Sources of the Organic Compounds.* Journal of Geophysical Research Atmospheres, 2004. **109**(D19).

98. Zheng, S., et al., *Blending Effects of Nitrogenous Bio-Fuel on Soot and NO_x Formation During Gasoline Combustion.* Energy & Fuels, 2023. **37**(6): p. 4596-4607.

99. Wang, B., et al., *Growth of Polycyclic Aromatic Hydrocarbon and Soot Inception by in Silico Simulation.* 2022.

100. Bond, T.C., et al., *Bounding the Role of Black Carbon in the Climate System: A Scientific Assessment.* Journal of Geophysical Research Atmospheres, 2013. **118**(11): p. 5380-5552.

101. Liu, J., et al., *Source Apportionment and Dynamic Changes of Carbonaceous Aerosols During the Haze Bloom-Decay Process in China Based on Radiocarbon and Organic Molecular Tracers.* Atmospheric Chemistry and Physics, 2016. **16**(5): p. 2985-2996.

102. Karacan, R., *Kurumsal Karbon Ayak İzi Çevre Kalitesi Ve Covid-19 Salgını Ile Mücadele (ABD Örneği).* Konuralp Tıp Dergisi, 2022. **14**(S1): p. 251-259.

103. Andersson, A., et al., *Regionally-Varying Combustion Sources of the January 2013 Severe Haze Events Over Eastern China.* Environmental Science & Technology, 2015. **49**(4): p. 2038-2043.

104. Kiran, D.V., N.M.G. Kumar, and S.M. Shashidhara, *A Comprehensive Review on Optimization Strategies for Combined Economic Emission Dispatch Problem.* Asian Journal of Electrical Sciences, 2018. **7**(1): p. 68-74.

105. Duan, M. and Y. Duan, *Research on Oil and Gas Energy Cooperation Between China and Central-North Asian Countries Under the "One Belt and One Road" Strategy.* Energies, 2023. **16**(21): p. 7326.

106. Johnsson, F., J. Kjärstad, and J. Rootzén, *The Threat to Climate Change Mitigation Posed by the Abundance of Fossil Fuels.* Climate Policy, 2018. **19**(2): p. 258-274.

107. Buchanan, E., *Russian Energy Strategy in the Asia-Pacific: Implications for Australia.* 2021.

108. Guo, Y., et al., *Establishment and Application of Prediction Model of Natural Gas Reserve and Production in Sichuan Basin.* Journal of Petroleum Exploration and Production Technology, 2021. **11**(6): p. 2679-2689.

109. Zhang, B. and L. Sun, *Artificial Photosynthesis: Opportunities and Challenges of Molecular Catalysts.* Chemical Society Reviews, 2019. **48**(7): p. 2216-2264.

110. Liu, C. and Q. Li, *Air Pollution, Global Warming and Difficulties to Replace Fossil Fuel With Renewable Energy.* Atmospheric and Climate Sciences, 2023. **13**(04): p. 526-538.

111. Paul, A., et al., *En Route to Artificial Photosynthesis: The Role of Polyoxometalate Based Photocatalysts.* Journal of Materials Chemistry A, 2022. **10**(25): p. 13152-13169.

112. Wang, H., et al., *Application Progress of NiMoO$_4$ Electrocatalyst in Basic Oxygen Evolution Reaction.* Catalysis Science & Technology, 2024. **14**(3): p. 533-554.

113. Menezes, F.S.R., et al., *Evaluation of Endoglucanase and B-Glucosidase Production by Bacteria and Yeasts Isolated From an Eucalyptus Plantation in the Cerrado of Minas Gerais.* Ambiente E Agua - An Interdisciplinary Journal of Applied Science, 2019. **14**(4): p. 1.

114. Oludaisi, A., A. Kayode, and O. Ayodeji, *Bridging Environmental Impact of Fossil Fuel Energy: The Contributing Role of Alternative Energy.* Journal of Engineering Studies and Research, 2017. **23**(2): p. 22-27.

115. Christy, E.J.S.B.A., et al., *Optimization of Pre-Treatment and Culture Conditions for Bioethanol Yield Enhancement From ≪em>Azolla Filiculoides</Em> Substrate Using ≪em>Saccharomyces Cerevisiae</Em&gt.* Journal of Dry Zone Agriculture, 2023. **9**(1): p. 1-25.

116. Skovgaard, J. and H.v. Asselt, *The Politics of Fossil Fuel Subsidies and Their Reform: Implications for Climate Change Mitigation.* Wiley Interdisciplinary Reviews Climate Change, 2019. **10**(4).

117. Piggot, G., et al., *Swimming Upstream: Addressing Fossil Fuel Supply Under the UNFCCC.* Climate Policy, 2018. **18**(9): p. 1189-1202.

118. Qubeissi, M.A., N. Al-Esawi, and R. Kolodnytska, *Atomization of Bio-Fossil Fuel Blends.* 2018.

119. Fæhn, T., et al., *Climate Policies in a Fossil Fuel Producing Country - Demand Versus Supply Side Policies.* SSRN Electronic Journal, 2014.

120. Zhang, W., et al., *Acid-etched Pt/Al-MCM-41 Catalysts for Fuel Production by One-step Hydrotreatment of Jatropha Oil.* GCB Bioenergy, 2021. **13**(4): p. 679-690.

121. Ploeg, F.v.d., *Untapped Fossil Fuel and the Green Paradox: A Classroom Calibration of the Optimal Carbon Tax.* Environmental Economics and Policy Studies, 2014. **17**(2): p. 185-210.

122. Wu, H., P. Wu, and Y. Chen, *Extraction Technology and Pillar Design Analysis of End-Wall Mining System in China.* International Journal of Mining Science, 2017. **3**(4).

123. Pagouni, C., et al., *Transitional and Post-Mining Land Uses: A Global Review of Regulatory Frameworks, Decision-Making Criteria, and Methods.* Land, 2024. **13**(7): p. 1051.

124. Lima, A.T., et al., *The Legacy of Surface Mining: Remediation, Restoration, Reclamation and Rehabilitation.* Environmental Science & Policy, 2016. **66**: p. 227-233.

125. Mukherjee, S. and D.P. Pahari, *Underground and Opencast Coal Mining Methods in India: A Comparative Assessment.* Space and Culture India, 2019. **7**(1).

126. Sasaoka, T., et al., *Punch Multi-Slice Longwall Mining System for Thick Coal Seam Under Weak Geological Conditions.* Journal of Geological Resource and Engineering, 2016. **4**(1).

127. Phuc, L.Q., В.П. Зубов, and P.M. Dac, *Improvement of the Loading Capacity of Narrow Coal Pillars and Control Roadway Deformation in the Longwall Mining System. A Case Study at Khe Cham Coal Mine (Vietnam).* Inżynieria Mineralna, 2020. **1**(2).

128. Zhang, X., *Simulation Study on the Safety Evaluation of Coal Mine General Mining Working Face Production System.* 2023.

129. Liu, Q., X. Li, and F. Guan, *Research on Effectiveness of Coal Mine Safety Supervision System Reform on Three Types of Collieries in China.* International Journal of Coal Science & Technology, 2014. **1**(3): p. 376-382.

130. Agostinelli, S., et al., *Mastering Robotic Process Automation With Process Mining.* 2022: p. 47-53.

131. Wang, J.X., *Research on Mobile Coal Mining Safety Monitoring System.* Applied Mechanics and Materials, 2013. **368-370**: p. 1995-2001.

132. Ma, T., P. Chen, and J. Zhao, *Overview on Vertical and Directional Drilling Technologies for the Exploration and Exploitation of Deep Petroleum Resources.* Geomechanics and Geophysics for Geo-Energy and Geo-Resources, 2016. **2**(4): p. 365-395.

133. Al-Jawad, D.M.S., D.A.A.A.L. Dabaj, and H.A.H.A. Hussien, *Prediction of Performance of Rotary Bottom Hole Assembly for Directional Well.* Journal of Petroleum Research and Studies, 2013. **4**(3): p. 1-15.

134. Onsarigo, L., S. Adamtey, and A. Atalah, *Analysis of Horizontal Directional Drilling Construction Risks Using the Probability-Impact Model: A Contractor's Perspective.* 2014: p. 1772-1783.

135. Kong, L., et al., *Simulation and Experimental Study on Cuttings -Carrying for Reverse Circulation Horizontal Directional Drilling With Dual Drill Pipes.* Advances in Civil Engineering, 2019. **2019**(1).

136. Daniel, J., et al., *Physicochemical Characterization of Horizontal Directional Drilling Residuals.* Sustainability, 2020. **12**(18): p. 7707.

137. Jin, L. and J. Wei, *Research on Horizontal Directional Drilling (HDD) Trajectory Design and Optimization Using Improved Radial Movement Optimization.* Applied Sciences, 2022. **12**(23): p. 12207.

138. Santoso, M.Y., M.D. Khairiansyah, and R.F. Hernasa, *Improving Safety Design for Gas Pipeline Installation via Horizontal Directional Drilling: A Pipe Stress Analysis Approach.* Jurnal Polimesin, 2024. **22**(3): p. 354.

139. Davis, C., *Substate Federalism and Fracking Policies: Does State Regulatory Authority Trump Local Land Use Autonomy?* Environmental Science & Technology, 2014. **48**(15): p. 8397-8403.

140. Murcott, M. and E. Webster, *Litigation and Regulatory Governance in the Age of the Anthropocene: The Case of Fracking in the Karoo.* Transnational Legal Theory, 2020. **11**(1-2): p. 144-164.

141. Stephenson, T., J.E. Valle, and X. Riera-Palou, *Modeling the Relative GHG Emissions of Conventional and Shale Gas Production.* Environmental Science & Technology, 2011. **45**(24): p. 10757-10764.

142. Meegoda, J.N., et al., *Can Fracking Be Environmentally Acceptable?* Journal of Hazardous Toxic and Radioactive Waste, 2017. **21**(2).

143. Osborn, S.G., et al., *Methane Contamination of Drinking Water Accompanying Gas-Well Drilling and Hydraulic Fracturing.* Proceedings of the National Academy of Sciences, 2011. **108**(20): p. 8172-8176.

144. *To What Extent Could Shale Gas "Fracking" Contaminate Ground Water.* Journal of Petroleum & Environmental Biotechnology, 2015. **06**(05).

145. Barkley, Z., et al., *Estimating Methane Emissions From Underground Coal and Natural Gas Production in Southwestern Pennsylvania.* Geophysical Research Letters, 2019. **46**(8): p. 4531-4540.

146. Schwartz, M.O., *Modelling the Hypothetical Methane-Leakage in a Shale-Gas Project and the Impact on Groundwater Quality.* Environmental Earth Sciences, 2014. **73**(8): p. 4619-4632.

147. Yudhowijoyo, A., et al., *Subsurface Methane Leakage in Unconventional Shale Gas Reservoirs: A Review of Leakage Pathways and Current Sealing Techniques.* Journal of Natural Gas Science and Engineering, 2018. **54**: p. 309-319.

148. Fisk, J.M., *The Right to Know? State Politics of Fracking Disclosure.* Review of Policy Research, 2013. **30**(4): p. 345-365.

149. Zacher, S., *The US Political Economy of Climate Change: Impacts of the "Fracking" Boom on State-Level Climate Policies.* State Politics & Policy Quarterly, 2023. **23**(2): p. 140-165.

150. Zhang, Y., et al., *Remarkable Spatial Disparity of Life Cycle Inventory for Coal Production in China.* Environmental Science & Technology, 2023. **57**(41): p. 15443-15453.

151. Yun, X., et al., *Coal Is Dirty, but Where It Is Burned Especially Matters.* Environmental Science & Technology, 2021. **55**(11): p. 7316-7326.

152. Kholod, N., et al., *Global Methane Emissions From Coal Mining to Continue Growing Even With Declining Coal Production.* Journal of Cleaner Production, 2020. **256**: p. 120489.

153. Zhuang, G., Z. Zhang, and M. Jaber, *Organoclays Used as Colloidal and Rheological Additives in Oil-Based Drilling Fluids: An Overview.* Applied Clay Science, 2019. **177**: p. 63-81.

154. Karabalayeva, A.B., et al., *Monitoring the Environment and Recycling Approaches for Managing Oil and Drilling Waste.* Instrumentation Mesure Métrologie, 2024. **23**(5): p. 355-361.

155. Chen, L., L. Pan, and K. Zhang, *The Dynamic Cointegration Relationship Between International Crude Oil, Natural Gas, and Coal Price.* Energies, 2024. **17**(13): p. 3126.

156. Blondeel, M., et al., *The Geopolitics of Energy System Transformation: A Review.* Geography Compass, 2021. **15**(7).

157. Mercure, J.-F., et al., *Macroeconomic Impact of Stranded Fossil Fuel Assets.* Nature Climate Change, 2018. **8**(7): p. 588-593.

158. Hou, Z., et al., *Spatiotemporal Evolution and Market Dynamics of the International Liquefied Natural Gas Trade: A Multilevel Network Analysis.* Energies, 2023. **17**(1): p. 228.

159. Daudu, C.D., et al., *LNG and Climate Change: Evaluating Its Carbon Footprint in Comparison to Other Fossil Fuels.* Engineering Science & Technology Journal, 2024. **5**(2): p. 412-426.

160. Peña-Ramos, J.A., M.d. Pino-García, and A. Sánchez-Bayón, *The Spanish Energy Transition Into the EU Green Deal: Alignments and Paradoxes.* Energies, 2021. **14**(9): p. 2535.

161. Ritchie, H. and P. Rosado, *Fossil fuels.* 2024.

162. Adam, N.M. and A. Shanableh, *Combined Production of Biofuels From Locally Grown Microalgae.* International Journal of Thermal and Environmental Engineering, 2016. **13**(1).

163. Driver, T., A.K. Bajhaiya, and J.K. Pittman, *Potential of Bioenergy Production From Microalgae.* Current Sustainable/Renewable Energy Reports, 2014. **1**(3): p. 94-103.

164. Sharma, P., et al., *Biofuel Production, Study &Amp; Characterisation From Macro-Algae (Azolla Pinnata).* Brazilian Journal of Science, 2023. **2**(3): p. 75-81.

165. Alam, F., S. Mobin, and H. Chowdhury, *Third Generation Biofuel From Algae.* Procedia Engineering, 2015. **105**: p. 763-768.

166. Mu, D., et al., *Life Cycle Environmental Impacts of Wastewater-Based Algal Biofuels.* Environmental Science & Technology, 2014. **48**(19): p. 11696-11704.

167. Villarreal, J.V., C. Burgués, and C. Rösch, *Acceptability of Genetically Engineered Algae Biofuels in Europe: Opinions of Experts and Stakeholders.* Biotechnology for Biofuels, 2020. **13**(1).

168. Zhao, C., T. Brück, and J.A. Lercher, *Catalytic Deoxygenation of Microalgae Oil to Green Hydrocarbons.* Green Chemistry, 2013. **15**(7): p. 1720.

169. Rösch, C., M. Roßmann, and S. Weickert, *Microalgae for Integrated Food and Fuel Production.* GCB Bioenergy, 2018. **11**(1): p. 326-334.

170. Sinambela, R. and I. Samanlangi, *Using Biotechnology to Make Biofuels From Algae as Renewable Energy.* 2024. **1**(3): p. 1-8.

171. Adeniyi, O., U. Azimov, and A.A. Burluka, *Algae Biofuel: Current Status and Future Applications.* Renewable and Sustainable Energy Reviews, 2018. **90**: p. 316-335.

172. Singh, A., P.S.N. Nigam, and J.D. Murphy, *Renewable Fuels From Algae: An Answer to Debatable Land Based Fuels.* Bioresource Technology, 2011. **102**(1): p. 10-16.

173. Sandalcı, T., et al., *An Experimental Investigation of Ethanol-Diesel Blends on Performance and Exhaust Emissions of Diesel Engines.* Advances in Mechanical Engineering, 2014. **6**: p. 409739.

174. Luo, M., et al., *Particulate Matter and Gaseous Emission of Hydrous Ethanol Gasoline Blends Fuel in a Port Injection Gasoline Engine.* Energies, 2017. **10**(9): p. 1263.

175. Norvell, K.L. and N.P. Nghiem, *Soaking in Aqueous Ammonia (SAA) Pretreatment of Whole Corn Kernels for Cellulosic Ethanol Production From the Fiber Fractions.* 2018.

176. Solomon, B.D., *Biofuels and Sustainability.* Annals of the New York Academy of Sciences, 2010. **1185**(1): p. 119-134.

177. Tsantili, I.C., M.N. Karim, and M.I. Klapa, *Quantifying the Metabolic Capabilities of Engineered Zymomonas Mobilis Using Linear Programming Analysis.* Microbial Cell Factories, 2007. **6**(1).

178. Handler, R.M., et al., *Life Cycle Assessments of Ethanol Production via Gas Fermentation: Anticipated Greenhouse Gas Emissions for Cellulosic and Waste Gas Feedstocks.* Industrial & Engineering Chemistry Research, 2015. **55**(12): p. 3253-3261.

179. Petrova, S., et al., *Optical Ratiometric Sensor for Alcohol Measurements.* Analytical Letters, 2007. **40**(4): p. 715-727.

180. Huang, J., et al., *Experimental Investigation on the Performance and Emissions of a Diesel Engine Fuelled With Ethanol–diesel Blends.* Applied Thermal Engineering, 2009. **29**(11-12): p. 2484-2490.

181. Freitas, E.S.d.C., et al., *Emission and Performance Evaluation of a Diesel Engine Using Addition of Ethanol to Diesel/Biodiesel Fuel Blend.* Energies, 2022. **15**(9): p. 2988.

182. Withers, J., H. Quesada, and R.A. Smith, *Internal and External Barriers Impacting Non-Food Cellulosic Biofuel Projects in the United States.* Bioresources, 2015. **10**(3).

183. Wahlen, B.D., et al., *Biodiesel From Microalgae, Yeast, and Bacteria: Engine Performance and Exhaust Emissions.* Energy & Fuels, 2012. **27**(1): p. 220-228.

184. Manurung, R. and A. Alexander, *Minor Component Extraction From Palm Methyl Ester Using Potassium Carbonate Glycerol Based Deep Eutectic Solvent (Des)*. Rasayan Journal of Chemistry, 2018. **11**(4): p. 1519-1524.

185. Lackey, L.G. and S.E. Paulson, *Influence of Feedstock: Air Pollution and Climate-Related Emissions From a Diesel Generator Operating on Soybean, Canola, and Yellow Grease Biodiesel*. Energy & Fuels, 2011. **26**(1): p. 686-700.

186. Mueller, C.J., A.L. Boehman, and G.C. Martin, *An Experimental Investigation of the Origin of Increased NO$_x$ Emissions When Fueling a Heavy-Duty Compression-Ignition Engine With Soy Biodiesel*. Sae International Journal of Fuels and Lubricants, 2009. **2**(1): p. 789-816.

187. Alptekin, E. and M. Çanakçı, *Determination of the Density and the Viscosities of Biodiesel–diesel Fuel Blends*. Renewable Energy, 2008. **33**(12): p. 2623-2630.

188. Bittle, J.A., B.M. Knight, and T.J. Jacobs, *Interesting Behavior of Biodiesel Ignition Delay and Combustion Duration*. Energy & Fuels, 2010. **24**(8): p. 4166-4177.

189. Fregolente, P.B.L., L.V. Fregolente, and M.R.W. Maciel, *Water Content in Biodiesel, Diesel, and Biodiesel–Diesel Blends*. Journal of Chemical & Engineering Data, 2012. **57**(6): p. 1817-1821.

190. Babulal, A.P.d.K.S., *Experimental Performance and Emission Analysis of C.I Engine Fuelled With Biodiesel*. International Journal of Innovative Research in Science Engineering and Technology, 2015. **04**(02): p. 133-138.

191. Çanakçı, M. and J.H.V. Gerpen, *Comparison of Engine Performance and Emissions for Petroleum Diesel Fuel, Yellow Grease Biodiesel, and Soybean Oil Biodiesel*. Transactions of the Asae, 2003. **46**(4).

192. Koçak, M.S., E. İleri, and Z. Utlu, *Experimental Study of Emission Parameters of Biodiesel Fuels Obtained From Canola, Hazelnut, and Waste Cooking Oils*. Energy & Fuels, 2007. **21**(6): p. 3622-3626.

193. Tompkins, B.T. and T.J. Jacobs, *Low-Temperature Combustion With Biodiesel: Its Enabling Features in Improving Efficiency and Emissions*. Energy & Fuels, 2013. **27**(5): p. 2794-2803.

194. Boehman, A.L., J. Song, and M. Alam, *Impact of Biodiesel Blending on Diesel Soot and the Regeneration of Particulate Filters*. Energy & Fuels, 2005. **19**(5): p. 1857-1864.

195. Pienaar, J.F. and A.C. Brent, *A Model for Evaluating the Economic Feasibility of Small-Scale Biodiesel Production Systems for on-Farm Fuel Usage*. Renewable Energy, 2012. **39**(1): p. 483-489.

196. Cecrle, E., et al., *Investigation of the Effects of Biodiesel Feedstock on the Performance and Emissions of a Single-Cylinder Diesel Engine*. Energy & Fuels, 2012. **26**(4): p. 2331-2341.

197. Hidalgo, P., et al., *Improving the FAME Yield of in Situ Transesterification From Microalgal Biomass Through Particle Size Reduction and Cosolvent Incorporation*. Energy & Fuels, 2015. **29**(2): p. 823-832.

198. Rostami, M., et al., *Liquid–Liquid Phase Equilibria of Systems of Palm and Soya Biodiesels: Experimental and Modeling*. Industrial & Engineering Chemistry Research, 2012. **51**(24): p. 8302-8307.

199. Swanson, K.J., et al., *Release of the Pro-Inflammatory Markers by BEAS-2B Cells Following in Vitro Exposure to Biodiesel Extracts*. The Open Toxicology Journal, 2009. **3**(1): p. 8-15.

200. Gümüş, M., C. Sayın, and M. Çanakçı, *Effect of Fuel Injection Timing on the Injection, Combustion, and Performance Characteristics of a Direct-Injection (DI) Diesel Engine Fueled With Canola Oil Methyl Ester–Diesel Fuel Blends*. Energy & Fuels, 2010. **24**(5): p. 3199-3213.

201. Peng, D.X., *The Effect on Diesel Injector Wear, and Exhaust Emissions by Using Ultralow Sulphur Diesel Blending With Biofuels*. Materials Transactions, 2015. **56**(5): p. 642-647.

202. Walsh, M.J., et al., *Algal Food and Fuel Coproduction Can Mitigate Greenhouse Gas Emissions While Improving Land and Water-Use Efficiency*. Environmental Research Letters, 2016. **11**(11): p. 114006.

203. Menetrez, M.Y., *An Overview of Algae Biofuel Production and Potential Environmental Impact*. Environmental Science & Technology, 2012. **46**(13): p. 7073-7085.

204. Naik, A.N., M. Singh, and Y. Qurishi, *Algal Biofuel: A Promising Perspective*. Annals of Plant Sciences, 2018. **7**(5): p. 2262.

205. Williams, C., et al., *Evaluating Algae as an Alternative Fuel for Chemical Looping Combustion*. Pam Review Energy Science & Technology, 2018. **5**: p. 37-55.

206. Jones, J.N., et al., *Resins That Reversibly Bind Algae for Harvesting and Concentration*. Environmental Progress & Sustainable Energy, 2012. **32**(4): p. 1143-1149.

207. Siddiqua, S., A. Al-Mamun, and S.M.E. Babar, *Production of Biodiesel From Coastal Macroalgae (Chara Vulgaris) and Optimization of Process Parameters Using Box-Behnken Design*. Springerplus, 2015. **4**(1).

208. Taparia, T., et al., *Developments and Challenges in Biodiesel Production From Microalgae: A Review*. Biotechnology and Applied Biochemistry, 2015. **63**(5): p. 715-726.

209. Jangiam, W., P. Tongtubtim, and M. Penjun, *Biological Hydrogen Production From Amphora Sp. Isolated From Eastern Coast of Thailand*. E3s Web of Conferences, 2019. **93**: p. 03004.

210. Kozhevnikov, Y.A., I. Knyazeva, and O. Vershinina, *Microalgae Is an Unclaimed Bioenergy Resource in Russia*. Bio Web of Conferences, 2021. **40**: p. 01007.

211. Morales, M., A. Hélias, and O. Bernard, *Optimal Integration of Microalgae Production With Photovoltaic Panels: Environmental Impacts and Energy Balance*. Biotechnology for Biofuels, 2019. **12**(1).

212. Hu, M., et al., *Systematic Review and Perspective on the Progress of Algal Biofuels*. E3s Web of Conferences, 2021. **257**: p. 03008.

213. Barreiro, D.L., et al., *Assessing Microalgae Biorefinery Routes for the Production of Biofuels via Hydrothermal Liquefaction*. Bioresource Technology, 2014. **174**: p. 256-265.

214. al, M.P.J.M.P.J.e. and et al., *An Evaluation of Algae Biofuel as the Next Generation Alternative Fuel and Its Effects on Engine Characteristics, a Review*. International Journal of Mechanical and Production Engineering Research and Development, 2019. **9**(1): p. 435-440.

215. Malik, K., et al., *Biofuels Production: A Review on Sustainable Alternatives to Traditional Fuels and Energy Sources*. Fuels, 2024. **5**(2): p. 157-175.

216. Jeswani, H.K., A. Chilvers, and A. Azapagic, *Environmental Sustainability of Biofuels: A Review*. Proceedings of the Royal Society a Mathematical Physical and Engineering Sciences, 2020. **476**(2243).

217. Mishra, P.K. and A. Dutta, *Biofuel: Upcoming Sources of Energy*. 2024: p. 202-216.

218. Ajanović, A. and R. Haas, *Renewable Energy Systems Implementation in Road Transport: Prospects and Impediments*. Renewable Energy and Environmental Sustainability, 2021. **6**: p. 39.

219. Skevis, G., *Liquid Biofuels: Bioalcohols, Biodiesel and Biogasoline and Algal Biofuels.* 2016: p. 1-43.

220. Culaba, A.B., et al., *Biofuel From Microalgae: Sustainable Pathways.* Sustainability, 2020. **12**(19): p. 8009.

221. Olabi, A.G., et al., *Emerging Technologies for Enhancing Microalgae Biofuel Production: Recent Progress, Barriers, and Limitations.* Fermentation, 2022. **8**(11): p. 649.

222. Allen, J.D., et al., *Integration of Biology, Ecology and Engineering for Sustainable Algal-Based Biofuel and Bioproduct Biorefinery.* Bioresources and Bioprocessing, 2018. **5**(1).

223. Hoyer, J., et al., *Bioenergy From Microalgae – Vision or Reality?* Chembioeng Reviews, 2018. **5**(4): p. 207-216.

224. Medipally, S.R., et al., *Microalgae as Sustainable Renewable Energy Feedstock for Biofuel Production.* Biomed Research International, 2015. **2015**: p. 1-13.

225. Park, H. and C.G. Lee, *Theoretical Calculations on the Feasibility of Microalgal Biofuels: Utilization of Marine Resources Could Help Realizing the Potential of Microalgae.* Biotechnology Journal, 2016. **11**(11): p. 1461-1470.

226. Li, Y., *Comparison and Alternative Feasibility of Biofuels With Other Fuels.* 2024. **1**(9).

227. World Nuclear Association, *Hydrogen Production and Uses.* 2024.

228. Elshafei, A.M. and R. Mansour, *Green Hydrogen as a Potential Solution for Reducing Carbon Emissions: A Review.* Journal of Energy Research and Reviews, 2023. **13**(2): p. 1-10.

229. Bauer, C., et al., *On the Climate Impacts of Blue Hydrogen Production.* Sustainable Energy & Fuels, 2022. **6**(1): p. 66-75.

230. Hamed, A.M., et al., *A Review on Blue and Green Hydrogen Production Process and Their Life Cycle Assessments.* Iop Conference Series Earth and Environmental Science, 2023. **1281**(1): p. 012034.

231. Zhao, C., et al., *Solutions for Decarbonising Urban Bus Transport: A Life Cycle Case Study in Saudi Arabia.* Communications Engineering, 2024. **3**(1).

232. Guo, P., et al., *Study of Hydrogen Internal Combustion Engine Vehicles Based on the Whole Life Cycle Evaluation Method.* Trends in Renewable Energy, 2022. **8**(1): p. 27-37.

233. Ingale, G.U., et al., *Assessment of Greenhouse Gas Emissions From Hydrogen Production Processes: Turquoise Hydrogen vs. Steam Methane Reforming.* Energies, 2022. **15**(22): p. 8679.

234. Racha, A., et al., *Sustainable Fuel Additives Derived From Renewable Resources: Promising Strategies for a Greener Future.* Acs Omega, 2025. **10**(19): p. 19256-19282.

235. Tan, E.C.D., et al., *Reduction of Greenhouse Gas and Criteria Pollutant Emissions by Direct Conversion of Associated Flare Gas to Synthetic Fuels at Oil Wellheads.* International Journal of Energy and Environmental Engineering, 2018. **9**(3): p. 305-321.

236. Towoju, O.A., *Fuels for Automobiles: The Sustainable Future.* Journal of Energy Research and Reviews, 2021: p. 8-13.

237. Brynolf, S., et al., *Review of Electrofuel Feasibility—prospects for Road, Ocean, and Air Transport.* Progress in Energy, 2022. **4**(4): p. 042007.

238. Khalili, S., et al., *Global Transportation Demand Development With Impacts on the Energy Demand and Greenhouse Gas Emissions in a Climate-Constrained World.* Energies, 2019. **12**(20): p. 3870.

239. Fasihi, M., D. Bogdanov, and C. Breyer, *Long-Term Hydrocarbon Trade Options for the Maghreb Region and Europe—Renewable Energy Based Synthetic Fuels for a Net Zero Emissions World.* Sustainability, 2017. **9**(2): p. 306.

240. Rozzi, E., et al., *Green Synthetic Fuels: Renewable Routes for the Conversion of Non-Fossil Feedstocks Into Gaseous Fuels and Their End Uses.* Energies, 2020. **13**(2): p. 420.

241. Breyer, C., S. Khalili, and D. Bogdanov, *Solar Photovoltaic Capacity Demand for a Sustainable Transport Sector to Fulfil the Paris Agreement by 2050.* Progress in Photovoltaics Research and Applications, 2019. **27**(11): p. 978-989.

242. Etukudoh, E.A., et al., *A Review of Sustainable Transportation Solutions: Innovations, Challenges, and Future Directions.* World Journal of Advanced Research and Reviews, 2024. **21**(1): p. 1440-1452.

243. Kofler, R., et al., *Techno-Economic Analysis of Dimethyl Ether Production From Different Biomass Resources and Off-Grid Renewable Electricity.* Energy & Fuels, 2024. **38**(10): p. 8777-8803.

244. Wątróbski, J. and A. Bączkiewicz, *Towards Sustainable Transport Assessment Considering Alternative Fuels Based on McDa Methods.* 2022.

245. Agrawal, R. and N.R. Singh, *Solar Energy to Biofuels.* Annual Review of Chemical and Biomolecular Engineering, 2010. **1**(1): p. 343-364.

246. Lehtveer, M., S. Brynolf, and M. Grahn, *What Future for Electrofuels in Transport? Analysis of Cost Competitiveness in Global Climate Mitigation.* Environmental Science & Technology, 2019. **53**(3): p. 1690-1697.

247. Ali, S.A., et al., *Review on the Role of Electrofuels in Decarbonizing Hard-to-Abate Transportation Sectors: Advances, Challenges, and Future Directions.* Energy & Fuels, 2025. **39**(11): p. 5051-5098.

248. Sherwin, E.D., *Electrofuel Synthesis From Variable Renewable Electricity: An Optimization-Based Techno-Economic Analysis.* Environmental Science & Technology, 2021. **55**(11): p. 7583-7594.

249. Willauer, H.D., et al., *The Feasibility and Current Estimated Capital Costs of Producing Jet Fuel at Sea Using Carbon Dioxide and Hydrogen.* Journal of Renewable and Sustainable Energy, 2012. **4**(3).

250. Bardow, A., et al., *Towards Carbon-Neutral and Clean Combustion With Hydroformylated Fischer-Tropsch (HyFiT) Fuels.* 2023.

251. Bellotti, D., et al., *Techno-Economic Analysis for the Integration of a Power to Fuel System With a CCS Coal Power Plant.* Journal of Co2 Utilization, 2019. **33**: p. 262-272.

252. Marques, L., et al., *Review of Power-to-Liquid (PtL) Technology for Renewable Methanol (E-MeOH): Recent Developments, Emerging Trends and Prospects for the Cement Plant Industry.* Energies, 2024. **17**(22): p. 5589.

253. Araya, S.S., et al., *A Review of the Methanol Economy: The Fuel Cell Route.* Energies, 2020. **13**(3): p. 596.

254. Cabrera, E. and J.M.M. Sousa, *Use of Sustainable Fuels in Aviation—A Review.* Energies, 2022. **15**(7): p. 2440.

255. Kilian, T. and M. Horn, *Verification of a Chamberless HLFC Design With an Outer Skin of Variable Porosity.* Ceas Aeronautical Journal, 2021. **12**(4): p. 835-845.

256. Speizer, S., et al., *Integrated Assessment Modeling of a Zero-Emissions Global Transportation Sector.* Nature Communications, 2024. **15**(1).

257. Dray, L., et al., *Cost and Emissions Pathways Towards Net-Zero Climate Impacts in Aviation.* Nature Climate Change, 2022. **12**(10): p. 956-962.

258. Yfantis, E.A., et al., *Shipping Decarbonization: E-Fuels as a Potential Solution and the Role of Resources.* E3s Web of Conferences, 2023. **436**: p. 11009.

259. Kanchiralla, F.M., S. Brynolf, and A. Mjelde, *Role of Biofuels, Electro-Fuels, and Blue Fuels for Shipping: Environmental and Economic Life Cycle Considerations.* Energy & Environmental Science, 2024. **17**(17): p. 6393-6418.

260. Shi, K., et al., *Perspectives and Outlook of E-Fuels: Production, Cost Effectiveness, and Applications.* Energy & Fuels, 2024. **38**(9): p. 7665-7692.

261. Muratori, M., et al., *Role of the Freight Sector in Future Climate Change Mitigation Scenarios.* Environmental Science & Technology, 2017. **51**(6): p. 3526-3533.

262. Oke, D., J.B. Dunn, and T.R. Hawkins, *Reducing Economy-Wide Greenhouse Gas Emissions With Electrofuels and Biofuels as the Grid Decarbonizes.* Energy & Fuels, 2024. **38**(7): p. 6048-6061.

263. Wang, Q., et al., *The Use of Alternative Fuels for Maritime Decarbonization: Special Marine Environmental Risks and Solutions From an International Law Perspective.* Frontiers in Marine Science, 2023. **9**.

264. Bullock, S., C. Hoolohan, and A. Bows-Larkin, *Accelerating Shipping Decarbonisation: A Case Study on UK Shore Power.* Heliyon, 2023. **9**(7): p. e17475.

265. Garrouste, M., et al., *Techno-Economic Analysis of Synthetic Fuel Production From Existing Nuclear Power Plants Across the United States.* 2025. **2**(1): p. 015018.

266. Zang, G., et al., *Synthetic Methanol/Fischer–Tropsch Fuel Production Capacity, Cost, and Carbon Intensity Utilizing CO_2 From Industrial and Power Plants in the United States.* Environmental Science & Technology, 2021. **55**(11): p. 7595-7604.

267. Subedi, A. and B.S. Thapa, *Parametric Modeling of Re-Electrification by Green Hydrogen as an Alternative to Backup Power.* Iop Conference Series Earth and Environmental Science, 2022. **1037**(1): p. 012057.

268. Pivovar, B.S., N. Rustagi, and S. Satyapal, *Hydrogen at Scale (H_2 @Scale): Key to a Clean, Economic, and Sustainable Energy System.* The Electrochemical Society Interface, 2018. **27**(1): p. 47-52.

269. Grim, R.G., et al., *Transforming the Carbon Economy: Challenges and Opportunities in the Convergence of Low-Cost Electricity and Reductive CO_2 Utilization.* Energy & Environmental Science, 2020. **13**(2): p. 472-494.

270. Isaacs, S., et al., *Environmental and Economic Performance of Hybrid Power-to-Liquid and Biomass-to-Liquid Fuel Production in the United States.* Environmental Science & Technology, 2021. **55**(12): p. 8247-8257.

271. Ibraheem, A., et al., *Innovative Technological Scheme of Iraq Oils Refining.* Chemistry & Chemical Technology, 2014. **8**(2): p. 219-224.

272. Alzahawy, A. and H. Issa, *The Iraqi Petroleum Refining Industry: A Review of Environmental Impact Assessment.* 2024.

273. Zhakirova, N.K., et al., *Increasing the Yield of Light Distillates by Wave Action on Oil Raw Materials.* Eurasian Chemico-Technological Journal, 2021. **23**(2): p. 125-132.

274. Primo, A. and H. García, *Zeolites as Catalysts in Oil Refining.* Chemical Society Reviews, 2014. **43**(22): p. 7548-7561.

275. Omarova, A., et al., *Processing of Model N-Alcanes and Diesel Fraction on the Catalyst La-Zn-Mn / Al2O3 + ZSM.* International Journal of Biology and Chemistry, 2019. **12**(1): p. 153-158.

276. Wang, M., et al., *Structure-Oriented Lumping Method: An Effective Tool for Molecular Refining.* Industrial & Engineering Chemistry Research, 2023. **62**(33): p. 12845-12854.

277. Saadi, M.A.S.R., et al., *Sustainable Valorization of Asphaltenes via Flash Joule Heating.* Science Advances, 2022. **8**(46).

278. Polanco-Martínez, J.M., L.M. Abadie, and J. Fernández-Macho, *A Multi-Resolution and Multivariate Analysis of the Dynamic Relationships Between Crude Oil and Petroleum-Product Prices.* Applied Energy, 2018. **228**: p. 1550-1560.

279. Vetere, A., D. Pröfrock, and W. Schräder, *Quantitative and Qualitative Analysis of Three Classes of Sulfur Compounds in Crude Oil.* Angewandte Chemie, 2017. **56**(36): p. 10933-10937.

280. Wang, S., F. Tao, and Y. Shi, *Optimization of Inventory Routing Problem in Refined Oil Logistics With the Perspective of Carbon Tax.* Energies, 2018. **11**(6): p. 1437.

281. Dyk, S.v., et al., *Potential Synergies of Drop-in Biofuel Production With Further Co-processing at Oil Refineries.* Biofuels Bioproducts and Biorefining, 2019. **13**(3): p. 760-775.

282. Balke, N.S., M. Plante, and M.K. Yücel, *Fuel Subsidies, the Oil Market and the World Economy.* The Energy Journal, 2015. **36**(1_suppl): p. 99-128.

283. Zhao, F., et al., *Research Status and Analysis of Stabilization Mechanisms and Demulsification Methods of Heavy Oil Emulsions.* Energy Science & Engineering, 2020. **8**(12): p. 4158-4177.

284. Jia, H., R. Ådland, and Y. Wang, *Global Oil Export Destination Prediction: A Machine Learning Approach.* The Energy Journal, 2021. **42**(4): p. 111-130.

285. Rajabi, H., et al., *Emissions of Volatile Organic Compounds From Crude Oil Processing – Global Emission Inventory and Environmental Release.* The Science of the Total Environment, 2020. **727**: p. 138654.

286. Young, B., et al., *Expansion of the Petroleum Refinery Life Cycle Inventory Model to Support Characterization of a Full Suite of Commonly Tracked Impact Potentials.* Environmental Science & Technology, 2019. **53**(4): p. 2238-2248.

287. Zaharova, E.A. and N.A. Lihacheva, *Environmental Performance Evaluation of Oil Refineries.* Chemistry and Technology of Fuels and Oils, 2021. **57**(3): p. 482-486.

288. Polvara, E., et al., *Estimation of Emission Factors for Hazardous Air Pollutants From Petroleum Refineries.* Atmosphere, 2021. **12**(11): p. 1531.

289. Al-Rubaye, A.H., et al., *The Side Effect of Oil Refineries on Environment: As a Mini Review.* Iop Conference Series Earth and Environmental Science, 2023. **1262**(2): p. 022024.

290. Kponee, K.Z., et al., *Petroleum contaminated water and health symptoms: a cross-sectional pilot study in a rural Nigerian community.* Environmental Health, 2015. **14**: p. 1-8.

291. Alsaqer, S., et al., *Biosurfactant-Facilitated Leaching of Metals From Spent Hydrodesulphurization Catalyst.* Journal of Applied Microbiology, 2018. **125**(5): p. 1358-1369.

292. Schaefer, M.E., *Hazardous Waste Management.* 2016: p. 161-194.

293. Hidayah, M., D. Fitasari, and A. Ma'ruf. *Synthesis and Characterization of Polyacrilonitrile-Fe2o3 Membrane Nano Composites for Oil Refinery Liquid Waste Separation.* in *International Conference on Science and Engineering (ICSE-UIN-SUKA 2021).* 2021. Atlantis Press.

294. Ma, Q., et al., *Benefit Analysis of Organic Rankine Cycle Power Generation by Using Waste Heat Recovery in Refinery.* E3s Web of Conferences, 2022. **352**: p. 02014.

295. Yu, H., X. Feng, and Y. Wang, *Working Fluid Selection for Organic Rankine Cycle (ORC) Considering the Characteristics of Waste Heat Sources.* Industrial & Engineering Chemistry Research, 2016. **55**(5): p. 1309-1321.

296. Yunusov, O., et al., *Development of Technology for Refining of Black Cotton Oil Miscella.* E3s Web of Conferences, 2024. **497**: p. 03034.

297. Stanford University. *Fast Facts About Energy for Transportation*. 2025 [cited 2025 28/5/2025]; Available from: https://understand-energy.stanford.edu/energy-services/energy-transportation.

298. Ali, M.M. and S.M. Salih, *Factors Affecting Performance of Dual Fuel Compression Ignition Engines.* Applied Mechanics and Materials, 2013. **388**: p. 217-222.

299. Khan, R., Y. Kong, and J. Xinxin, *Influence of Environmental Degradation and Economic Growth on CO2 Emissions: Evidence From Developing Countries.* Asian Journal of Economics and Empirical Research, 2020. **7**(2): p. 193-206.

300. Cheah, W.Y., et al., *Pretreatment Methods for Lignocellulosic Biofuels Production: Current Advances, Challenges and Future Prospects.* Biofuel Research Journal, 2020. **7**(1): p. 1115-1127.

301. Grange, S.K., et al., *Post-Dieselgate: Evidence of NO$_x$ Emission Reductions Using on-Road Remote Sensing.* Environmental Science & Technology Letters, 2020. **7**(6): p. 382-387.

302. Tanaka, K., et al., *Climate Effects of Non-Compliant Volkswagen Diesel Cars.* Environmental Research Letters, 2018. **13**(4): p. 044020.

303. Holland, S.P., et al., *Damages and Expected Deaths Due to Excess NO$_x$ Emissions From 2009 to 2015 Volkswagen Diesel Vehicles.* Environmental Science & Technology, 2016. **50**(3): p. 1111-1117.

304. Mews, A., *The Impact of Scandalous News in the Automobile Manufacture on Companies From the Same Industry: A Comparative Study on the Chinese and European Markets.* European Scientific Journal Esj, 2021. **17**(23).

305. Topal, İ., et al., *The Effect of Greenwashing on Online Consumer Engagement: A Comparative Study in France, Germany, Turkey, and the United Kingdom.* Business Strategy and the Environment, 2019. **29**(2): p. 465-480.

306. Lim, J., *Volkswagen's Crisis Communication: Twitter Use During #Dieselgate.* 2021.

307. An, Q., et al., *Volkswagen's Diesel Emission Scandal: Analysis of Facebook Engagement and Financial Outcomes.* 2018: p. 260-276.

308. Hou, L., et al., *Public Health Impact and Economic Costs of Volkswagen's Lack of Compliance With the United States' Emission Standards.* International Journal of Environmental Research and Public Health, 2016. **13**(9): p. 891.

309. Barrett, S.R.H., et al., *Impact of the Volkswagen Emissions Control Defeat Device on US Public Health.* Environmental Research Letters, 2015. **10**(11): p. 114005.

310. Zhang, M., G. Atwal, and M. Kaiser, *Corporate Social Irresponsibility and Stakeholder Ecosystems: The Case of <scp>Volkswagen Dieselgate</Scp> Scandal.* Strategic Change, 2021. **30**(1): p. 79-85.

311. Law, A., R. Martinez-Botas, and P.T. Blythe, *Current Vehicle Emission Standards Will Not Mitigate Climate Change or Improve Air Quality.* Scientific Reports, 2023. **13**(1).

312. Strittmatter, A. and M. Lechner, *Sorting in the Used-Car Market After the Volkswagen Emission Scandal.* Journal of Environmental Economics and Management, 2020. **101**: p. 102305.

313. Ameen, K., *Failure of Ethical Compliance: The Case of Volkswagen.* International Journal of Science and Management Studies (Ijsms), 2020: p. 7-13.

314. Bachmann, R., et al., *Firms and Collective Reputation: A Study of the Volkswagen Emissions Scandal*. Journal of the European Economic Association, 2022. **21**(2): p. 484-525.

315. Boiral, O., et al., *Through the Smokescreen of the Dieselgate Disclosure: Neutralizing the Impacts of a Major Sustainability Scandal*. Organization & Environment, 2021. **35**(2): p. 175-201.

316. Painter, C. and J.T. Martins, *Organisational Communication Management During the Volkswagen Diesel Emissions Scandal: A Hermeneutic Study in Attribution, Crisis Management, and Information Orientation*. Knowledge and Process Management, 2017. **24**(3): p. 204-218.

317. Che, X., H. Katayama, and P. Lee, *Product-Harm Crises and Spillover Effects: A Case Study of the Volkswagen Diesel Emissions Scandal in eBay Used Car Auction Markets*. Journal of Marketing Research, 2022. **60**(2): p. 409-424.

318. Rahman, M.M., et al., *Energy Demand of the Road Transport Sector of Saudi Arabia—Application of a Causality-Based Machine Learning Model to Ensure Sustainable Environment*. Sustainability, 2022. **14**(23): p. 16064.

319. Skar, A., et al., *Internet-of-Things (IoT) Platform for Road Energy Efficiency Monitoring*. Sensors, 2023. **23**(5): p. 2756.

320. Gao, Y.X., et al., *Road Transport Energy Consumption in the G7 and BRICS: 1973-2010*. International Journal of Global Energy Issues, 2015. **38**(4/5/6): p. 342.

321. Nurjani, E., et al., *Carbon Emissions From the Transportation Sector During the Covid-19 Pandemic in the Special Region of Yogyakarta, Indonesia*. Iop Conference Series Earth and Environmental Science, 2021. **940**(1): p. 012039.

322. Chen, S.L., J. Wu, and Y. Zong, *The Impact of the Freight Transport Modal Shift Policy on China's Carbon Emissions Reduction*. Sustainability, 2020. **12**(2): p. 583.

323. Curran, S., et al., *In-Cylinder Fuel Blending of Gasoline/Diesel for Improved Efficiency and Lowest Possible Emissions on a Multi-Cylinder Light-Duty Diesel Engine*. 2010.

324. Timilsina, R.R., et al., *Are Households Shifting Toward Cleaner Cooking Fuel? Empirical Evidence From India During 2005–2021*. Frontiers in Environmental Economics, 2023. **2**.

325. Zhang, D., et al., *Review and Outlook of Global Energy Use Under the Impact of <scp>COVID</Scp>-19*. Engineering Reports, 2022. **5**(3).

326. Şahin, Z. and O. Durgun, *Improving of Diesel Combustion-Pollution-Fuel Economy and Performance by Gasoline Fumigation*. Energy Conversion and Management, 2013. **76**: p. 620-633.

327. Durdağ, C. and E. Şahin, *The Effect of Energy Policies in Turkey on Transportation Sector: The Analysis of Energy Related Price and Cost in Road Transportation*. Marmara Fen Bilimleri Dergisi, 2016. **28**(5).

328. Park, S.H., et al., *Influence of the Mixture of Gasoline and Diesel Fuels on Droplet Atomization, Combustion, and Exhaust Emission Characteristics in a Compression Ignition Engine*. Fuel Processing Technology, 2013. **106**: p. 392-401.

329. Ghosh, C. and R.H. Chaudhury, *A Comparative Study of Saving Behaviour Between India and China*. Millennial Asia, 2022. **14**(4): p. 461-479.

330. Rivera-González, L., et al., *Long-Term Forecast of Energy and Fuels Demand Towards a Sustainable Road Transport Sector in Ecuador (2016–2035): A LEAP Model Application*. Sustainability, 2020. **12**(2): p. 472.

331. Rehman, S., et al., *A Multicriteria Decision-Making Approach in Exploring the Nexus Between Wind and Solar Energy Generation, Economic Development, Fossil Fuel Consumption, and CO2 Emissions.* Frontiers in Environmental Science, 2022. **9**.

332. Zou, Z., et al., *Dynamic Nexus Between Non-Renewable Energy Consumption, Economic Growth and CO_2 Emission: A Comparison Analysis Between Major Emitters.* Energy & Environment, 2023. **35**(8): p. 4339-4360.

333. Ershov, M.A., et al., *An Overview of the Global Market, Fleet, and Components in the Field of Aviation Gasoline.* Aerospace, 2023. **10**(10): p. 863.

334. Corporan, E., et al., *Chemical, Thermal Stability, Seal Swell, and Emissions Studies of Alternative Jet Fuels.* Energy & Fuels, 2011. **25**(3): p. 955-966.

335. Mickevičius, T., S. Slavinskas, and G. Labeckas, *Impact of Aviation Fuel on Durability of Diesel Common Rail Injection System.* 2019.

336. Dhamo, D., et al., *Coupling the High-Temperature Fischer–Tropsch Synthesis and the Skeletal Isomerization Reaction at Optimal Operation Conditions in the Power-to-Fuels Process Route for the Production of Sustainable Aviation Gasoline.* Sustainable Energy & Fuels, 2024. **8**(9): p. 2094-2103.

337. Kantaroğlu, E. and A. Doğan, *Experimental Investigation of the Effects of JP8 and Amorphous Elemental Boron Additives on Combustion Characteristics for I-Dsi Engine.* International Journal of Engine Research, 2025. **26**(6): p. 885-902.

338. Labeckas, G. and S. Slavinskas, *Combustion Phenomenon, Performance and Emissions of a Diesel Engine With Aviation Turbine JP-8 Fuel and Rapeseed Biodiesel Blends.* Energy Conversion and Management, 2015. **105**: p. 216-229.

339. Doğan, İ., M.S. GÖKmen, and H. Aydoğan, *Effect of Gasoline-Avgas Blends on Engine Performance of Engine With Direct Injection.* Bioenergy Studies Black Sea Agricultural Research Institute, 2021. **1**(1): p. 1-6.

340. Jiménez-Díaz, L., et al., *Microbial Alkane Production for Jet Fuel Industry: Motivation, State of the Art and Perspectives.* Microbial Biotechnology, 2016. **10**(1): p. 103-124.

341. Hisham, S., et al., *Catalytic Hydrocracking of Jatropha Oil for Bio-Jetfuel Production Using Natural Clay.* 2023.

342. Sharma, V., et al., *Microalgal Biodiesel: A Challenging Route Toward a Sustainable Aviation Fuel.* Fermentation, 2023. **9**(10): p. 907.

343. Mallette, N., et al., *Evaluation of Cellulose as a Substrate for Hydrocarbon Fuel Production by &Amp;Lt;i>Ascocoryne Sarcoides</I> (NRRL 50072).* Journal of Sustainable Bioenergy Systems, 2014. **04**(01): p. 33-49.

344. Wang, E., et al., *Knock Suppression of a Spark-Ignition Aviation Piston Engine Fuelled With Kerosene.* 2020.

345. Windom, B., T.M. Lovestead, and T.J. Bruno, *Application of the Advanced Distillation Curve Method to the Development of Unleaded Aviation Gasoline.* Energy & Fuels, 2010. **24**(5): p. 3275-3284.

346. Vozka, P., P. Šimáček, and G. Kılaz, *Impact of HEFA Feedstocks on Fuel Composition and Properties in Blends With Jet A.* Energy & Fuels, 2018. **32**(11): p. 11595-11606.

347. Franz, S.M., M. Rottoli, and C. Bertram, *The Wide Range of Possible Aviation Demand Futures After the COVID-19 Pandemic.* Environmental Research Letters, 2022. **17**(6): p. 064009.

348. Gößling, S., *Risks, Resilience, and Pathways to Sustainable Aviation: A COVID-19 Perspective.* Journal of Air Transport Management, 2020. **89**: p. 101933.

349. Adekitan, A.I., *Sustainable Supply of Aviation Fuel in Nigeria: The Status Quo and the Challenges.* International Journal of Advances in Applied Sciences, 2022. **11**(1): p. 38.

350. Balogun, S.A., et al., *Renewable Aviation Fuel: Review of Bio-Jet Fuel for Aviation Industry.* Engineering Science Letter, 2022. **1**(01): p. 7-11.

351. Singh, J., et al., *Who Should Hold the Baton of Aviation Sustainability?* Social Responsibility Journal, 2022. **19**(7): p. 1161-1177.

352. Nasrullah, M.N.C.H., et al., *Advancements and Challenges in the Search for Lead-Free Aviation Fuel: A Review.* 2024. **1**(1): p. 123-128.

353. Singh, J., S.K. Sharma, and R. Srivastava, *Managing Fuel Efficiency in the Aviation Sector: Challenges, Accomplishments and Opportunities.* Fiib Business Review, 2018. **7**(4): p. 244-251.

354. Kieckhäfer, K., et al., *Simulation-Based Analysis of the Potential of Alternative Fuels Towards Reducing CO2 Emissions From Aviation.* Energies, 2018. **11**(1): p. 186.

355. Sulung, S.D., et al., *Impact of the Fuel Mixture Ratio of AVGAS 100LL and RON 92 Fuel on Combustion Characteristics.* Journal of Science Technology (Jostec), 2023. **5**(1): p. 07-13.

356. Diniz, A.P.M.M., R. Sargeant, and G.J. Millar, *Stochastic Techno-Economic Analysis of the Production of Aviation Biofuel From Oilseeds.* Biotechnology for Biofuels, 2018. **11**(1).

357. Hermann, R.R. and K. Wigger, *Eco-Innovation Drivers in Value-Creating Networks: A Case Study of Ship Retrofitting Services.* Sustainability, 2017. **9**(5): p. 733.

358. Li, S., et al., *Techno-Economic Analysis of Sustainable Biofuels for Marine Transportation.* Environmental Science & Technology, 2022. **56**(23): p. 17206-17214.

359. Tan, E.C.D., et al., *Adoption of Biofuels for Marine Shipping Decarbonization: A <scp>long-term</Scp> Price and Scalability Assessment.* Biofuels Bioproducts and Biorefining, 2022. **16**(4): p. 942-961.

360. Chountalas, T., M.A. Founti, and D.T. Hountalas, *Review of Biofuel Effect on Emissions of Various Types of Marine Propulsion and Auxiliary Engines.* Energies, 2023. **16**(12): p. 4647.

361. Moshiul, A.M., R. Mohammad, and F.A. Hira, *Alternative Fuel Selection Framework Toward Decarbonizing Maritime Deep-Sea Shipping.* Sustainability, 2023. **15**(6): p. 5571.

362. Gerlitz, L., E. Mildenstrey, and G. Prause, *Ammonia as Clean Shipping Fuel for the Baltic Sea Region.* Transport and Telecommunication Journal, 2022. **23**(1): p. 102-112.

363. Mollaoğlu, M., et al., *Evaluation of Various Fuel Alternatives in Terms of Sustainability for the Ship Investment Decision Using Single Valued Neutrosophic Numbers With TOPSIS Methods.* Proceedings of the Institution of Mechanical Engineers Part M Journal of Engineering for the Maritime Environment, 2022. **237**(1): p. 215-226.

364. Nasser, M.A., et al., *Environmental and Economic Performance Investigation of Natural Gas and Methanol as a Marine Alternative Fuel.* 2022.

365. Tan, E.C.D., et al., *Biofuel Options for Marine Applications: Technoeconomic and Life-Cycle Analyses.* Environmental Science & Technology, 2021. **55**(11): p. 7561-7570.

366. Curran, S., et al., *The Future of Ship Engines: Renewable Fuels and Enabling Technologies for Decarbonization.* International Journal of Engine Research, 2023. **25**(1): p. 85-110.

367. Reusser, C.A. and J.R.P. Osses, *Challenges for Zero-Emissions Ship.* Journal of Marine Science and Engineering, 2021. **9**(10): p. 1042.

368. Gallucci, M., *The Ammonia Solution: Ammonia Engines and Fuel Cells in Cargo Ships Could Slash Their Carbon Emissions*. Ieee Spectrum, 2021. **58**(3): p. 44-50.

369. Elkafas, A.G., *Advanced Operational Measure for Reducing Fuel Consumption Onboard Ships*. Environmental Science and Pollution Research, 2022. **29**(60): p. 90509-90519.

370. Shim, H., et al., *Marine Demonstration of Alternative Fuels on the Basis of Propulsion Load Sharing for Sustainable Ship Design*. Journal of Marine Science and Engineering, 2023. **11**(3): p. 567.

371. Awoyomi, A., K. Patchigolla, and E.J. Anthony, *Process and Economic Evaluation of an Onboard Capture System for LNG-Fueled CO$_2$ Carriers*. Industrial & Engineering Chemistry Research, 2019. **59**(15): p. 6951-6960.

372. Ni, P., X. Wang, and H. Li, *A Review on Regulations, Current Status, Effects and Reduction Strategies of Emissions for Marine Diesel Engines*. Fuel, 2020. **279**: p. 118477.

373. Nielsen, J.B., K.K. Yum, and E. Pedersen, *Improving Pre-Turbine Selective Catalytic Reduction Systems in Marine Two-Stroke Diesel Engines Using Hybrid Turbocharging: A Numerical Study of Selective Catalytic Reduction Operation Range and System Fuel Efficiency*. Proceedings of the Institution of Mechanical Engineers Part M Journal of Engineering for the Maritime Environment, 2019. **234**(2): p. 463-474.

374. Elkafas, A.G., et al., *Fuel Cell Systems for Maritime: A Review of Research Development, Commercial Products, Applications, and Perspectives*. Processes, 2022. **11**(1): p. 97.

375. Kim, B. and K.-I. Hwang, *Numerical Analysis of the Effects of Ship Motion on Hydrogen Release and Dispersion in an Enclosed Area*. Applied Sciences, 2022. **12**(3): p. 1259.

376. Özbey, S. and İ. Tıkız, *Renewable Energy Solutions for Commercial Ships*. 2024.

377. Kim, M.-K., et al., *A Study on Characteristic Emission Factors of Exhaust Gas From Diesel Locomotives*. International Journal of Environmental Research and Public Health, 2020. **17**(11): p. 3788.

378. Malozyomov, B.V., et al., *Study of Supercapacitors Built in the Start-Up System of the Main Diesel Locomotive*. Energies, 2023. **16**(9): p. 3909.

379. Merchan, A.L., S. Belboom, and A. Léonard, *Life Cycle Assessment of Rail Freight Transport in Belgium*. Clean Technologies and Environmental Policy, 2020. **22**(5): p. 1109-1131.

380. Dolinayová, A., V. Zitrický, and L. Černá, *Decision-Making Process in the Case of Insufficient Rail Capacity*. Sustainability, 2020. **12**(12): p. 5023.

381. Folęga, P. and M. Irlik, *Effect of Train Position Reporting on Railway Line Capacity*. Transport Problems, 2021. **16**(2): p. 59-72.

382. Soofastaei, A., *Energy-Efficiency Improvement in Mine-Railway Operation Using AI*. Journal of Energy and Power Engineering, 2019. **13**(9).

383. Barbosa, F.C., *Battery Only Electric Traction for Freight Trains - A Technical and Operational Assessment*. Proceedings of the Institution of Mechanical Engineers Part F Journal of Rail and Rapid Transit, 2023. **238**(3): p. 322-337.

384. Goryunov, A., et al., *Numerical Simulation of Aerodynamic Processes in the Fan in the Diesel Locomotive Cooling Module*. E3s Web of Conferences, 2024. **471**: p. 02018.

385. Feng, X., et al., *A Review Study on Traction Energy Saving of Rail Transport*. Discrete Dynamics in Nature and Society, 2013. **2013**: p. 1-9.

386. Fullerton, G.A., G.C. DiDomenico, and C.T. Dick, *Sensitivity of Freight and Passenger Rail Fuel Efficiency to Infrastructure, Equipment, and Operating Factors*. Transportation Research Record Journal of the Transportation Research Board, 2015. **2476**(1): p. 59-66.

387. Liu, L., et al., *Emission Projections for Long-Haul Freight Trucks and Rail in the United States Through 2050.* Environmental Science & Technology, 2015. **49**(19): p. 11569-11576.

388. Andrzejewski, M., et al., *Impact of a Locomotive Engine Modernization on Fuel Consumption.* Matec Web of Conferences, 2021. **338**: p. 01001.

389. Mitrofanov, A.N., et al., *Indirect Fuel Rationing for a Special Self-Propelled Rolling Stock.* Energies, 2022. **15**(3): p. 836.

390. Sebeşan, I. and T. Bogdan, *The Impact of Aerodynamics on Fuel Consumption in Railway Applications.* Incas Bulletin, 2012. **4**(1): p. 93-102.

391. Dordal, O.B., et al., *Strong Reduction in Fuel Consumption Driving Trains in Bi-Directional Single Line Using Crossing Loops.* 2011: p. 1597-1602.

392. Fullerton, G., et al., *Congestion as a Source of Variation in Passenger and Freight Railway Fuel Efficiency.* 2014.

393. Su, S., T. Tang, and Y. Wang, *Evaluation of Strategies to Reducing Traction Energy Consumption of Metro Systems Using an Optimal Train Control Simulation Model.* Energies, 2016. **9**(2): p. 105.

394. Elkhoury, N., et al., *Degradation Prediction of Rail Tracks: A Review of the Existing Literature.* The Open Transportation Journal, 2018. **12**(1): p. 88-104.

395. Mahamudin, N.S. and F.J. Darsivan, *The Effect of Coefficient of Friction Between Rail Vehicle Wheels and Rail Track on Operation Power Consumption.* 2023: p. 1-6.

396. Maheras, S., E. Lahti, and S. Ross, *Transportation Shock and Vibration Literature Review.* 2013.

397. Pineda-Jaramillo, J., R. Insa, and P.M. Fernández, *Modeling the Energy Consumption of Trains by Applying Neural Networks.* Proceedings of the Institution of Mechanical Engineers Part F Journal of Rail and Rapid Transit, 2017. **232**(3): p. 816-823.

398. Brenna, M., et al., *A Review on Energy Efficiency in Three Transportation Sectors: Railways, Electrical Vehicles and Marine.* Energies, 2020. **13**(9): p. 2378.

399. Gołębiowski, P., M. Jacyna, and A. Stańczak, *The Assessment of Energy Efficiency Versus Planning of Rail Freight Traffic: A Case Study on the Example of Poland.* Energies, 2021. **14**(18): p. 5629.

400. Saputra, M.N., et al., *Modeling Fuel Consumption in Various External Vehicle Conditions for Military Vehicle Using Mixed Linear Models.* 2023. **1**(2): p. 67-74.

401. Department of Climate Change, E., the Environment and Water,. *Rail Transport.* 2024 [cited 2025 28/5/2025]; Available from: https://www.energy.gov.au/business/sector-guides/transport/rail-transport.

402. Sotnyk, I.M., et al., *Development of the US Electric Car Market: Macroeconomic Determinants and Forecasts.* Polityka Energetyczna – Energy Policy Journal, 2020. **23**(3): p. 147-164.

403. Niu, D.X., et al., *Research on Evaluation System of China's Energy Security Considering Energy Transportation.* Advanced Materials Research, 2012. **535-537**: p. 2049-2052.

404. Lyulyov, O., et al., *Comprehensive Assessment of Smart Grids: Is There a Universal Approach?* Energies, 2021. **14**(12): p. 3497.

405. Боркова, Е.А., et al., *Economic Security in the Context of the Transport Economics Digitalization.* Iop Conference Series Materials Science and Engineering, 2020. **918**(1): p. 012182.

406. Jeong, E.-K., et al., *Carbon and Nitrogen Isotope Characterization of Imported Coals in South Korea.* Frontiers in Environmental Science, 2023. **11**.

407. Debiagi, P., et al., *Iron as a Sustainable Chemical Carrier of Renewable Energy: Analysis of Opportunities and Challenges for Retrofitting Coal-Fired Power Plants.* 2022.

408. Kartal, M.T., et al., *Asymmetric Nexus of Coal Consumption With Environmental Quality and Economic Growth: Evidence From BRICS, E7, and Fragile Five Countries by Novel Quantile Approaches.* Energy & Environment, 2023. **35**(4): p. 2228-2247.

409. Sun, W., et al., *Spatial and Temporal Characterization of Mine Water Inrush Accidents in China, 2014–2022.* Water, 2024. **16**(5): p. 656.

410. Vinichenko, V., et al., *Phasing Out Coal for 2 °C Target Requires Worldwide Replication of Most Ambitious National Plans Despite Security and Fairness Concerns.* Environmental Research Letters, 2023. **18**(1): p. 014031.

411. Nacke, L., et al., *Compensating Affected Parties Necessary for Rapid Coal Phase-Out but Expensive if Extended to Major Emitters.* Nature Communications, 2024. **15**(1).

412. Yang, D., et al., *Understanding Institutional and Policy Challenges in China's Coal Transition: Insights From the Case of Chongqing.* 2023. **4**(1).

413. Guo-dang, Z., et al., *Using a Novel Fractal-Time-Series Prediction Model to Predict Coal Consumption.* Discrete Dynamics in Nature and Society, 2023. **2023**: p. 1-11.

414. Rah, S.M.T., A.M. Hashan, and S.M.M. Rah, *Numerical Analysis of the Equilibrium Composition and Structural Features of Coal Combustion Products.* 2024.

415. Deetjen, T.A. and I.L. Azevedo, *Climate and Health Benefits of Rapid Coal-to-Gas Fuel Switching in the U.S. Power Sector Offset Methane Leakage and Production Cost Increases.* Environmental Science & Technology, 2020. **54**(18): p. 11494-11505.

416. Goforth, T. and D. Nock, *Air Pollution Disparities and Equality Assessments of US National Decarbonization Strategies.* Nature Communications, 2022. **13**(1).

417. Peng, W. and Y. Ou, *Integrating Air Quality and Health Considerations Into Power Sector Decarbonization Strategies.* Environmental Research Letters, 2022. **17**(8): p. 081002.

418. Vega-Araújo, J., et al., *Navigating a Just Energy Transition From Coal in the Colombian Caribbean.* 2023.

419. Pai, S., et al., *Solar Has Greater Techno-Economic Resource Suitability Than Wind for Replacing Coal Mining Jobs.* Environmental Research Letters, 2020. **15**(3): p. 034065.

420. Nie, B., et al., *Coal Dilemma: Is in Situ Clean Power Plant a Way Out?* 2024. **2**(2): p. 220-228.

421. Wang, D., X. Xue, and Y. Wang, *Overcapacity Risk of China's Coal Power Industry: A Comprehensive Assessment and Driving Factors.* Sustainability, 2021. **13**(3): p. 1426.

422. Campos, C.M.M., et al., *Designing Retirement Strategies for Coal-Fired Power Plants to Mitigate Air Pollution and Health Impacts.* 2024.

423. Mayfield, E., *Phasing Out Coal Power Plants Based on Cumulative Air Pollution Impact and Equity Objectives in Net Zero Energy System Transitions.* Environmental Research Infrastructure and Sustainability, 2022. **2**(2): p. 021004.

424. Ladage, S., et al., *On the Climate Benefit of a Coal-to-Gas Shift in Germany's Electric Power Sector.* Scientific Reports, 2021. **11**(1).

425. Burns, D. and E. Grubert, *Contribution of Regionalized Methane Emissions to Greenhouse Gas Intensity of Natural Gas-Fired Electricity and Carbon Capture in the United States.* Environmental Science & Technology Letters, 2021. **8**(9): p. 811-817.

426. Kemfert, C., et al., *The Expansion of Natural Gas Infrastructure Puts Energy Transitions at Risk.* Nature Energy, 2022. **7**(7): p. 582-587.

427. Gordon, D., et al., *Evaluating Net Life-Cycle Greenhouse Gas Emissions Intensities From Gas and Coal at Varying Methane Leakage Rates*. Environmental Research Letters, 2023. **18**(8): p. 084008.

428. Dvinin, D., et al., *Integral Assessment of Low-Carbon Energy Sources Effectiveness Based on Multi-Criteria Analysis*. E3s Web of Conferences, 2024. **498**: p. 01002.

429. Burns, D. and E. Grubert, *Attribution of Production-Stage Methane Emissions to Assess Spatial Variability in the Climate Intensity of US Natural Gas Consumption*. Environmental Research Letters, 2021. **16**(4): p. 044059.

430. Avellino, N. and J.J. Bonello, *From Heavy Fuel Oil to Liquified Natural Gas: Electricity Generation Transition in Malta*. 2021.

431. Zhang, X., et al., *Life Cycle Assessment of Power-to-Gas With Biogas as the Carbon Source*. Sustainable Energy & Fuels, 2020. **4**(3): p. 1427-1436.

432. Wang, G., et al., *Key Problems of Gas-fired Power Plants Participating in Peak Load Regulation: A Review*. Iet Cyber-Physical Systems Theory & Applications, 2022. **8**(4): p. 222-235.

433. Bauer, C., et al., *On the Climate Impacts of Blue Hydrogen Production*. 2021.

434. Xin-gang, Z., et al., *Impacts of Government Policies on the Adoption of Biomass Power: A System Dynamic Perspective*. Sustainability, 2023. **15**(2): p. 1723.

435. Watanabe, T., *The Development of Straw-Based Biomass Power Generation in Rural Area in Northeast China—An Institutional Analysis Grounded in a Risk Management Perspective*. Sustainability, 2020. **12**(5): p. 1973.

436. Menéndez, J. and J. Loredo, *An Economic Assessment of Lignocellulosic Biomass Power Plants*. E3s Web of Conferences, 2020. **191**: p. 02003.

437. Ilari, A., et al., *Carbon Footprint and Feedstock Quality of a Real Biomass Power Plant Fed With Forestry and Agricultural Residues*. Resources, 2022. **11**(2): p. 7.

438. Zheng, Y., et al., *Carbon Footprint Analysis for Biomass-Fueled Combined Heat and Power Station: A Case Study*. Agriculture, 2022. **12**(8): p. 1146.

439. Sutrisno, Z.P., A.A. Meiritza, and A. Raksajati, *Understanding the Potential of Bio-Carbon Capture and Storage From Biomass Power Plant in Indonesia*. Indonesian Journal of Energy, 2021. **4**(1): p. 36-56.

440. Chen, H., et al., *Main Problems Faced by Biomass Coupling Power Generation Metering*. Highlights in Science Engineering and Technology, 2023. **30**: p. 138-142.

441. Eswaran, U., *Integrating Renewable Energy Sources Into Existing Energy Systems for Achieving a Low-Carbon Energy Transition*. 2024: p. 262-277.

442. Pasupuleti, M.K., *Advancement of Renewable Energy Integration Into National Grids*. 2025. **05**(04): p. 259-266.

443. Lukashevych, Y.P., et al., *Innovation in the Energy Sector: The Transition to Renewable Sources as a Strategic Step Towards Sustainable Development*. African Journal of Applied Research, 2024. **10**(1): p. 43-56.

444. Barrie, I., et al., *Leveraging Machine Learning to Optimize Renewable Energy Integration in Developing Economies*. Global Journal of Engineering and Technology Advances, 2024. **20**(3): p. 080-093.

445. Meegahapola, L., et al., *Role of Fault Ride-through Strategies for Power Grids With 100% Power Electronic-interfaced Distributed Renewable Energy Resources*. Wiley Interdisciplinary Reviews Energy and Environment, 2018. **7**(4).

446. Prianka, Y.M., A. Sharma, and C. Biswas, *Integration of Renewable Energy, Microgrids, and EV Charging Infrastructure: Challenges and Solutions*. 2024. **2**(3): p. 317-326.

447. Roesdi, H., R. Lestari, and R. Amini, *Konflik Sosial Dan Lingkungan Di Sektor Energi Terbarukan: Tinjauan Pada Skala Global*. 2024. **1**(1).

448. Ohiri, P.C., et al., *Impact of Renewable Energy Integration on Grid Stability: A Case Study of Nigeria's Power System*. World Journal of Advanced Engineering Technology and Sciences, 2025. **14**(2): p. 199-204.

449. Varanasi, S.S., *Global Energy Consumption: Challenges in Transitioning to Renewable Sources*. 2025.

450. Akpan, J., H. Kumba, and O.A. Olanrewaju, *Sustainable Energy in Zimbabwe - Status, Challenges and Solutions*. 2024. **2**(2).

451. Brown, M.A., S. Zhou, and M. Ahmadi, *Smart Grid Governance: An International Review of Evolving Policy Issues and Innovations*. Wiley Interdisciplinary Reviews Energy and Environment, 2018. **7**(5).

452. Chhaya, L., *IoT-Based Management of Smart Microgrid*. 2019: p. 1-13.

453. Wang, Y., C. Xu, and Y. Ping-hong, *Is There a Grid-Connected Effect of Grid Infrastructure on Renewable Energy Generation? Evidence From China's Upgrading Transmission Lines*. Energy & Environment, 2021. **33**(5): p. 975-995.

454. Ritchie, H. and P. Rosado, *Energy Mix*. 2024.

455. Achuo, K.E., M. Seraj, and H. Özdeşer, *Modelling the Determinants of Coal Consumption in China*. Regional Economic Development Research, 2024: p. 1-19.

456. Qi, Y., et al., *China's Post-Coal Growth*. Nature Geoscience, 2016. **9**(8): p. 564-566.

457. Liu, F., et al., *Is Renewable Energy a Path Towards Sustainable Development?* Sustainable Development, 2023. **31**(5): p. 3869-3880.

458. Sofian, A.D.A.B.A., et al., *Machine Learning and the Renewable Energy Revolution: Exploring Solar and Wind Energy Solutions for a Sustainable Future Including Innovations in Energy Storage*. Sustainable Development, 2024. **32**(4): p. 3953-3978.

459. Shahzad, Q. and K. Aruga, *Does the Environmental Kuznets Curve Hold for Coal Consumption? Evidence From South and East Asian Countries*. Sustainability, 2023. **15**(6): p. 5532.

460. Striełkowski, W., E.V. Lutsenko, and D. Pavlov, *Fossil Fuel Industry Development in the 21st Century: A Case of Coal*. SHS Web of Conferences, 2021. **128**: p. 02004.

461. Chen, G.C. and C. Lees, *Growing China's Renewables Sector: A Developmental State Approach*. New Political Economy, 2016. **21**(6): p. 574-586.

462. Steckel, J.C., O. Edenhofer, and M. Jakob, *Drivers for the Renaissance of Coal*. Proceedings of the National Academy of Sciences, 2015. **112**(29).

463. Mastoi, M.S., et al., *A Comprehensive Analysis of the Power Demand–Supply Situation, Electricity Usage Patterns, and the Recent Development of Renewable Energy in China*. Sustainability, 2022. **14**(6): p. 3391.

464. Xu, B., et al., *How Does Coal Consumption Constraint Policy Affect Electrical Energy Efficiency? Evidence From 30 Chinese Provinces*. Energy Efficiency, 2022. **15**(4).

465. Zhao, W., et al., *Will Coal Price Fluctuations Affect Renewable Energy Substitution and Carbon Emission? A Computable General Equilibrium-Based Study of China*. Energy Engineering, 2021. **118**(4): p. 1009-1026.

466. Destek, M.A. and A. Sinha, *Renewable, Non-Renewable Energy Consumption, Economic Growth, Trade Openness and Ecological Footprint: Evidence From Organisation for Economic Co-Operation and Development Countries*. Journal of Cleaner Production, 2020. **242**: p. 118537.

467. Yinnan, H., R. Qin, and B. Wang, *On the Club Convergence in China's Provincial Coal Consumptions: Evidence From a Nonlinear Time-Varying Factor Model.* Sustainability, 2023. **15**(3): p. 1881.

468. Wang, Y. and Q. Yan, *Energy Consumption and Saving of China's Electric Power Industry.* Destech Transactions on Computer Science and Engineering, 2018(mmsta).

469. Akbal, G.F., *Sustainable Energy and China: The Importance of Institutional Authority.* China Report, 2024. **60**(2): p. 148-166.

470. She, K.Y. and W.J. Chen, *Spatial Pattern Evolution of Coal Consumption at Provincial Level in China.* Advanced Materials Research, 2013. **734-737**: p. 1752-1756.

471. Luo, Z., H. Yu, and L. Jin, *Effects of a Global Monopole on the Thermodynamic Phase Transition of a Charged AdS Black Hole*.* Chinese Physics C, 2022. **46**(12): p. 125101.

472. Hao, Y., et al., *China's Farewell to Coal: A Forecast of Coal Consumption Through 2020.* Energy Policy, 2015. **86**: p. 444-455.

473. Li, R., K. Zhang, and J. Wang, *Thermal Dynamic Phase Transition of Reissner-Nordström Anti-De Sitter Black Holes on Free Energy Landscape.* Journal of High Energy Physics, 2020. **2020**(10).

474. Demirbaş, A., *Biomass Co-Firing for Coal-Fired Boilers.* Energy Exploration & Exploitation, 2003. **21**(3): p. 269-278.

475. Baker, L., P. Newell, and J. Phillips, *The Political Economy of Energy Transitions: The Case of South Africa.* New Political Economy, 2014. **19**(6): p. 791-818.

476. Liu, J., et al., *Life Cycle Assessment of the Coal-to-Synthetic Natural Gas/Methanol Coproduction Process in Series.* Acs Sustainable Chemistry & Engineering, 2024.

477. Coenen, L., P. Benneworth, and B. Truffer, *Toward a Spatial Perspective on Sustainability Transitions.* Research Policy, 2012. **41**(6): p. 968-979.

478. Theiventhran, G.M., *Emerging Frontiers of Energy Transition in Sri Lanka.* 2021: p. 185-210.

479. Broto, V.C., *Innovation Territories and Energy Transitions: Energy, Water and Modernity in Spain, 1939–1975.* Journal of Environmental Policy & Planning, 2015. **18**(5): p. 712-729.

480. Hurlbert, M., et al., *Diverse Community Energy Futures in Saskatchewan, Canada.* Clean Technologies and Environmental Policy, 2020. **22**(5): p. 1157-1172.

481. Paul, F.C., *Deep Entanglements: History, Space and (Energy) Struggle in the German Energiewende.* Geoforum, 2018. **91**: p. 1-9.

482. Geels, F.W., *Ontologies, Socio-Technical Transitions (To Sustainability), and the Multi-Level Perspective.* Research Policy, 2010. **39**(4): p. 495-510.

483. Biserčić, A.Z. and U. Bugarić, *Reliability of Baseload Electricity Generation From Fossil and Renewable Energy Sources.* Energy and Power Engineering, 2021. **13**(05): p. 190-206.

484. Sekar, L.K. and E.R. Okoroafor, *Maximizing Geothermal Energy Recovery From Enhanced Geothermal Systems Through Huff-and-Puff: A Comprehensive Simulation Study.* 2024.

485. Ball, P., *A Review of Geothermal Technologies and Their Role in Reducing Greenhouse Gas Emissions in the USA.* Journal of Energy Resources Technology, 2020. **143**(1).

486. Yoshidome, D., et al., *Actual Measurement and Evaluation of the Balance Between Electricity Supply and Demand in Waste-Treatment Facilities and Development of Adjustment Methods.* Applied Sciences, 2021. **11**(22): p. 10747.

487. Clegg, J. and S. Krase, *New-Generation Geothermal Wells – How the Upstream Oil and Gas Industry Can Bring Its Experience to Bear.* 2022.

488. Matek, B. and K. Gawell, *The Benefits of Baseload Renewables: A Misunderstood Energy Technology.* The Electricity Journal, 2015. **28**(2): p. 101-112.

489. Bailey, J.R., et al., *A Grand Challenge Update: Geothermal Energy.* Journal of Petroleum Technology, 2024. **76**(04): p. 52-57.

490. Nardini, I., *Geothermal Power Generation.* 2022: p. 183-194.

491. Vargas, C.A., L. Caracciolo, and P. Ball, *Geothermal Energy as a Means to Decarbonize the Energy Mix of Megacities.* Communications Earth & Environment, 2022. **3**(1).

492. Wilson, A., *Oil's Wisdom Heats Up Drive for Geothermal Energy.* Journal of Petroleum Technology, 2020. **72**(09): p. 41-42.

493. Ayzit, T., et al., *A Sustainable Clean Energy Source for Mitigating CO2 Emissions: Numerical Simulation of Hamit Granitoid, Central Anatolian Massif.* Geomechanics and Geophysics for Geo-Energy and Geo-Resources, 2024. **10**(1).

494. Kaven, J.O., D.C. Templeton, and A.P. Bathija, *Introduction to This Special Section: Geothermal Energy.* The Leading Edge, 2020. **39**(12): p. 855-856.

495. Jabir, H.J., et al., *Impacts of Demand-Side Management on Electrical Power Systems: A Review.* Energies, 2018. **11**(5): p. 1050.

496. Vardakas, J.S., N. Zorba, and C. Verikoukis, *A Survey on Demand Response Programs in Smart Grids: Pricing Methods and Optimization Algorithms.* Ieee Communications Surveys & Tutorials, 2015. **17**(1): p. 152-178.

497. Nikoukar, J., *Unit Commitment Considering the Emergency Demand Response Programs and Interruptible/Curtailable Loads.* Turkish Journal of Electrical Engineering & Computer Sciences, 2018. **26**(2): p. 1069-1080.

498. Sosnovsky, I.K., A.H. Козлов, and F. Liu, *Study of Operation Modes of Hybrid Microgrid With Using Solar Energy.* E3s Web of Conferences, 2023. **470**: p. 01003.

499. Pálenský, P. and D. Dietrich, *Demand Side Management: Demand Response, Intelligent Energy Systems, and Smart Loads.* Ieee Transactions on Industrial Informatics, 2011. **7**(3): p. 381-388.

500. Yang, J., G. Zhang, and K. Ma, *A Nonlinear Control Method for Price-Based Demand Response Program in Smart Grid.* International Journal of Electrical Power & Energy Systems, 2016. **74**: p. 322-328.

501. Cappers, P., C. Goldman, and D. Kathan, *Demand Response in U.S. Electricity Markets: Empirical Evidence.* Energy, 2010. **35**(4): p. 1526-1535.

502. Faria, P., Z. Vale, and J. Baptista, *Constrained Consumption Shifting Management in the Distributed Energy Resources Scheduling Considering Demand Response.* Energy Conversion and Management, 2015. **93**: p. 309-320.

503. Abrishambaf, O., P. Faria, and Z. Vale, *Ramping of Demand Response Event With Deploying Distinct Programs by an Aggregator.* Energies, 2020. **13**(6): p. 1389.

504. Liu, D., et al., *Analysis and Accurate Prediction of User's Response Behavior in Incentive-Based Demand Response.* Ieee Access, 2019. **7**: p. 3170-3180.

505. Goswami, G.G., U. Rahman, and M. Chowdhury, *Estimating the Economic Cost of Setting Up a Nuclear Power Plant at Rooppur in Bangladesh.* Environmental Science and Pollution Research, 2022. **29**(23): p. 35073-35095.

506. Hong, S., C.J.A. Bradshaw, and B.W. Brook, *Global Zero-Carbon Energy Pathways Using Viable Mixes of Nuclear and Renewables.* Applied Energy, 2015. **143**: p. 451-459.

507. Wasti, A., et al., *Climate Change and the Hydropower Sector: A Global Review.* Wiley Interdisciplinary Reviews Climate Change, 2022. **13**(2).

508.   Norbeck, J. and T. Latimer, *Commercial-Scale Demonstration of a First-of-a-Kind Enhanced Geothermal System.* 2023.

509.   Parra, C.R., et al., *Prospects for Bioenergy Development Potential From Dedicated Energy Crops in Ecuador: An Agroecological Zoning Study.* Agriculture, 2023. **13**(1): p. 186.

510.   Eser, P., N. Chokani, and R.S. Abhari, *Operational and Financial Performance of Fossil Fuel Power Plants Within a High Renewable Energy Mix.* 2017. **1**: p. 2BIOTO.

511.   Johnson, N., et al., *A Reduced-Form Approach for Representing the Impacts of Wind and Solar PV Deployment on the Structure and Operation of the Electricity System.* Energy Economics, 2017. **64**: p. 651-664.

512.   Bloomfield, H., et al., *Quantifying the Increasing Sensitivity of Power Systems to Climate Variability.* Environmental Research Letters, 2016. **11**(12): p. 124025.

513.   Slusarewicz, J.H. and D.S. Cohan, *Assessing Solar and Wind Complementarity in Texas.* Renewables Wind Water and Solar, 2018. **5**(1).

514.   Surve, S., et al., *Impact Assessment of Different Battery Energy Storage Technologies in Distribution Grids With High Penetration of Renewable Energies.* Renewable Energy and Power Quality Journal, 2024. **20**(5).

515.   Uzondu, N.C. and D.D. Lele, *Comprehensive Analysis of Integrating Smart Grids With Renewable Energy Sources: Technological Advancements, Economic Impacts, and Policy Frameworks.* Engineering Science & Technology Journal, 2024. **5**(7): p. 2334-2363.

516.   Liu, Y., et al., *Study on the Stability of Power Market Regulation Under the Multi-User Coupling Demand Response.* 2023.

517.   Acquah, M.A., D. Kodaira, and S. Han, *Real-Time Demand Side Management Algorithm Using Stochastic Optimization.* Energies, 2018. **11**(5): p. 1166.

518.   Ratshitanga, M., et al., *Demand-Side Management as a Network Planning Tool: Review of Drivers, Benefits and Opportunities for South Africa.* Energies, 2023. **17**(1): p. 116.

519.   Papurello, D., et al., *Strategies for Demand-Side Management in an Office Building Integrated With Rooftop Façade PV Installations.* Tecnica Italiana-Italian Journal of Engineering Science, 2019. **63**(2-4): p. 311-314.

520.   Paul, W.U.H., A.S. Siddiqui, and S. Kirmani, *Demand Side Management Strategies in Residential Load With Renewable Energy Integration: A Brief Overview.* 2022: p. 66-69.

521.   Luo, X., X. Zhu, and E.G. Lim, *Electrical Vehicle-Assisted Demand Side Energy Management.* 2019.

522.   Areed, E.F., et al., *Electric Spring Average Model Development and Dynamic Analysis for Demand-side Management.* Iet Smart Grid, 2020. **3**(2): p. 226-236.

523.   Avwioroko, A., et al., *Smart Grid Integration of Solar and Biomass Energy Sources.* European Journal of Computer Science and Information Technology, 2024. **12**(3): p. 1-14.

524.   Forouli, A., et al., *Assessment of Demand Side Flexibility in European Electricity Markets: A Country Level Review.* Energies, 2021. **14**(8): p. 2324.

525.   Perez-Moscote, D.A. and M. Tyagunov, *Modeling of a Distributed Energy System With Renewable Generation and Demand-Side Flexibility.* 2022.

526.   Usman, R., et al., *Systematic Review of Demand-Side Management Strategies in Power Systems of Developed and Developing Countries.* Energies, 2022. **15**(21): p. 7858.

527. Escobar, M., et al., *The Contribution of Bottom-Up Energy Models to Support Policy Design of Electricity End-Use Efficiency for Residential Buildings and the Residential Sector: A Systematic Review.* Energies, 2021. **14**(20): p. 6466.

528. Aquino, C.C.C.B.d. and C. Unsihuay-Vila, *Optimal Dispatch and Technical and Economic Feasibility Analysis of Residential Microgrids Considering Demand Side Management.* Brazilian Archives of Biology and Technology, 2021. **64**(spe).

529. Rocca, R., et al., *Design Trade-Offs and Feasibility Assessment of a Novel One-Body, Laminated-Rotor Flywheel Switched Reluctance Machine.* Energies, 2020. **13**(22): p. 5857.

530. Ciabattoni, L., et al., *A Novel Open-Source Simulator of Electric Vehicles in a Demand-Side Management Scenario.* Energies, 2021. **14**(6): p. 1558.

531. Borst, F., et al., *Investigating the Electrical Demand-Side Management Potential of Industrial Steam Supply Systems Using Dynamic Simulation.* Energies, 2021. **14**(6): p. 1533.

532. Jayachandran, M., et al., *Challenges and Opportunities in Green Hydrogen Adoption for Decarbonizing Hard-to-Abate Industries: A Comprehensive Review.* Ieee Access, 2024. **12**: p. 23363-23388.

533. Borup, R.L., T. Krause, and J. Brouwer, *Hydrogen Is Essential for Industry and Transportation Decarbonization.* The Electrochemical Society Interface, 2021. **30**(4): p. 79-84.

534. Plank, F., et al., *External Hydrogen Relations of the European Union: Framing Processes in the Public Discourse Towards and Within Partner Countries.* Sustainability, 2023. **15**(20): p. 14757.

535. Ram, K.E.S. and H. Zhang, *Hydrogen in Decarbonization Strategies in Asia and the Pacific.* 2023.

536. Locke, K., *The Urgency of Hydrogen: Environmental Issues and the Need for Change.* 2024. **2**(2): p. 46-58.

537. Watari, T., et al., *Feasible Supply of Steel and Cement Within a Carbon Budget Is Likely to Fall Short of Expected Global Demand.* Nature Communications, 2023. **14**(1).

538. Leicher, J., et al., *The Impact of Hydrogen Admixture Into Natural Gas on Residential and Commercial Gas Appliances.* Energies, 2022. **15**(3): p. 777.

539. Alia, S.M., et al., *Chalkboard 2 - How to Make Clean Hydrogen.* The Electrochemical Society Interface, 2021. **30**(4): p. 49-56.

540. Yuan, M., *Accelerating the Net Zero Transition in Asia and the Pacific: Low-Carbon Hydrogen for Industrial Decarbonization.* 2023.

541. Fan, Z.H., *The Role of Hydrogen in Achieving Sustainable Energy Systems: Challenges and Opportunities.* Highlights in Science Engineering and Technology, 2024. **116**: p. 237-240.

542. Wang, C., et al., *Proton Exchange Membrane (PEM) Water Electrolysis: Cell-Level Considerations for Gigawatt-Scale Deployment.* Chemical Reviews, 2025. **125**(3): p. 1257-1302.

543. Keshk, W.F., E. Hekal, and M. Mohamed, *Mitigation of Greenhouse Gases Emissions and Reduction of Energy Consumption in Cement Industry in Egypt Case Study: Arabian Cement Plant.* Journal of Environmental Science, 2023. **52**(11): p. 1-17.

544. El-Hassan, H., *Accelerated Carbonation Curing as a Means of Reducing Carbon Dioxide Emissions.* 2021.

545. Mais, F. and H. Kolasch, *Tools and Frameworks for Sustainable Business Model Innovation in the Hydrogen Context for German Steel, Cement, and Chemical Industries.* 2024.

546. Fernández, L.M.G., et al., *Evaluation of Different Pretreatment Systems for the Energy Recovery of Greenhouse Agricultural Wastes in a Cement Plant.* Acs Sustainable Chemistry & Engineering, 2019. **7**(20): p. 17137-17144.

547. An, J., R.S. Middleton, and Y. Li, *Environmental Performance Analysis of Cement Production With CO2 Capture and Storage Technology in a Life-Cycle Perspective.* Sustainability, 2019. **11**(9): p. 2626.

548. Fennell, P.S., S.J. Davis, and A.M. Mohammed, *Decarbonizing Cement Production.* Joule, 2021. **5**(6): p. 1305-1311.

549. Cormoş, A.-M., et al., *Techno-Economic and Environmental Evaluations of Decarbonized Fossil-Intensive Industrial Processes by Reactive Absorption &Amp; Adsorption CO2 Capture Systems.* Energies, 2020. **13**(5): p. 1268.

550. Arens, M., et al., *Pathways to a Low-Carbon Iron and Steel Industry in the Medium-Term – The Case of Germany.* Journal of Cleaner Production, 2017. **163**: p. 84-98.

551. Mallapragada, D.S., et al., *Decarbonization of the Chemical Industry Through Electrification: Barriers and Opportunities.* 2022.

552. Davis, S.J., et al., *Net-Zero Emissions Energy Systems.* Science, 2018. **360**(6396).

553. Petroche, D.M. and Á.D. Ramírez, *The Environmental Profile of Clinker, Cement, and Concrete: A Life Cycle Perspective Study Based on Ecuadorian Data.* Buildings, 2022. **12**(3): p. 311.

554. Suer, J., M. Traverso, and N. Jäger, *Review of Life Cycle Assessments for Steel and Environmental Analysis of Future Steel Production Scenarios.* Sustainability, 2022. **14**(21): p. 14131.

555. Bhaskar, A., M. Assadi, and H.N. Somehsaraei, *Decarbonization of the Iron and Steel Industry With Direct Reduction of Iron Ore With Green Hydrogen.* Energies, 2020. **13**(3): p. 758.

556. Gajdzik, B., R. Wolniak, and W. Grebski, *Process of Transformation to Net Zero Steelmaking: Decarbonisation Scenarios Based on the Analysis of the Polish Steel Industry.* Energies, 2023. **16**(8): p. 3384.

557. Jordan, K., et al., *The Role of Hydrogen in Decarbonizing U.S. Iron and Steel Production.* Environmental Science & Technology, 2025. **59**(10): p. 4915-4925.

558. wang, c., S.D.C. Walsh, and A. Feitz, *A Techno-Economic Analysis of Australian Green Steel Production From Hydrogen.* 2022.

559. Digiesi, S., G. Mummolo, and M. Vitti, *Minimum Emissions Configuration of a Green Energy–Steel System: An Analytical Model.* Energies, 2022. **15**(9): p. 3324.

560. Bampaou, M., et al., *Integration of Renewable Hydrogen Production in Steelworks Off-Gases for the Synthesis of Methanol and Methane.* Energies, 2021. **14**(10): p. 2904.

561. Quitzow, R. and L. Eicke, *Toward a Renewables-Driven Industrial Landscape: Evidence on Investment Decisions in the Chemical and Steel Sectors.* 2025.

562. Abu-Elyazeed, O.S.M., et al., *Co-Combustion of RDF and Biomass Mixture With Bituminous Coal: A Case Study of Clinker Production Plant in Egypt.* Waste Disposal & Sustainable Energy, 2021. **3**(4): p. 257-266.

563. Sasso, O.R., et al., *Valorization of Biomass and Industrial Wastes as Alternative Fuels for Sustainable Cement Production.* Clean Technologies, 2024. **6**(2): p. 814-825.

564. Manalac, E.J., M.S. Berana, and S. Kim, *Performance Analysis of Cement Plant Waste-Heat Recovery for Power Generation Based on Partial Evaporating Cycle With Ejector.* Energy Technology, 2024. **13**(1).

565. Giddey, S., et al., *Ammonia as a Renewable Energy Transportation Media.* Acs Sustainable Chemistry & Engineering, 2017. **5**(11): p. 10231-10239.

566. Shaker, I., et al., *Process Simulation and Design of Green Ammonia Production Plant for Carbon Neutrality.* International Journal of Industry and Sustainable Development, 2024. **5**(1): p. 79-103.

567. Zhang, Y., et al., *Efficient Interlayer Confined Nitrate Reduction Reaction and Oxygen Generation Enabled by Interlayer Expansion.* Nanoscale, 2023. **15**(1): p. 204-214.

568. Mallouppas, G., C. Ioannou, and E.A. Yfantis, *A Review of the Latest Trends in the Use of Green Ammonia as an Energy Carrier in Maritime Industry.* Energies, 2022. **15**(4): p. 1453.

569. Tnifasse, K., et al., *Green Ammonia Unleashed: Cutting-Edge Electrochemical Synthesis, Catalyst Innovations, and SOFC Integration for a Sustainable Energy Future–A Review and Perspectives.* Energy & Fuels, 2025. **39**(4): p. 1793-1812.

570. Nayak-Luke, R., R. Bañares-Alcántara, and I. Wilkinson, *"Green" Ammonia: Impact of Renewable Energy Intermittency on Plant Sizing and Levelized Cost of Ammonia.* Industrial & Engineering Chemistry Research, 2018. **57**(43): p. 14607-14616.

571. Gezerman, A.O., *A Critical Assessment of Green Ammonia Production and Ammonia Production Technologies.* Kemija U Industriji, 2022(1-2).

572. Verschuur, J., et al., *Optimal Fuel Supply of Green Ammonia to Decarbonise Global Shipping.* Environmental Research Infrastructure and Sustainability, 2024. **4**(1): p. 015001.

573. Xue, M., et al., *Assessment of Ammonia as an Energy Carrier From the Perspective of Carbon and Nitrogen Footprints.* Acs Sustainable Chemistry & Engineering, 2019.

574. Vinardell, S., et al., *Sustainability Assessment of Green Ammonia Production to Promote Industrial Decarbonization in Spain.* Acs Sustainable Chemistry & Engineering, 2023. **11**(44): p. 15975-15983.

575. Valera-Medina, A., et al., *Review on Ammonia as a Potential Fuel: From Synthesis to Economics.* Energy & Fuels, 2021. **35**(9): p. 6964-7029.

576. Edomah, N., *Deindustrialization and Effects of Poorly Implemented Energy Policies on Sustainable Industrial Growth.* 2017: p. 1-12.

577. González, n.K., *Agencies of Transition: Why German Coal Workers Are Not Accepting an Energy Transition Despite Social Provisions.* 2025. **2**(1): p. 015005.

578. Levinson, A., *Energy Intensity: Deindustrialization, Composition, Prices, and Policies in U.S. States.* Resource and Energy Economics, 2021. **65**: p. 101243.

579. Gasim, A., *The Embodied Energy in Trade: What Role Does Specialization Play?* Energy Policy, 2015. **86**: p. 186-197.

580. Lin, B. and K. Liu, *Using LMDI to Analyze the Decoupling of Carbon Dioxide Emissions From China’s Heavy Industry.* 2017.

581. Sterzinger, G., *The Economic Promise of Renewable Energy.* New Labor Forum, 2007. **16**(3): p. 80-91.

582. Tomassetti, P., *From Treadmill of Production to Just Transition and Beyond.* European Journal of Industrial Relations, 2020. **26**(4): p. 439-457.

583. Paterson, M. and X. P-Laberge, *Political Economies of Climate Change.* Wiley Interdisciplinary Reviews Climate Change, 2018. **9**(2).

584. Karasoy, A., *Assessing the Impacts of Industrialization, Deindustrialization and Financialization on Turkey's Energy Security: Evidence From the Augmented NARDL Method.* International Journal of Energy Sector Management, 2023. **17**(6): p. 1053-1073.

585. Babiker, M., *Climate Change Policy, Market Structure, and Carbon Leakage.* Journal of International Economics, 2005. **65**(2): p. 421-445.

586. Fragkos, P., K. Fragkiadakis, and L. Paroussos, *Reducing the Decarbonisation Cost Burden for EU Energy-Intensive Industries.* Energies, 2021. **14**(1): p. 236.

587. Thompson, S., *Strategic Analysis of the Renewable Electricity Transition: Power to the World Without Carbon Emissions?* Energies, 2023. **16**(17): p. 6183.

588. Wright, C., et al., *'We're in the Coal Business': Maintaining Fossil Fuel Hegemony in the Face of Climate Change.* Journal of Industrial Relations, 2022. **64**(4): p. 544-563.

589. Janardhanan, N., et al., *How Can Japan Help Create a Sustainable Hydrogen Society in Asia?* 2023.

590. Jones, N., et al., *Tapping the Potential of NDCs and LT-LEDS to Address Fossil Fuel Production.* 2021.

591. Green, F. and A. Gambhir, *Transitional Assistance Policies for Just, Equitable and Smooth Low-Carbon Transitions: Who, What and How?* Climate Policy, 2019. **20**(8): p. 902-921.

592. Muttitt, G. and S. Kartha, *Equity, Climate Justice and Fossil Fuel Extraction: Principles for a Managed Phase Out.* Climate Policy, 2020. **20**(8): p. 1024-1042.

593. Tuncer, A.D., et al., *Efficient Energy Systems Models for Sustainable Food Processing.* Turkish Journal of Agriculture - Food Science and Technology, 2019. **7**(8): p. 1138-1145.

594. Tadini, C.C. and J.A.W. Gut, *The Importance of Heating Unit Operations in the Food Industry to Obtain Safe and High-Quality Products.* Frontiers in Nutrition, 2022. **9**.

595. Dziki, D., *Recent Trends in Pretreatment of Food Before Freeze-Drying.* Processes, 2020. **8**(12): p. 1661.

596. Araújo, M.V., J.M.P.Q. Delgado, and A.G.B.d. Lima, *On the Use of CFD in Thermal Analysis of Industrial Hollow Ceramic Brick.* Diffusion Foundations, 2017. **10**: p. 70-82.

597. Martynenko, A. and T. Kudra, *Non-Isothermal Drying of Medicinal Plants.* Drying Technology, 2015. **33**(13): p. 1550-1559.

598. Mujumdar, A.S. and C.L. Law, *Drying Technology: Trends and Applications in Postharvest Processing.* Food and Bioprocess Technology, 2010. **3**(6): p. 843-852.

599. Monteiro, C.A., et al., *A New Classification of Foods Based on the Extent and Purpose of Their Processing.* Cadernos De Saúde Pública, 2010. **26**(11): p. 2039-2049.

600. Dolgun, G.K., et al., *Photovoltaic-Thermal Solar Energy System Design for Dairy Industry.* Journal of Energy Systems, 2019. **3**(2): p. 86-95.

601. Gruzdev, N., R. Pinto, and S. Sela, *Effect of Desiccation on Tolerance Of Salmonella Enterica to Multiple Stresses.* Applied and Environmental Microbiology, 2011. **77**(5): p. 1667-1673.

602. Nkem, O.M., A.O. Oladejo, and A.F. Alonge, *Influence of Ultrasound Pretreatment on Drying Characteristics of Cocoyam (Xanthosoma Sagittifolium) Slices During Convective Hot Air Drying.* Journal of the Science of Food and Agriculture, 2023. **104**(5): p. 3047-3056.

603. Guanco, J. and L.F. Casinillo, *Performance Evaluation of a Rice Hull-Fueled Cabinet Food Dryer.* Recoletos Multidisciplinary Research Journal, 2021. **9**(1): p. 23-38.

604. Bajan, B., A. Mrówczyńska-Kamińska, and W. Poczta, *Economic Energy Efficiency of Food Production Systems.* Energies, 2020. **13**(21): p. 5826.

605. Clairand, J.-M., et al., *Review of Energy Efficiency Technologies in the Food Industry: Trends, Barriers, and Opportunities.* Ieee Access, 2020. **8**: p. 48015-48029.

606. Vora, N., et al., *Food–Energy–Water Nexus: Quantifying Embodied Energy and GHG Emissions From Irrigation Through Virtual Water Transfers in Food Trade.* Acs Sustainable Chemistry & Engineering, 2017. **5**(3): p. 2119-2128.

607. Laso, J., et al., *Assessing Energy and Environmental Efficiency of the Spanish Agri-Food System Using the LCA/DEA Methodology.* Energies, 2018. **11**(12): p. 3395.

608. Vittuari, M., F.D. Menna, and M. Pagani, *The Hidden Burden of Food Waste: The Double Energy Waste in Italy.* Energies, 2016. **9**(8): p. 660.

609. Tannous, H., V. Stojceska, and S. Tassou, *The Use of Solar Thermal Heating in SPIRE and Non-Spire Industrial Processes.* Sustainability, 2023. **15**(10): p. 7807.

610. Johnson, N., *Exploring Integrated Biogas Solutions for Industrial Processes: An Advanced Approach Towards Renewable Energy.* 2024.

611. Nunes, L.J.R., R. Godina, and J.C.O. Matias, *Technological Innovation in Biomass Energy for the Sustainable Growth of Textile Industry.* Sustainability, 2019. **11**(2): p. 528.

612. Hua, Y., M. Oliphant, and E. Hu, *Development of Renewable Energy in Australia and China: A Comparison of Policies and Status.* Renewable Energy, 2016. **85**: p. 1044-1051.

613. Wu, X., et al., *Industrial Characteristics of Renewable Energy and Spatial Aggregation Correlations in Beijing–Tianjin–Hebei.* Science and Technology for Energy Transition, 2022. **77**: p. 3.

614. Song, D., et al., *Overview of the Policy Instruments for Renewable Energy Development in China.* Energies, 2022. **15**(18): p. 6513.

615. Boix, M., et al., *Benefits Analysis of Optimal Design of Eco-Industrial Parks Through Life Cycle Indicators.* 2017: p. 1951-1956.

616. Farel, R., et al., *Sustainable Manufacturing Through Creation and Governance Of Eco-Industrial Parks.* Journal of Manufacturing Science and Engineering, 2016. **138**(10).

617. Boix, M., et al., *Optimization Methods Applied to the Design of Eco-Industrial Parks: A Literature Review.* Journal of Cleaner Production, 2015. **87**: p. 303-317.

618. Blakers, A., et al., *Pathway to 100% Renewable Electricity.* Ieee Journal of Photovoltaics, 2019. **9**(6): p. 1828-1833.

619. Wahyuni, W., I. Meutia, and S. Syamsurijal, *The Effect of Green Accounting Implementation on Improving the Environmental Performance of Mining and Energy Companies in Indonesia.* Binus Business Review, 2019. **10**(2): p. 131-137.

620. Harianto, B.B. and M. Karjadi, *Planning of Photovoltaic (PV) Type Solar Power Plant as an Alternative Energy of the Future in Indonesia.* Endless International Journal of Future Studies, 2022. **5**(2): p. 182-195.

621. Adrian, M.M., et al., *Energy Transition Towards Renewable Energy in Indonesia.* Heritage and Sustainable Development Issn 2712-0554, 2023. **5**(1): p. 107-118.

622. Azhar, M., et al., *The New Renewable Energy Consumption Policy of Rare Earth Metals to Build Indonesia's National Energy Security.* E3s Web of Conferences, 2018. **68**: p. 03008.

623. Sambodo, M.T., et al., *Breaking Barriers to Low-Carbon Development in Indonesia: Deployment of Renewable Energy.* Heliyon, 2022. **8**(4): p. e09304.

624. Maharani, Y.N., I. Ikhsan, and T. Wibowo, *Renewable Energy Transition in Facing Climate Change in Indonesia.* Iop Conference Series Earth and Environmental Science, 2024. **1339**(1): p. 012036.

625. Chakraborty, D., N.K. Mondal, and J.K. Datta, *Indoor Pollution From Solid Biomass Fuel and Rural Health Damage: A Micro-Environmental Study in Rural Area of Burdwan, West Bengal.* International Journal of Sustainable Built Environment, 2014. **3**(2): p. 262-271.

626. Bisu, D.Y., A. Kuhe, and H.A. Iortyer, *Urban Household Cooking Energy Choice: An Example of Bauchi Metropolis, Nigeria.* Energy Sustainability and Society, 2016. **6**(1).

627. Cameron, C., et al., *Policy Trade-Offs Between Climate Mitigation and Clean Cook-Stove Access in South Asia.* Nature Energy, 2016. **1**(1).

628. Mhagama, G., *Determinants of Household Energy Use for Cooking in Tanzania.* 2024. **25**(2): p. 47-60.

629. Yang, Z., et al., *The Relationship Between Cooking Fuel and Health Status From the Perspective of Income Heterogeneity: Evidence From China.* Energy & Environment, 2023. **35**(6): p. 3210-3231.

630. Nie, P., A. Sousa-Poza, and J. Xue, *Fuel for Life: Domestic Cooking Fuels and Women's Health in Rural China.* International Journal of Environmental Research and Public Health, 2016. **13**(8): p. 810.

631. Savvakis, N., et al., *Environmental Effects From the Use of Traditional Biomass for Heating in Rural Areas: A Case Study of Anogeia, Crete.* Environment Development and Sustainability, 2021. **24**(4): p. 5473-5495.

632. Khavari, B., et al., *A Geospatial Approach to Understanding Clean Cooking Challenges in Sub-Saharan Africa.* Nature Sustainability, 2023. **6**(4): p. 447-457.

633. Amadi, A.H., et al., *Cooking Gas (LPG) Distribution to Rivers State Homes, Case Study: Choba Community.* European Journal of Engineering and Technology Research, 2021. **6**(4): p. 77-84.

634. Rosenthal, J., et al., *Clean Cooking and the SDGs: Integrated Analytical Approaches to Guide Energy Interventions for Health and Environment Goals.* Energy for Sustainable Development, 2018. **42**: p. 152-159.

635. Dagnachew, A.G., et al., *Scenario Analysis for Promoting Clean Cooking in Sub-Saharan Africa: Costs and Benefits.* Energy, 2020. **192**: p. 116641.

636. Vásquez, M.A.D., et al., *Adoption and Impact of Improved Cookstoves in Lambayeque, Peru, 2017.* Global Health Promotion, 2020. **27**(4): p. 123-130.

637. Heltberg, R., *Factors Determining Household Fuel Choice in Guatemala.* Environment and Development Economics, 2005. **10**(3): p. 337-361.

638. Campbell, C., et al., *Investigating Cooking Activity Patterns and Perceptions of Air Quality Interventions Among Women in Urban Rwanda.* International Journal of Environmental Research and Public Health, 2021. **18**(11): p. 5984.

639. Al-Rubaye, H., et al., *Advances in Energy Hybridization for Resilient Supply: A Sustainable Approach to the Growing World Demand.* Energies, 2022. **15**(16): p. 5903.

640. Famewo, A.S. and V.A. Uwala, *Socio-Economic Impacts of Rural Energy Poverty on Women and Students in Esa-Oke, Nigeria.* Journal of Sustainability and Environmental Management, 2022. **1**(2): p. 84-93.

641. Destaw, F., A. Birhanu, and A. Gurmessa, *Carbon Emission Reduction Potentials of Improved Biomass Cookstoves Used in Gambella Refugee Camps, Southwest Ethiopia* 2024.

642. Gould, C. and J. Urpelainen, *LPG as a Clean Cooking Fuel: Adoption, Use, and Impact in Rural India.* Energy Policy, 2018. **122**: p. 395-408.

643. Abdulai, M.A., et al., *Experiences With the Mass Distribution of LPG Stoves in Rural Communities of Ghana.* Ecohealth, 2018. **15**(4): p. 757-767.

644. Makonese, T., A.P. Ifegbesan, and I.T. Rampedi, *Household Cooking Fuel Use Patterns and Determinants Across Southern Africa: Evidence From the Demographic and Health Survey Data.* Energy & Environment, 2017. **29**(1): p. 29-48.

645. Ibraimo, M., J. Robinson, and H.J. Annegarn, *Household Energy Use and Emission.* 2017.

646. Junejo, A., et al., *Comparative Analysis of Fuel Wood Consumption of Improved Mud Stove and Three Stone Fire Stove: A case Study.* Comptes Rendus Chimie, 2024. **26**(S1): p. 25-35.

647. Ahmed, N. and M.Z. Iqbal, *Benefits of Improved Cook Stoves: Evidence From Rural Bangladesh.* The Bangladesh Development Studies, 2019. **XLI**(4): p. 01-27.

648. Wei, S., X. Zhang, and Q. Cheng, *The Temperature and Pressure Remote Intelligent Control System for Biogas Digester Based on STM32.* SHS Web of Conferences, 2023. **166**: p. 01060.

649. Li, X., I.S. Chang, and J. Wu, *Assessment on Biogas Utilizations in Rural Areas in Urumqi, Xinjiang.* Advanced Materials Research, 2012. **512-515**: p. 343-346.

650. Oluwatosin, D., K. Ola, and A.O. Olayinka, *The Impacts of the Use of Biomass Solid Fuels for Household Cooking in Sub-Saharan Africa - A Review.* 2022.

651. Agbokey, F., et al., *Determining the Enablers and Barriers for the Adoption of Clean Cookstoves in the Middle Belt of Ghana—A Qualitative Study.* International Journal of Environmental Research and Public Health, 2019. **16**(7): p. 1207.

652. Cui, Y., *Mechanization's Impact on Agricultural Total Factor Productivity.* Agricultural Economics (Zemědělská Ekonomika), 2023. **69**(11): p. 446-457.

653. Lampridi, M., et al., *Energy Footprint of Mechanized Agricultural Operations.* Energies, 2020. **13**(3): p. 769.

654. Han, G.-G., et al., *Analysis of Air Pollutant Emissions for Mechanized Rice Cultivation in Korea.* Agriculture, 2021. **11**(12): p. 1208.

655. Silveira, F.d., et al., *Fuel Consumption by Agricultural Machinery: A Review of Pollutant Emission Control Technologies.* Ciência Rural, 2023. **53**(5).

656. Masset, E., et al., *PROTOCOL: The Impact of Agricultural Mechanisation on Women's Economic Empowerment: A Mixed-methods Systematic Review.* Campbell Systematic Reviews, 2023. **19**(3).

657. Chen, H.C., et al., *Electrification of Agricultural Machinery: One Design Case of a 4 kW Air Compressor.* Energies, 2024. **17**(15): p. 3647.

658. Scolaro, E., et al., *Electrification of Agricultural Machinery: A Review.* Ieee Access, 2021. **9**: p. 164520-164541.

659. Vahdanjoo, M., R. Gislum, and C.A.G. Sørensen, *Estimating Fuel Consumption of an Agricultural Robot by Applying Machine Learning Techniques During Seeding Operation.* Agriengineering, 2024. **6**(1): p. 754-772.

660. Chen, P. and L.I. Jing-long, *Sustainable Agricultural Management: How to Achieve Carbon Neutrality in Agriculture – Evidence From China Agricultural Sustainable Development Plan.* Sustainable Development, 2023. **32**(3): p. 2846-2857.

661. Amanor, K. and A.Y. Iddrisu, *Old Tractors, New Policies and Induced Technological Transformation: Agricultural Mechanisation, Class Formation, and Market Liberalisation in Ghana.* The Journal of Peasant Studies, 2021. **49**(1): p. 158-178.

662. Daum, T. and R. Birner, *The Neglected Governance Challenges of Agricultural Mechanisation in Africa – Insights From Ghana.* Food Security, 2017. **9**(5): p. 959-979.

663. Pirdashti, H., et al., *Efficient Use of Energy Through Organic Rice–duck Mutualism System.* Agronomy for Sustainable Development, 2015. **35**(4): p. 1489-1497.

664. Li, N., et al., *Diesel Consumption of Agriculture in China*. Energies, 2012. **5**(12): p. 5126-5149.

665. Rajaeifar, M.A., et al., *Energy Consumption and Greenhouse Gas Emissions of Biodiesel Production From Rapeseed in Iran*. Journal of Renewable and Sustainable Energy, 2013. **5**(6).

666. Catão, L. and R. Chang, *World Food Prices and Monetary Policy*. 2010.

667. Azizpanah, A., R. Fathi, and M. Taki, *Eco-Energy and Environmental Evaluation of Cantaloupe Production by Life Cycle Assessment Method*. Environmental Science and Pollution Research, 2022. **30**(1): p. 1854-1870.

668. Li, W. and P. Zhang, *Relationship and Integrated Development of Low-Carbon Economy, Food Safety, and Agricultural Mechanization*. Environmental Science and Pollution Research, 2021. **28**(48): p. 68679-68689.

669. Flammini, A., et al., *Emissions of Greenhouse Gases From Energy Use in Agriculture, Forestry and Fisheries: 1970–2019*. Earth System Science Data, 2022. **14**(2): p. 811-821.

670. Patra, A.K., *Trends and Projected Estimates of GHG Emissions From Indian Livestock in Comparisons With GHG Emissions From World and Developing Countries*. Asian-Australasian Journal of Animal Sciences, 2014. **27**(4): p. 592-599.

671. Haider, A., et al., *Nexus Between Nitrous Oxide Emissions and Agricultural Land Use in Agrarian Economy: An ARDL Bounds Testing Approach*. Sustainability, 2021. **13**(5): p. 2808.

672. Wysocka-Czubaszek, A., et al., *Methane and Nitrous Oxide Emissions From Agriculture on a Regional Scale*. Journal of Ecological Engineering, 2018. **19**(3): p. 206-217.

673. Brankov, T., *Factors Affecting Agricultural Emissions in the Western Balkans: Panel Data Analysis*. Serbian Journal of Management, 2023. **18**(2): p. 275-283.

674. Strand, S.E., L. Zhang, and M. Flury, *Theoretical Analysis of Engineered Plants for Control of Atmospheric Nitrous Oxide and Methane by Modification of the Mitochondrial Proteome*. Acs Sustainable Chemistry & Engineering, 2022. **10**(17): p. 5441-5452.

675. Ji, E.S. and K.-H. Park, *Methane and Nitrous Oxide Emissions From Livestock Agriculture in 16 Local Administrative Districts of Korea*. Asian-Australasian Journal of Animal Sciences, 2012. **25**(12): p. 1768-1774.

676. Prather, M.J., et al., *Measuring and Modeling the Lifetime of Nitrous Oxide Including Its Variability*. Journal of Geophysical Research Atmospheres, 2015. **120**(11): p. 5693-5705.

677. Sun, Q., et al., *Nitrous Oxide Emissions From a Waterbody in the Nenjiang Basin, China*. Hydrology Research, 2012. **43**(6): p. 862-869.

678. Bu, M., et al., *Scenario-Based Modeling of Agricultural Nitrous Oxide Emissions in China*. 2024.

679. Hunt, K.A., et al., *Contribution of Microorganisms With the Clade II Nitrous Oxide Reductase to Suppression of Surface Emissions of Nitrous Oxide*. Environmental Science & Technology, 2024. **58**(16): p. 7056-7065.

680. Daelman, M.R.J., et al., *Methane and Nitrous Oxide Emissions From Municipal Wastewater Treatment – Results From a Long-Term Study*. Water Science & Technology, 2013. **67**(10): p. 2350-2355.

681. Nagata, O., T. Yazaki, and Y. Yanai, *Nitrous Oxide Emissions From Drained and Mineral Soil-Dressed Peatland in Central Hokkaido, Japan*. Journal of Agricultural Meteorology, 2010. **66**(1): p. 23-30.

682. Khan-Perez, J., T. MacCarrick, and F. Martin, *The Use of Nitrous Oxide 'Cracking' Technology in the Labour Ward: A Case Report and Patient Account.* Anaesthesia Reports, 2022. **10**(2).

683. Suenaga, T., et al., *Biokinetic Characterization and Activities of N2o-Reducing Bacteria in Response to Various Oxygen Levels.* Frontiers in Microbiology, 2018. **9**.

684. Kołoszko-Chomentowska, Z., L. Sieczko, and R. Trochimczuk, *Production Profile of Farms and Methane and Nitrous Oxide Emissions.* Energies, 2021. **14**(16): p. 4904.

685. Xin, C., et al., *Precipitation Patterns and Their Role in Modulating Nitrous Oxide Emissions From Arid Desert Soil.* Land, 2024. **13**(11): p. 1920.

686. Law, Y., et al., *Nitrous Oxide Emissions From Wastewater Treatment Processes.* Philosophical Transactions of the Royal Society B Biological Sciences, 2012. **367**(1593): p. 1265-1277.

687. Wang, Z., et al., *Clarifying Microbial Nitrous Oxide Reduction Under Aerobic Conditions: Tolerant, Intolerant, and Sensitive.* Microbiology Spectrum, 2023. **11**(2).

688. Bergquist, P., M. Mildenberger, and L. Stokes, *Combining Climate, Economic, and Social Policy Builds Public Support for Climate Action in the US.* Environmental Research Letters, 2020. **15**(5): p. 054019.

689. Mulholland, M., *A Moment of Intersecting Crises: Climate Justice in the Era of Coronavirus.* Development, 2020. **63**(2-4): p. 257-261.

690. Ukoha, A.H., *Assessment of the Cost-Benefits of Climate Change Adaptation Strategies of Cassava-Based Farmers in Southern Nigeria.* International Journal of Environment and Climate Change, 2020: p. 99-110.

691. Nolan, A. and J.P. Bohoslavsky, *Human Rights and Economic Policy Reforms.* The International Journal of Human Rights, 2020. **24**(9): p. 1247-1267.

692. Soemanto, A. and R.H.S. Koestoer, *Scenario Insight of Energy Transition.* Indonesian Journal of Energy, 2023. **6**(1).

693. Redon, F., et al., *Meeting Stringent 2025 Emissions and Fuel Efficiency Regulations With an Opposed-Piston, Light-Duty Diesel Engine.* 2014.

694. Gupta, M., M. Mohan, and S. Bhati, *Assessment of Air Pollution Mitigation Measures on Secondary Pollutants PM10 and Ozone Using Chemical Transport Modelling Over Megacity Delhi, India.* Urban Science, 2022. **6**(2): p. 27.

695. Khodke, A., A. Watabe, and N. Mehdi, *Implementation of Accelerated Policy-Driven Sustainability Transitions: Case of Bharat Stage 4 to 6 Leapfrogs in India.* Sustainability, 2021. **13**(8): p. 4339.

696. Rogelj, J., et al., *Paris Agreement Climate Proposals Need a Boost to Keep Warming Well Below 2 °C.* Nature, 2016. **534**(7609): p. 631-639.

697. Maris, G. and F. Flouros, *The Green Deal, National Energy and Climate Plans in Europe: Member States' Compliance and Strategies.* Administrative Sciences, 2021. **11**(3): p. 75.

698. Roelfsema, M., et al., *Integrated Assessment of International Climate Mitigation Commitments Outside the UNFCCC.* Global Environmental Change, 2018. **48**: p. 67-75.

699. Ronzon, T., S. Iost, and G. Philippidis, *Has the European Union Entered a Bioeconomy Transition? Combining an Output-Based Approach With a Shift-Share Analysis.* Environment Development and Sustainability, 2022. **24**(6): p. 8195-8217.

700. Eckert, S., *The European Green Deal and the EU's Regulatory Power in Times of Crisis.* JCMS Journal of Common Market Studies, 2021. **59**(S1): p. 81-91.

701. Aznar, E.T., A. Sánchez-Bayón, and J.M. Vindel, *The European Union Green Deal: Clean Energy Wellbeing Opportunities and the Risk of the Jevons Paradox.* Energies, 2021. **14**(14): p. 4148.

702. Heriyanti, S.I. and S. Ariyanto, *Epilog: Sprint With Renewable Energy Towards Net Zero Emissions 2060.* 2023.

703. Bahuguna, P., *The Role on International Law in Addressing Climate Change in India: An Analytical Perspective.* 2023. **55**(01).

704. Pianta, M. and M. Lucchese, *Rethinking the European Green Deal.* Review of Radical Political Economics, 2020. **52**(4): p. 633-641.

705. Yurdakul, M. and H. Kazan, *Effects of Eco-Innovation on Economic and Environmental Performance: Evidence From Turkey's Manufacturing Companies.* Sustainability, 2020. **12**(8): p. 3167.

706. Biresselioğlu, M.E., et al., *Legal Provisions and Market Conditions for Energy Communities in Austria, Germany, Greece, Italy, Spain, and Turkey: A Comparative Assessment.* Sustainability, 2021. **13**(20): p. 11212.

707. Wang, F., et al., *Technologies and Perspectives for Achieving Carbon Neutrality.* The Innovation, 2021. **2**(4): p. 100180.

708. Zhou, H., *Air Passenger Carbon Offset and Carbon Neutrality Strategies: Implementation Mechanism by Convolutional Neural Network.* Heliyon, 2024. **10**(19): p. e37495.

709. Chen, L., et al., *Strategies to Achieve a Carbon Neutral Society: A Review.* Environmental Chemistry Letters, 2022. **20**(4): p. 2277-2310.

710. Le, A.T.H., et al., *Policy Mapping for Net-Zero-Carbon Buildings: Insights From Leading Countries.* Buildings, 2023. **13**(11): p. 2766.

711. Shaik, A.S., et al., *Artificial Intelligence (AI)-driven Strategic Business Model Innovations in Small- and Medium-sized Enterprises. Insights on Technological and Strategic Enablers for Carbon Neutral Businesses.* Business Strategy and the Environment, 2023. **33**(4): p. 2731-2751.

712. Loveday, J., G.M. Morrison, and D.A. Martin, *Identifying Knowledge and Process Gaps From a Systematic Literature Review of Net-Zero Definitions.* Sustainability, 2022. **14**(5): p. 3057.

713. Luqman, A., et al., *Artificial Intelligence and Corporate Carbon Neutrality: A Qualitative Exploration.* Business Strategy and the Environment, 2024. **33**(5): p. 3986-4003.

714. Guo, Y., et al., *Forecasting and Scenario Analysis of Carbon Emissions in Key Industries: A Case Study in Henan Province, China.* Energies, 2023. **16**(20): p. 7103.

715. Dong, X., J.K. Kim, and S. Son, *Corporate Carbon Neutrality Strategy and Sustainable Development Capabilities: A Resource-Based Approach To Key Dynamics and Impacts.* Corporate Social Responsibility and Environmental Management, 2025. **32**(3): p. 3965-3977.

716. Mondal, S.R., S. Das, and V. Vrana, *Exploring the Role of Artificial Intelligence in Achieving a Net Zero Carbon Economy in Emerging Economies: A Combination of PLS-SEM and fsQCA Approaches to Digital Inclusion and Climate Resilience.* Sustainability, 2024. **16**(23): p. 10299.

717. Bortoli, A.d., et al., *Planning Sustainable Carbon Neutrality Pathways: Accounting Challenges Experienced by Organizations and Solutions From Industrial Ecology.* The International Journal of Life Cycle Assessment, 2023. **28**(7): p. 746-770.

718. Achakulwisut, P., et al., *Global Fossil Fuel Reduction Pathways Under Different Climate Mitigation Strategies and Ambitions.* Nature Communications, 2023. **14**(1).

719. Heuberger, C.F., et al., *Quantifying the Value of CCS for the Future Electricity System.* Energy & Environmental Science, 2016. **9**(8): p. 2497-2510.

720. Gabrielli, P., M. Gazzani, and M. Mazzotti, *The Role of Carbon Capture and Utilization, Carbon Capture and Storage, and Biomass to Enable a Net-Zero-CO_2 Emissions Chemical Industry.* Industrial & Engineering Chemistry Research, 2020. **59**(15): p. 7033-7045.

721. Cuéllar-Franca, R.M. and A. Azapagic, *Carbon Capture, Storage and Utilisation Technologies: A Critical Analysis and Comparison of Their Life Cycle Environmental Impacts.* Journal of Co2 Utilization, 2015. **9**: p. 82-102.

722. Muratori, M., et al., *Global Economic Consequences of Deploying Bioenergy With Carbon Capture and Storage (BECCS).* Environmental Research Letters, 2016. **11**(9): p. 095004.

723. Fan, J.L., et al., *A Net-Zero Emissions Strategy for China's Power Sector Using Carbon-Capture Utilization and Storage.* Nature Communications, 2023. **14**(1).

724. Asayama, S., *The Oxymoron of Carbon Dioxide Removal: Escaping Carbon Lock-in and Yet Perpetuating the Fossil Status Quo?* Frontiers in Climate, 2021. **3**.

725. Agyekum, E.B., et al., *A Critical Review of Renewable Hydrogen Production Methods: Factors Affecting Their Scale-Up and Its Role in Future Energy Generation.* Membranes, 2022. **12**(2): p. 173.

726. Griffiths, S., et al., *Bridging the Gap Between Defossilization and Decarbonization to Achieve Net-Zero Industry.* Environmental Research Letters, 2025. **20**(2): p. 024063.

727. Agarwala, N., *Is Hydrogen a Decarbonizing Fuel for Maritime Shipping?* Maritime Technology and Research, 2024. **6**(4): p. 271244.

728. Yang, C., et al., *Promoting Economic and Environmental Resilience in the Post-Covid-19 Era Through the City and Regional on-Road Fuel Sustainability Development.* NPJ Urban Sustainability, 2022. **2**(1).

729. Hirunsit, P., et al., *From CO_2 to Sustainable Aviation Fuel: Navigating the Technology Landscape.* Acs Sustainable Chemistry & Engineering, 2024. **12**(32): p. 12143-12160.

730. Yalamati, H.P.S., R.K. Vij, and R. Srivastava, *Green Hydrogen Revolution: Mapping G20 Nations' Policies and Geopolitical Strategies in the Net-zero Emission Era.* Wiley Interdisciplinary Reviews Energy and Environment, 2024. **13**(5).

731. Soest, H.v., M.d. Elzen, and D.P.v. Vuuren, *Net-Zero Emission Targets for Major Emitting Countries Consistent With the Paris Agreement.* Nature Communications, 2021. **12**(1).

732. Fankhauser, S., et al., *The Meaning of Net Zero and How to Get It Right.* Nature Climate Change, 2021. **12**(1): p. 15-21.

733. Averchenkova, A., S. Fankhauser, and J. Finnegan, *The Impact of Strategic Climate Legislation: Evidence From Expert Interviews on the UK Climate Change Act.* Climate Policy, 2020. **21**(2): p. 251-263.

734. Sakrabani, R., et al., *Towards Net Zero in Agriculture: Future Challenges and Opportunities for Arable, Livestock and Protected Cropping Systems in the UK.* Outlook on Agriculture, 2023. **52**(2): p. 116-125.

735. Armstrong, C. and D. McLaren, *Which Net Zero? Climate Justice and Net Zero Emissions.* Ethics & International Affairs, 2022. **36**(4): p. 505-526.

736. Gangadhari, R.K., S. Karadayi-Usta, and W.M. Lim, *Breaking Barriers Toward a Net-zero Economy.* Natural Resources Forum, 2023. **49**(1): p. 138-159.

737. Sousa, T.J.C., et al., *Unified Traction and Battery Charging Systems for Electric Vehicles: A Sustainability Perspective.* 2020: p. 58-69.

738. Khan, M., *Innovations in Battery Technology: Enabling the Revolution in Electric Vehicles and Energy Storage.* British Journal of Multidisciplinary and Advanced Studies, 2024. **5**(1): p. 23-41.

739. Chen, Y., *Enhance the Market Competitiveness of Electric Vehicles by Improving the Design of Lithium-Ion Batteries.* Highlights in Science Engineering and Technology, 2023. **32**: p. 245-251.

740. Liu, X., K. Li, and X. Li, *The Electrochemical Performance and Applications of Several Popular Lithium-Ion Batteries for Electric Vehicles - A Review.* 2018: p. 201-213.

741. Li, L., *Lithium-Ion Battery Management System for Electric Vehicles.* 2018.

742. Wu, X., et al., *Robust Online Estimation of State of Health for Lithium-Ion Batteries Based on Capacities Under Dynamical Operation Conditions.* Batteries, 2024. **10**(7): p. 219.

743. Zhang, Y., et al., *Characterization of Interaction Between Electric Vehicles and Smart Grid.* E3s Web of Conferences, 2021. **237**: p. 02004.

744. Unterweger, A., et al., *Low-Risk Privacy-Preserving Electric Vehicle Charging With Payments.* 2021.

745. Bezha, M. and N. Nagaoka, *Improved ANN for Estimation of Power Consumption of EV for Real-Time Battery Diagnosis.* Ieej Journal of Industry Applications, 2019. **8**(3): p. 532-538.

746. Soares, L.O., et al., *Electric Vehicle Supply Chain Management: A Bibliometric and Systematic Review.* Energies, 2023. **16**(4): p. 1563.

747. Shen, W.X., et al., *Research on Life Cycle Energy Consumption and Environmental Emissions of Light-Duty Battery Electric Vehicles.* Materials Science Forum, 2015. **814**: p. 447-457.

748. Tamayao, M.-A.M., et al., *Regional Variability and Uncertainty of Electric Vehicle Life Cycle CO_2 Emissions Across the United States.* Environmental Science & Technology, 2015. **49**(14): p. 8844-8855.

749. Yüksel, T., et al., *Effect of Regional Grid Mix, Driving Patterns and Climate on the Comparative Carbon Footprint of Gasoline and Plug-in Electric Vehicles in the United States.* Environmental Research Letters, 2016. **11**(4): p. 044007.

750. Archsmith, J., A. Kendall, and D. Rapson, *From Cradle to Junkyard: Assessing the Life Cycle Greenhouse Gas Benefits of Electric Vehicles.* Research in Transportation Economics, 2015. **52**: p. 72-90.

751. Raman, V., J. Venkatesan, and R.A. Subramanian, *Comparative Analysis of CO2 Emissions of Electric Ride-Hailing Vehicles Over Conventional Gasoline Personal Vehicles.* 2024.

752. Dai, Q., et al., *Life Cycle Analysis of Lithium-Ion Batteries for Automotive Applications.* Batteries, 2019. **5**(2): p. 48.

753. Gan, Y., et al., *Cradle-to-Grave Lifecycle Analysis of Greenhouse Gas Emissions of Light-Duty Passenger Vehicles in China: Towards a Carbon-Neutral Future.* Sustainability, 2023. **15**(3): p. 2627.

754. Wang, B., et al., *Assessing the Supply Risk of Geopolitics on Critical Minerals for Energy Storage Technology in China.* Frontiers in Energy Research, 2023. **10**.

755. Jursová, S., D. Burchart-Korol, and P. Pustějovská, *Carbon Footprint and Water Footprint of Electric Vehicles and Batteries Charging in View of Various Sources of Power Supply in the Czech Republic.* Environments, 2019. **6**(3): p. 38.

756. Elgowainy, A., et al., *Current and Future United States Light-Duty Vehicle Pathways: Cradle-to-Grave Lifecycle Greenhouse Gas Emissions and Economic Assessment.* Environmental Science & Technology, 2018. **52**(4): p. 2392-2399.

757. Hoehne, C. and M. Chester, *Optimizing Plug-in Electric Vehicle and Vehicle-to-Grid Charge Scheduling to Minimize Carbon Emissions.* Energy, 2016. **115**: p. 646-657.

758. Muha, R. and A. PeroŠA, *Energy Consumption and Carbon Footprint of an Electric Vehicle and a Vehicle With an Internal Combustion Engine.* Transport Problems, 2018. **13**(2): p. 49-58.

759. Sheth, R.P., et al., *The Lithium-Ion Battery Recycling Process From a Circular Economy Perspective—A Review and Future Directions.* Energies, 2023. **16**(7): p. 3228.

760. Mrozik, W., et al., *Environmental Impacts, Pollution Sources and Pathways of Spent Lithium-Ion Batteries.* Energy & Environmental Science, 2021. **14**(12): p. 6099-6121.

761. Liu, Y., et al., *Current and Future Lithium-Ion Battery Manufacturing.* Iscience, 2021. **24**(4): p. 102332.

762. Ayerbe, E., et al., *Digitalization of Battery Manufacturing: Current Status, Challenges, and Opportunities.* Advanced Energy Materials, 2021. **12**(17).

763. Gastol, D., et al., *Reclaimed and Up-Cycled Cathodes for Lithium-Ion Batteries.* Global Challenges, 2022. **6**(12).

764. Neumann, J., et al., *Recycling of Lithium-Ion Batteries—Current State of the Art, Circular Economy, and Next Generation Recycling.* Advanced Energy Materials, 2022. **12**(17).

765. Rachmadhani, D.R. and B. Priyono, *Techno-Economic Analysis of the Business Potential of Recycling Lithium-Ion Batteries Using Hydrometallurgical Methods.* International Journal of Engineering Business and Social Science, 2024. **2**(02): p. 938-948.

766. Du, H., et al., *Easily Recyclable Lithium-ion Batteries: Recycling-oriented Cathode Design Using Highly Soluble LiFeMnPO$_4$ With a Water-soluble Binder.* Battery Energy, 2023. **2**(4).

767. Abdelbaky, M., J. Peeters, and W. Dewulf, *On the Influence of Second Use, Future Battery Technologies, and Battery Lifetime on the Maximum Recycled Content of Future Electric Vehicle Batteries in Europe.* Waste Management, 2021. **125**: p. 1-9.

768. Misch, F., Y. Camara, and B. Holtsmark, *Electric Vehicles, Tax Incentives and Emissions: Evidence From Norway.* Imf Working Paper, 2021. **2021**(162): p. 1.

769. Fridstrøm, L., *The Norwegian Vehicle Electrification Policy and Its Implicit Price of Carbon.* Sustainability, 2021. **13**(3): p. 1346.

770. Mersky, A.C., et al., *Effectiveness of Incentives on Electric Vehicle Adoption in Norway.* Transportation Research Part D Transport and Environment, 2016. **46**: p. 56-68.

771. Aasness, M.A. and J. Odeck, *The Increase of Electric Vehicle Usage in Norway—incentives and Adverse Effects.* European Transport Research Review, 2015. **7**(4).

772. Li, M., et al., *How Shenzhen, China Pioneered the Widespread Adoption of Electric Vehicles in a Major City: Implications for Global Implementation.* Wiley Interdisciplinary Reviews Energy and Environment, 2020. **9**(4).

773. Li, W., M. Yang, and S. Sandu, *Electric Vehicles in China: A Review of Current Policies.* Energy & Environment, 2018. **29**(8): p. 1512-1524.

774. Santos, G. and H. Davies, *Incentives for Quick Penetration of Electric Vehicles in Five European Countries: Perceptions From Experts and Stakeholders.* Transportation Research Part a Policy and Practice, 2020. **137**: p. 326-342.

775. Zhang, X., et al., *Policy Incentives for the Adoption of Electric Vehicles Across Countries.* Sustainability, 2014. **6**(11): p. 8056-8078.

776. Palmer, K., et al., *Total Cost of Ownership and Market Share for Hybrid and Electric Vehicles in the UK, US and Japan.* Applied Energy, 2018. **209**: p. 108-119.

777. Shi, X., et al., *Electric Vehicle Transformation in Beijing and the Comparative Eco-Environmental Impacts: A Case Study of Electric and Gasoline Powered Taxis.* Journal of Cleaner Production, 2016. **137**: p. 449-460.

778. Kanz, O., et al., *Environmental Impacts of Integrated Photovoltaic Modules in Light Utility Electric Vehicles.* Energies, 2020. **13**(19): p. 5120.

779. Xue, Y., et al., *Understanding the Barriers to Consumer Purchasing of Electric Vehicles: The Innovation Resistance Theory.* Sustainability, 2024. **16**(6): p. 2420.

780. Ramadani, V., et al., *Antecedents of Electric Vehicle Purchasing Behaviors: Evidence From Türkiye.* Business Ethics the Environment & Responsibility, 2024. **34**(2): p. 456-472.

781. Gao, Z., Y. Hu, and S. Tang, *Investigating the Impact of New Energy Policy on the Market for New Energy Vehicles.* Advances in Economics Management and Political Sciences, 2023. **64**(1): p. 213-221.

782. Setiawan, I.C., *Policy Simulation of Electricity-Based Vehicle Utilization in Indonesia (Electrified Vehicle - HEV, PHEV, BEV and FCEV).* Automotive Experiences, 2019. **2**(1): p. 1-8.

783. Neubauer, J., E. Wood, and A. Pesaran, *A Second Life for Electric Vehicle Batteries: Answering Questions on Battery Degradation and Value.* Sae International Journal of Materials and Manufacturing, 2015. **8**(2): p. 544-553.

784. Senol, M., et al., *Electric Vehicles Under Low Temperatures: A Review on Battery Performance, Charging Needs, and Power Grid Impacts.* Ieee Access, 2023. **11**: p. 39879-39912.

785. Neaimeh, M., et al., *Charge Control of Second Life EV Batteries on the DC Link of a Back-to-Back Converter.* Cired - Open Access Proceedings Journal, 2017. **2017**(1): p. 1625-1628.

786. Suárez-Bertoa, R., et al., *On-Road Emissions of Passenger Cars Beyond the Boundary Conditions of the Real-Driving Emissions Test.* Environmental Research, 2019. **176**: p. 108572.

787. Kawamoto, R., et al., *Estimation of CO2 Emissions of Internal Combustion Engine Vehicle and Battery Electric Vehicle Using LCA.* Sustainability, 2019. **11**(9): p. 2690.

788. Simons, A., *Road Transport: New Life Cycle Inventories for Fossil-Fuelled Passenger Cars and Non-Exhaust Emissions in Ecoinvent V3.* The International Journal of Life Cycle Assessment, 2013. **21**(9): p. 1299-1313.

789. Müller, D., T. Dufaux, and K.P. Birke, *Model-Based Investigation of Porosity Profiles in Graphite Anodes Regarding Sudden-Death and Second-Life of Lithium Ion Cells.* Batteries, 2019. **5**(2): p. 49.

790. Machuca, E., et al., *Availability of Lithium Ion Batteries From Hybrid and Electric Cars for Second Use: How to Forecast for Germany Until 2030.* J of Electrical Engineering, 2018. **6**(3).

791. Stanojević, N., M. Vasić, and V. Popović, *The Contribution of CNG Powered Vehicles in the Transition to Zero Emission Mobility - Example of the Light Commercial Vehicles Fleet.* Thermal Science, 2021. **25**(3 Part A): p. 1867-1878.

792. Hagedorn, T. and G. Sieg, *Emissions and External Environmental Costs From the Perspective of Differing Travel Purposes.* Sustainability, 2019. **11**(24): p. 7233.

793. Reşitoğlu, İ.A., K. Altınışık, and A. Keskin, *The Pollutant Emissions From Diesel-Engine Vehicles and Exhaust Aftertreatment Systems.* Clean Technologies and Environmental Policy, 2014. **17**(1): p. 15-27.

794. Li, Y., N. Ha, and T. Li, *Research on Carbon Emissions of Electric Vehicles Throughout the Life Cycle Assessment Taking Into Vehicle Weight and Grid Mix Composition.* Energies, 2019. **12**(19): p. 3612.

795. Uliasz-Misiak, B., et al., *Prospects for the Implementation of Underground Hydrogen Storage in the EU.* Energies, 2022. **15**(24): p. 9535.

796. Staffell, I., et al., *The Role of Hydrogen and Fuel Cells in the Global Energy System.* Energy & Environmental Science, 2019. **12**(2): p. 463-491.

797. Yap, J. and B. McLellan, *A Historical Analysis of Hydrogen Economy Research, Development, and Expectations, 1972 to 2020.* Environments, 2023. **10**(1): p. 11.

798. Azzeh, S.E., M.M. Sarshar, and R. Fayaz, *Hydrogen Economy and the Built Environment.* 2011. **57**: p. 1986-1995.

799. Bockris, J.O.M., *A Hydrogen Economy.* Science, 1972. **176**(4041): p. 1323-1323.

800. Bockris, J.O.M., *The Hydrogen Economy: Its History.* International Journal of Hydrogen Energy, 2013. **38**(6): p. 2579-2588.

801. Lo, S.L.Y., et al., *Synthesis of Bio-Hydrogen Supply Network via Graph-Theoretic Approach Coupled With Monte Carlo Simulation Model.* Iop Conference Series Materials Science and Engineering, 2022. **1257**(1): p. 012011.

802. Seo, B. and S.H. Joo, *Recent Advances in Unveiling Active Sites in Molybdenum Sulfide-Based Electrocatalysts for the Hydrogen Evolution Reaction.* Nano Convergence, 2017. **4**(1).

803. Barba, D., et al., *Mass Transfer Coefficient in Multi-Stage Reformer/Membrane Modules for Hydrogen Production.* 2018.

804. Shabani, B. and J. Andrews, *Hydrogen and Fuel Cells.* 2015: p. 453-491.

805. Veenstra, A.T., et al., *Green Hydrogen Plant: Optimal Control Strategies for Integrated Hydrogen Storage and Power Generation With Wind Energy.* 2021.

806. Wang, N., Z. Feng, and X. Guo, *Coordinated Operation Strategy for Hydrogen Energy Storage in the Incremental Distribution Network.* International Journal of Energy Research, 2022. **46**(15): p. 24158-24178.

807. Xie, P., et al., *Resilience Enhancement Strategies for Power Distribution Network Based on Hydrogen Storage and Hydrogen Vehicle.* Iet Energy Systems Integration, 2024. **6**(S1): p. 789-798.

808. Cullen, D.A., et al., *New Roads and Challenges for Fuel Cells in Heavy-Duty Transportation.* Nature Energy, 2021. **6**(5): p. 462-474.

809. Griffiths, S., et al., *Industrial Decarbonization via Hydrogen: A Critical and Systematic Review of Developments, Socio-Technical Systems and Policy Options.* Energy Research & Social Science, 2021. **80**: p. 102208.

810. Yang, M., L. Liu, and J. Du, *Design of Hydrogen Supply Chain Networks for Cross-Regional Distribution.* Industrial & Engineering Chemistry Research, 2025. **64**(8): p. 4498-4515.

811. Ibrahim, A., M. Paskevicius, and C.E. Buckley, *Chemical Compression and Transport of Hydrogen Using Sodium Borohydride.* Sustainable Energy & Fuels, 2023. **7**(5): p. 1196-1203.

812. Crandall, B.S., et al., *Techno-Economic Assessment of Green H$_2$ Carrier Supply Chains.* Energy & Fuels, 2022. **37**(2): p. 1441-1450.

813. Kar, S.K., et al., *Hydrogen Economy in India: A Status Review.* Wiley Interdisciplinary Reviews Energy and Environment, 2022. **12**(1).

814. Huang, J., W. Li, and X. Wu, *Forecasting the Hydrogen Demand in China: A System Dynamics Approach.* Mathematics, 2022. **10**(2): p. 205.

815. Berstad, D., et al., *Low-Temperature Phase Separation of CO_2 From Syngas Mixtures–Experimental Results.* Industrial & Engineering Chemistry Research, 2024. **63**(50): p. 21960-21973.

816. Claesen, J., D. Valkenborg, and T. Burzykowski, *A "Refined Hydrogen Rule" and a "Refined Hydrogen and Halogen Rule" for Organic Molecules.* Journal of the American Society for Mass Spectrometry, 2019. **31**(1): p. 132-136.

817. Shi, X., et al., *A Study on the Promoting Role of Renewable Hydrogen in the Transformation of Petroleum Refining Pathways.* Processes, 2024. **12**(7): p. 1317.

818. Kar, S.K., et al., *Overview of Hydrogen Economy in Australia.* Wiley Interdisciplinary Reviews Energy and Environment, 2022. **12**(1).

819. Kim, Y., et al., *An Analysis of Greenhouse Gas Emissions in Electrolysis for Certifying Clean Hydrogen.* Energies, 2024. **17**(15): p. 3698.

820. Ghezelbash, A., et al., *Impacts of Green Energy Expansion and Gas Import Reduction on South Korea's Economic Growth: A System Dynamics Approach.* Sustainability, 2023. **15**(12): p. 9281.

821. He, G., et al., *Sector Coupling via Hydrogen to Lower the Cost of Energy System Decarbonization.* Energy & Environmental Science, 2021. **14**(9): p. 4635-4646.

822. Bian, H., et al., *Coordinated Planning of Electricity-Hydrogen Integrated Energy System Considering Lifecycle Carbon Emissions.* Ieee Access, 2024. **12**: p. 33889-33909.

823. Skiba, R., *Insight into Carbon Capture and Storage (CCS) Technology.* Organic and Medicinal Chemistry International Journal, 2020. **9**(5): p. 129-132.

824. Lau, H.C., *Evaluation of Decarbonization Technologies for ASEAN Countries via an Integrated Assessment Tool.* Sustainability, 2022. **14**(10): p. 5827.

825. Lau, H.C., et al., *A Review of the Status of Fossil and Renewable Energies in Southeast Asia and Its Implications on the Decarbonization of ASEAN.* Energies, 2022. **15**(6): p. 2152.

826. Esiri, A.E., D.D. Jambol, and C. Ozowe, *Best Practices and Innovations in Carbon Capture and Storage (CCS) for Effective CO2 Storage.* International Journal of Applied Research in Social Sciences, 2024. **6**(6): p. 1227-1243.

827. Chauvy, R. and G.D. Weireld, *CO_2 Utilization Technologies in Europe: A Short Review.* Energy Technology, 2020. **8**(12).

828. Rumayor, M., et al., *Deep Decarbonization of the Cement Sector: A Prospective Environmental Assessment of CO_2 Recycling to Methanol.* Acs Sustainable Chemistry & Engineering, 2021. **10**(1): p. 267-278.

829. Gangotra, A., E.D. Gado, and J.I. Lewis, *3D Printing Has Untapped Potential for Climate Mitigation in the Cement Sector.* Communications Engineering, 2023. **2**(1).

830. Lau, H.C. and S. Tsai, *Net Zero by 2050: A Tale of Three Economies.* Energy & Fuels, 2024. **38**(21): p. 21079-21094.

831. Lau, H.C., et al., *The Role of Carbon Capture and Storage in the Energy Transition.* Energy & Fuels, 2021. **35**(9): p. 7364-7386.

832. Lau, H.C., *The Contribution of Carbon Capture and Storage to the Decarbonization of Coal-Fired Power Plants in Selected Asian Countries.* Energy & Fuels, 2023. **37**(20): p. 15919-15934.

833. Franco, A. and C. Giovannini, *Routes for Hydrogen Introduction in the Industrial Hard-to-Abate Sectors for Promoting Energy Transition.* Energies, 2023. **16**(16): p. 6098.

834. Rumayor, M., A. Domínguez-Ramos, and Á. Irabien, *Toward the Decarbonization of Hard-to-Abate Sectors: A Case Study of the Soda Ash Production.* Acs Sustainable Chemistry & Engineering, 2020. **8**(32): p. 11956-11966.

835. Franco, A. and C. Giovannini, *Integration of Hydrogen for Decarbonisation: The Possible Contribution in "Hard-to-Abate" Sectors.* Journal of Physics Conference Series, 2024. **2893**(1): p. 012069.

836. Ringrose, P., et al., *Storage of Carbon Dioxide in Saline Aquifers: Physicochemical Processes, Key Constraints, and Scale-Up Potential.* Annual Review of Chemical and Biomolecular Engineering, 2021. **12**(1): p. 471-494.

837. Aminu, M.D., et al., *A Review of Developments in Carbon Dioxide Storage.* Applied Energy, 2017. **208**: p. 1389-1419.

838. Keerthana, K.B., et al., *The United States Energy Consumption and Carbon Dioxide Emissions: A Comprehensive Forecast Using a Regression Model.* Sustainability, 2023. **15**(10): p. 7932.

839. Sanchez, D.L. and D.S. Callaway, *Optimal Scale of Carbon-Negative Energy Facilities.* Applied Energy, 2016. **170**: p. 437-444.

840. Lin, N. and L. Xu, *Navigating the Implementation of Tax Credits for Natural-Gas-Based Low-Carbon-Intensity Hydrogen Projects.* Energies, 2024. **17**(7): p. 1604.

841. Akinola, T.E., et al., *Non-Linear System Identification of Solvent-Based Post-Combustion CO2 Capture Process.* Fuel, 2019. **239**: p. 1213-1223.

842. Osazuwa-Peters, M. and M. Hurlbert, *Analyzing Regulatory Framework for Carbon Capture and Storage (CCS) Technology Development: A Case Study Approach.* The Central European Review of Economics and Management, 2020. **4**(1): p. 107-148.

843. Grubert, E. and F. Sawyer, *US Power Sector Carbon Capture and Storage Under the Inflation Reduction Act Could Be Costly With Limited or Negative Abatement Potential.* Environmental Research Infrastructure and Sustainability, 2023. **3**(1): p. 015008.

844. Bui, M., et al., *Carbon Capture and Storage (CCS): The Way Forward.* Energy & Environmental Science, 2018. **11**(5): p. 1062-1176.

845. Massol, O., S. Tchung-Ming, and A. Banal-Estañol, *Joining the CCS Club! The Economics of CO 2 Pipeline Projects.* European Journal of Operational Research, 2015. **247**(1): p. 259-275.

846. Nimmala, R., *Evaluating Financial Risks of Carbon Capture in the Oil &Amp; Gas Industry Using Advanced Machine Learning Techniques.* Design of Single Chip Microcomputer Control System for Stepping Motor, 2024: p. 1-5.

847. Chyong, C.K., E. Italiani, and N. Kazantzis, *Implications of the Inflation Reduction Act on Deployment of Low-Carbon Ammonia Technologies.* 2023.

848. Wang, X. and H. Zhang, *Valuation of CCS Investment in China's Coal-fired Power Plants Based on a Compound Real Options Model.* Greenhouse Gases Science and Technology, 2018. **8**(5): p. 978-988.

849. Wang, X. and H. Zhang, *Optimal Design of Carbon Tax to Stimulate CCS Investment in China's Coal-fired Power Plants: A Real Options Analysis.* Greenhouse Gases Science and Technology, 2018. **8**(5): p. 863-875.

850. Supekar, S.D. and S.J. Skerlos, *Sourcing of Steam and Electricity for Carbon Capture Retrofits.* Environmental Science & Technology, 2017. **51**(21): p. 12908-12917.

851. Zhang, W. and T. Qing, *The Effects of Policy Subsidy on the Investment Decisions of Carbon Capture and Storage—A Real-options Approach.* Greenhouse Gases Science and Technology, 2022. **12**(6): p. 698-711.

852. Shakirova, R., *An Investigation of Government Employees' Support for Public-Private Partnerships.* International Journal of Public Sector Management, 2017. **30**(5): p. 467-486.

853. Setyawati, C.E.N. and S. Wibawa, *Investigating the Impacts of Carbon Pricing Mechanism on CCS Development in ASEAN Countries.* Iop Conference Series Earth and Environmental Science, 2024. **1395**(1): p. 012034.

854. Naseeb, A., A. Ramadan, and S.M. Al–Salem, *Economic Feasibility Study of a Carbon Capture and Storage (CCS) Integration Project in an Oil-Driven Economy: The Case of the State of Kuwait.* International Journal of Environmental Research and Public Health, 2022. **19**(11): p. 6490.

855. Wang, Y., et al., *The Long-Term Impact of Carbon Emission Trading and Renewable Energy Support Policy on China's Power Sector Under the Context of Electricity Marketed Reform: An Analysis Based on System Dynamics.* 2023.

856. Woodall, C.M. and C. McCormick, *Assessing the Optimal Uses of Biomass: Carbon and Energy Price Conditions for the Aines Principle to Apply.* Frontiers in Climate, 2022. **4**.

857. Antonakakis, N., et al., *Geopolitical Risks and the Oil-Stock Nexus Over 1899–2016.* Finance Research Letters, 2017. **23**: p. 165-173.

858. Li, F., et al., *Do Tense Geopolitical Factors Drive Crude Oil Prices?* Energies, 2020. **13**(16): p. 4277.

859. Plakandaras, V., R. Gupta, and W.K. Wong, *Point and Density Forecasts of Oil Returns: The Role of Geopolitical Risks.* Resources Policy, 2019. **62**: p. 580-587.

860. Li, B., et al., *Oil Prices and Geopolitical Risks: What Implications Are Offered via Multi-Domain Investigations?* Energy & Environment, 2019. **31**(3): p. 492-516.

861. Shahbaz, M., et al., *Oil Prices and Geopolitical Risk: Fresh Insights Based on <scp>G</Scp>ranger-causality in Quantiles Analysis.* International Journal of Finance & Economics, 2023. **29**(3): p. 2865-2881.

862. Abdel-Latif, H., M.A. El-Gamal, and A.M. Jaffe, *The Ephemeral Brent Geopolitical Risk Premium.* Economics of Energy and Environmental Policy, 2020. **9**(2).

863. Hoque, M.E., L.S. Wah, and M.A.S. Zaidi, *Oil Price Shocks, Global Economic Policy Uncertainty, Geopolitical Risk, and Stock Price in Malaysia: Factor Augmented VAR Approach.* Economic Research-Ekonomska Istraživanja, 2019. **32**(1): p. 3700-3732.

864. Omar, A., T.P. Wisniewski, and S. Nolte, *Diversifying Away the Risk of War and Cross-Border Political Crisis.* Energy Economics, 2017. **64**: p. 494-510.

865. Considine, J.I. and M.L. Barcella, *Foreword.* 2018.

866. Abdel-Latif, H. and M.A. El-Gamal, *Antecedents of War: The Geopolitics of Low Oil Prices and Decelerating Financial Liquidity.* Applied Economics Letters, 2018. **26**(9): p. 765-769.

867. Gamso, J., A.C. Inkpen, and K. Ramaswamy, *Managing Geopolitical Risks: The Global Oil and Gas Industry Plays a Winning Game*. Journal of Business Strategy, 2023. **45**(3): p. 190-198.

868. Aslam, F., et al., *Application of Multifractal Analysis in Estimating the Reaction of Energy Markets to Geopolitical Acts and Threats*. Sustainability, 2022. **14**(10): p. 5828.

869. Γκίλλας, K., R. Gupta, and C. Pierdzioch, *Forecasting Realized Gold Volatility: Is There a Role of Geopolitical Risks?* Finance Research Letters, 2020. **35**: p. 101280.

870. Ekwurzel, B., et al., *The Rise in Global Atmospheric CO2, Surface Temperature, and Sea Level From Emissions Traced to Major Carbon Producers*. Climatic Change, 2017. **144**(4): p. 579-590.

871. Piggot, G., et al., *Curbing Fossil Fuel Supply to Achieve Climate Goals*. Climate Policy, 2020. **20**(8): p. 881-887.

872. Piggot, G., *The Influence of Social Movements on Policies That Constrain Fossil Fuel Supply*. Climate Policy, 2017. **18**(7): p. 942-954.

873. Ross, M.L., C. Hazlett, and P. Mahdavi, *Global Progress and Backsliding on Gasoline Taxes and Subsidies*. Nature Energy, 2017. **2**(1).

874. Ploeg, F.v.d. and A. Rezai, *Stranded Assets in the Transition to a Carbon-Free Economy*. Annual Review of Resource Economics, 2020. **12**(1): p. 281-298.

875. Asheim, G.B., et al., *The Case for a Supply-Side Climate Treaty*. Science, 2019. **365**(6451): p. 325-327.

876. McDonnell, C., *Pension Funds and Fossil Fuel Phase-Out: Historical Developments and Limitations of Pension Climate Strategies*. International Environmental Agreements Politics Law and Economics, 2024. **24**(1): p. 169-191.

877. Bauer, N., et al., *Global Fossil Energy Markets and Climate Change Mitigation – An Analysis With REMIND*. Climatic Change, 2013. **136**(1): p. 69-82.

878. Ye, Z., et al., *The Dynamic Time-Frequency Relationship Between International Oil Prices and Investor Sentiment in China: A Wavelet Coherence Analysis*. The Energy Journal, 2020. **41**(5): p. 251-270.

879. Pierru, A., J. Smith, and T. Zamrik, *OPEC's Impact on Oil Price Volatility: The Role of Spare Capacity*. The Energy Journal, 2018. **39**(2): p. 173-196.

880. Almutairi, H., A. Pierru, and L. Smith, *Oil Market Stabilization: The Performance of OPEC and Its Allies*. The Energy Journal, 2023. **44**(6): p. 1-22.

881. Economou, A. and B. Fattouh, *OPEC at 60: The World With and Without OPEC*. Opec Energy Review, 2021. **45**(1): p. 3-28.

882. Mukhtarov, S., et al., *Do High Oil Prices Obstruct the Transition to Renewable Energy Consumption?* Sustainability, 2020. **12**(11): p. 4689.

883. Sanati, R., *OPEC and the International System: A Political History of Decisions and Behavior*.

884. Colgan, J.D., *The Emperor Has No Clothes: The Limits of OPEC in the Global Oil Market*. International Organization, 2014. **68**(3): p. 599-632.

885. Graaf, T.V.d., *Is OPEC Dead? Oil Exporters, the Paris Agreement and the Transition to a Post-Carbon World*. Energy Research & Social Science, 2017. **23**: p. 182-188.

886. Behar, A. and R.A. Ritz, *OPEC vs US Shale: Analyzing the Shift to a Market-Share Strategy*. Energy Economics, 2017. **63**: p. 185-198.

887. Kalu, E.U., et al., *Green or Gas in OPEC Member Countries: A Linear and Asymmetric Investigation of Energy–growth Nexus*. Opec Energy Review, 2020. **44**(4): p. 451-485.

888. Okullo, S.J. and F. Reynès, *Can Reserve Additions in Mature Crude Oil Provinces Attenuate Peak Oil?* Energy, 2011. **36**(9): p. 5755-5764.

889. Johansson, D., et al., *OPEC Strategies and Oil Rent in a Climate Conscious World.* The Energy Journal, 2009. **30**(3): p. 23-50.

890. Goldthau, A. and J.M. Witte, *Assessing OPEC's Performance in Global Energy.* Global Policy, 2011. **2**(s1): p. 31-39.

891. Alabi, O., I. Ackah, and A. Lartey, *Re-Visiting the Renewable Energy–economic Growth Nexus.* International Journal of Energy Sector Management, 2017. **11**(3): p. 387-403.

892. Omri, A. and D.K. Nguyen, *On the Determinants of Renewable Energy Consumption: International Evidence.* Energy, 2014. **72**: p. 554-560.

893. Ang, B.W., W.L. Choong, and T.S. Ng, *Energy Security: Definitions, Dimensions and Indexes.* Renewable and Sustainable Energy Reviews, 2015. **42**: p. 1077-1093.

894. Cohen, G., F.L. Joutz, and P. Loungani, *Measuring Energy Security: Trends in the Diversification of Oil and Natural Gas Supplies.* Energy Policy, 2011. **39**(9): p. 4860-4869.

895. Bradshaw, M., *Global Energy Dilemmas: A Geographical Perspective.* Geographical Journal, 2010. **176**(4): p. 275-290.

896. Guta, D.D. and J. Börner, *Energy Security, Uncertainty and Energy Resource Use Options in Ethiopia.* International Journal of Energy Sector Management, 2017. **11**(1): p. 91-117.

897. Chalvatzis, K. and A. Ioannidis, *Energy Supply Security in the EU: Benchmarking Diversity and Dependence of Primary Energy.* Applied Energy, 2017. **207**: p. 465-476.

898. Aslani, A., E. Antila, and K.V. Wong, *Comparative Analysis of Energy Security in the Nordic Countries: The Role of Renewable Energy Resources in Diversification.* Journal of Renewable and Sustainable Energy, 2012. **4**(6).

899. Weiner, C., *Revisiting the Management of Stationary Fuel Supply Security and Gas Diversification in Hungary.* SSRN Electronic Journal, 2019.

900. Yakoviyk, I. and M. Tsvelikh, *Energy Security of the European Union in the Context of Russian Aggression Against Ukraine.* Problems of Legality, 2023(160): p. 170-191.

901. Stetsiuk, M., *Germany's Position Concerning the Nord Stream-2 Pipeline in the Context of the European Union's Energy Security.* Mediaforum Analytics Forecasts Information Management, 2021(9): p. 108-127.

902. Limi, F.B., *Effect of Financial Development on Energy Diversification in Sub-Saharan Africa.* Journal of Energy Research and Reviews, 2020: p. 29-38.

903. Štreimikienė, D., I. Šikšnelytė-Butkienė, and V. Lekavičius, *Energy Diversification and Security in the EU: Comparative Assessment in Different EU Regions.* Economies, 2023. **11**(3): p. 83.

904. Anwar, J., *The Role of Renewable Energy Supply and Carbon Tax in the Improvement of Energy Security: A Case Study of Pakistan.* The Pakistan Development Review, 2022: p. 347-370.

905. Seriño, M.N.V., *Energy Security Through Diversification of Non-Hydro Renewable Energy Sources in Developing Countries.* Energy & Environment, 2021. **33**(3): p. 546-561.

906. Pavlović, B. and D. Ivezić, *Availability as a Dimension of Energy Security in the Republic of Serbia.* Thermal Science, 2017. **21**(1 Part A): p. 323-333.

907. Cavus, M., *Integration Smart Grids, Distributed Generation, and Cybersecurity: Strategies for Securing and Optimizing Future Energy Systems.* 2024.

908. Alimi, O.A., K. Ouahada, and A.M. Abu-Mahfouz, *A Review of Machine Learning Approaches to Power System Security and Stability.* Ieee Access, 2020. **8**: p. 113512-113531.

909.   Solaymani, S., *Energy Security and Its Determinants in New Zealand.* 2024.

910.   Owusu, P.A. and S.A. Sarkodie, *A Review of Renewable Energy Sources, Sustainability Issues and Climate Change Mitigation.* Cogent Engineering, 2016. **3**(1): p. 1167990.

911.   Laldjebaev, M., et al., *Rethinking Energy Security and Services in Practice: National Vulnerability and Three Energy Pathways in Tajikistan.* Energy Policy, 2018. **114**: p. 39-50.

912.   Kuzior, A., et al., *Applying Energy Taxes to Promote a Clean, Sustainable and Secure Energy System: Finding the Preferable Approaches.* Energies, 2023. **16**(10): p. 4203.

913.   Chowaniak, M., et al., *The RES in the Countries of the Commonwealth of Independent States: Potential and Production From 2015 to 2019.* Energies, 2021. **14**(7): p. 1856.

914.   Kim, S.E. and J. Urpelainen, *Democracy, Autocracy and the Urban Bias: Evidence From Petroleum Subsidies.* Political Studies, 2015. **64**(3): p. 552-572.

915.   Coady, D., V. Flamini, and L.M. Sears, *The Unequal Benefits of Fuel Subsidies Revisited: Evidence for Developing Countries.* Imf Working Paper, 2015. **15**(250): p. 1.

916.   Amin, S.B., L. Marsiliani, and T.I. Renström, *The Impacts of Fossil Fuel Subsidy Removal on Bangladesh Economy.* The Bangladesh Development Studies, 2018. **XLI**(2): p. 65-86.

917.   Kyle, J., *Local Corruption and Popular Support for Fuel Subsidy Reform in Indonesia.* Comparative Political Studies, 2018. **51**(11): p. 1472-1503.

918.   Jara, H.X., et al., *Fuel Subsidies and Income Redistribution in Ecuador.* 2018.

919.   Anand, R., et al., *The Fiscal and Welfare Impacts of Reforming Fuel Subsidies in India.* Imf Working Paper, 2013. **13**(128): p. 1.

920.   Couharde, C. and S. Mouhoud, *Fossil Fuel Subsidies, Income Inequality, and Poverty: Evidence From Developing Countries.* Journal of Economic Surveys, 2020. **34**(5): p. 981-1006.

921.   Park, K., Y. Lee, and H. Joon, *Economic Perspective on Discontinuing Fossil Fuel Subsidies and Moving Toward a Low-Carbon Society.* Sustainability, 2021. **13**(3): p. 1217.

922.   Schaffitzel, F., et al., *Can Government Transfers Make Energy Subsidy Reform Socially Acceptable? A Case Study on Ecuador.* Energy Policy, 2020. **137**: p. 111120.

923.   Schmidt, T.S., T. Matsuo, and A. Michaelowa, *Renewable Energy Policy as an Enabler of Fossil Fuel Subsidy Reform? Applying a Socio-Technical Perspective to the Cases of South Africa and Tunisia.* Global Environmental Change, 2017. **45**: p. 99-110.

924.   Berument, H., N.B. Ceylan, and N. Doğan, *The Impact of Oil Price Shocks on the Economic Growth of Selected MENA1 Countries.* The Energy Journal, 2010. **31**(1): p. 149-176.

925.   Moshiri, S. and E. Kheirandish, *Global Impacts of Oil Price Shocks: The Trade Effect.* Journal of Economic Studies, 2023. **51**(1): p. 126-144.

926.   Szira, Z., H. Alghamdi, and E. Varga, *Examining the Impact of Oil Price Change on the Economy Through GDP Change.* Acta Carolus Robertus, 2019. **9**(2): p. 149-159.

927.   Filis, G. and I. Chatziantoniou, *Financial and Monetary Policy Responses to Oil Price Shocks: Evidence From Oil-Importing and Oil-Exporting Countries.* Review of Quantitative Finance and Accounting, 2013. **42**(4): p. 709-729.

928.   Aliyu, S.U.R., *Oil Price Shocks and the Macroeconomy of Nigeria: A Non-Linear Approach.* J for International Business and Entrepreneurship Development, 2011. **5**(3): p. 179.

929.   Ito, K., *The Impact of Oil Price Volatility on the Macroeconomy in Russia.* The Annals of Regional Science, 2010. **48**(3): p. 695-702.

930. Sadeghi, A., *Oil Price Shocks and Economic Growth in Oil-Exporting Countries: Does the Size of Government Matter?* Imf Working Paper, 2017. **17**(287): p. 1.

931. Joseph, A., et al., *Oil Prices and Exchange Rates Causality: New Evidences From Decomposed Oil Prices Shocks and Parametric in Quantile Analysis.* International Journal of Advanced Economics, 2024. **6**(11): p. 678-698.

932. McCulloch, N., T. Moerenhout, and J. Yang, *Fuel Subsidy Reform and the Social Contract in Nigeria: A Micro-Economic Analysis.* Energy Policy, 2021. **156**: p. 112336.

933. Kpodar, K. and P. Imam, *To Pass (Or Not to Pass) Through International Fuel Price Changes to Domestic Fuel Prices in Developing Countries.* Imf Working Paper, 2020. **20**(194).

934. Gil-Alana, L.A. and M. Monge, *Crude Oil Prices and COVID-19: Persistence of the Shock.* Energy Research Letters, 2020. **1**(1).

935. Regnier, E., *Oil and Energy Price Volatility.* Energy Economics, 2007. **29**(3): p. 405-427.

936. Abraham, J.O., A. Vonoyauyau, and S. Narayan, *Price Controlled Petroleum and LPG Prices and COVID-19: Some Evidence From Fiji.* Energy Research Letters, 2023. **4**(4).

937. Omotosho, B.S., *Oil Price Shocks, Fuel Subsidies and Macroeconomic (In)stability in Nigeria.* Central Bank of Nigeria Journal of Applied Statistics, 2020(Vol. 10 No. 2): p. 1-38.

938. Onogbosele, D.O. and M. Adejoh, *Fuel Subsidy Payments and Food Inflation in Nigeria.* 2024.

939. Abimanyu, A. and M.H. Imansyah, *The Impact of Fuel Subsidy to The Income Distribution: The Case of Indonesia.* Indonesian Treasury Review: Jurnal Perbendaharaan, Keuangan Negara dan Kebijakan Publik, 2023. **8**(3): p. 189-203.

940. Pereira, A.M. and R.M. Pereira, *On the Environmental, Economic and Budgetary Impacts of Fossil Fuel Prices: A Dynamic General Equilibrium Analysis of the Portuguese Case.* Energy Economics, 2014. **42**: p. 248-261.

941. Hamilton, J.D., *Understanding Crude Oil Prices.* The Energy Journal, 2009. **30**(2): p. 179-206.

942. Jayasekara, M., *An Analysis of Exchange Rate Pass-Through and Impact on Domestic Inflation in Sri Lanka.* Indian Journal of Economics and Financial Issues, 2022. **3**(1): p. 1-31.

943. Zhang, Y., *Towards Greener Skies: Exploring the Potential of Hydrogen Energy in Aviation.* Theoretical and Natural Science, 2024. **56**(1): p. 105-111.

944. Zhang, C., et al., *Recent Development in Studies of Alternative Jet Fuel Combustion: Progress, Challenges, and Opportunities.* Renewable and Sustainable Energy Reviews, 2016. **54**: p. 120-138.

945. Barke, A., et al., *Are Sustainable Aviation Fuels a Viable Option for Decarbonizing Air Transport in Europe? An Environmental and Economic Sustainability Assessment.* Applied Sciences, 2022. **12**(2): p. 597.

946. Gegg, P., L. Budd, and S. Ison, *The Market Development of Aviation Biofuel: Drivers and Constraints.* Journal of Air Transport Management, 2014. **39**: p. 34-40.

947. Zhang, X., *Green Hydrogen Policies in the Clean Energy Transition- A Comparative Analysis of Russia and China.* Highlights in Science Engineering and Technology, 2023. **80**: p. 356-361.

948. Braga, H.C., R. Gouvea, and M. Gutierrez, *The Green Hydrogen Economy and the Global Net Zero Agenda: A Green Powershoring &Amp;amp; Green Watershoring Perspective.* Modern Economy, 2024. **15**(01): p. 23-48.

949. AlZohbi, G., *Green Hydrogen Generation: Recent Advances and Challenges.* Iop Conference Series Earth and Environmental Science, 2022. **1050**(1): p. 012003.

950. Yang, L., et al., *Current Development Status, Policy Support and Promotion Path of China's Green Hydrogen Industries Under the Target of Carbon Emission Peaking and Carbon Neutrality.* Sustainability, 2023. **15**(13): p. 10118.

951. Chen, K., D. Seo, and P. Canteenwalla, *The Effect of High-Temperature Water Vapour on Degradation and Failure of Hot Section Components of Gas Turbine Engines.* Coatings, 2021. **11**(9): p. 1061.

952. Marszałek, N. and T. Lis, *The Future of Sustainable Aviation Fuels.* Combustion Engines, 2022.

953. Gao, Y., *Sustainable Aviation Fuel as a Pathway to Mitigate Global Warming in the Aviation Industry.* Theoretical and Natural Science, 2023. **26**(1): p. 60-67.

954. Schreyer, F., et al., *From Net-Zero to Zero-Fossil in Transforming the EU Energy System.* 2025.

955. Salvarli, M.S., *Some Views on Renewable Energy Marketing Due to Climate Change Impacts.* Dünya Multidisipliner Araştırmalar Dergisi, 2023. **6**(1): p. 161-180.

956. Iranzo, E.C., et al., *Current Knowledge on the Novel Semiarid Photovoltaic Ecosystems and Their Impacts on Biodiversity.* 2024.

957. Eljack, F. and M.-K. Kazi, *Prospects and Challenges of Green Hydrogen Economy via Multi-Sector Global Symbiosis in Qatar.* Frontiers in Sustainability, 2021. **1**.

958. Franz, S.M. and R. Bramstoft, *Impact of Endogenous Learning Curves on Maritime Transition Pathways.* Environmental Research Letters, 2024. **19**(5): p. 054014.

959. Bullerdiek, N., et al., *Scaling Sustainable Aviation Fuel for Net-Zero CO2 Emissions by 2050 – Pathways, Requirements and Challenges.* 2025.

960. Nikolaidis, P. and A. Poullikkas, *A Thorough Emission-Cost Analysis of the Gradual Replacement of Carbon-Rich Fuels With Carbon-Free Energy Carriers in Modern Power Plants: The Case of Cyprus.* Sustainability, 2022. **14**(17): p. 10800.

961. Loy, A.C.M., et al., *Technical, Economic, and Environmental Potential of Glycerol Hydrogenolysis: A Roadmap Towards Sustainable Green Chemistry Future.* Sustainable Energy & Fuels, 2023. **7**(11): p. 2653-2669.

962. Al-Habaibeh, B., *Challenges and Opportunities of the Implementation of NetZero Strategy: Analysis of Key Issues That Affect the Adoption of Domestic Renewable Energy in the UK From Consumers' Perspective.* 2024.

963. Lelieveld, J., et al., *Air Pollution Deaths Attributable to Fossil Fuels: Observational and Modelling Study.* BMJ, 2023: p. e077784.

964. Ampah, J.D., et al., *Prioritizing Non-Carbon Dioxide Removal Mitigation Strategies Could Reduce the Negative Impacts Associated With Large-Scale Reliance on Negative Emissions.* Environmental Science & Technology, 2024. **58**(8): p. 3755-3765.

965. Lindstad, E., et al., *Smart Use of Renewable Electricity and Carbon Capture in the Transport Sector.* 2023.

966. Radovanović, M., et al., *Decarbonisation of Eastern European Economies: Monitoring, Economic, Social and Security Concerns.* Energy Sustainability and Society, 2022. **12**(1).

967. Wójcik-Jurkiewicz, M., et al., *Determinants of Decarbonisation in the Transformation of the Energy Sector: The Case of Poland.* Energies, 2021. **14**(5): p. 1217.

968. Carmona-Martínez, A.A., et al., *Renewable Power and Heat for the Decarbonisation of Energy-Intensive Industries.* Processes, 2022. **11**(1): p. 18.

969. Pushkar, T.I., et al., *Decarbonisation of Cities in the Context of Their Sustainable Inclusive Development.* Iop Conference Series Earth and Environmental Science, 2024. **1376**(1): p. 012028.

970. Svensson, O., J. Khan, and R. Hildingsson, *Studying Industrial Decarbonisation: Developing an Interdisciplinary Understanding of the Conditions for Transformation in Energy-Intensive Natural Resource-Based Industry.* Sustainability, 2020. **12**(5): p. 2129.

971. Holder, D., S. Percy, and A. Yavari, *A Review of Port Decarbonisation Options: Identified Opportunities for Deploying Hydrogen Technologies.* Sustainability, 2024. **16**(8): p. 3299.

972. Urrutia-Azcona, K., et al., *ENER-BI: Integrating Energy and Spatial Data for Cities' Decarbonisation Planning.* Sustainability, 2021. **13**(1): p. 383.

973. Fotiou, T., P. Fragkos, and E. Zisarou, *Decarbonising the EU Buildings|Model-Based Insights From European Countries.* Climate, 2024. **12**(6): p. 85.

974. Yin, J., et al., *The Ventilation Efficiency of Urban Built Intensity and Ventilation Path Identification: A Case Study of Wuhan.* International Journal of Environmental Research and Public Health, 2021. **18**(21): p. 11684.

975. Rawesat, A. and P. Pilidis, *'Greening' an Oil Exporting Country: A Hydrogen, Wind and Gas Turbine Case Study.* Energies, 2024. **17**(5): p. 1032.

976. Komorowska, A., *Can Decarbonisation and Capacity Market Go Together? The Case Study of Poland.* Energies, 2021. **14**(16): p. 5151.

977. Boscherini, M., et al., *New Perspectives on Catalytic Hydrogen Production by the Reforming, Partial Oxidation and Decomposition of Methane and Biogas.* Energies, 2023. **16**(17): p. 6375.

978. Huang, B., et al., *Exploring Carbon Neutral Potential in Urban Densification: A Precinct Perspective and Scenario Analysis.* Sustainability, 2020. **12**(12): p. 4814.

979. Arogundade, S., M. Dulaimi, and S. Ajayi, *Holistic Review of Construction Process Carbon-Reduction Measures: A Systematic Literature Review Approach.* Buildings, 2023. **13**(7): p. 1780.

980. Urrutia-Azcona, K., et al., *Cities4ZERO: Overcoming Carbon Lock-in in Municipalities Through Smart Urban Transformation Processes.* Sustainability, 2020. **12**(9): p. 3590.

981. Wakefield, J.C. and M. Kassem, *Improving Site Access Verification and Operator Safety in Smart and Sustainable Cities: A Pilot Study in a UK Decarbonisation Project.* 2020.

982. Ballesty, S. and A. Sawhney, *Decarbonisation of the Built Environment: Using Integrated Life Cycle and Carbon Emissions Reporting.* Iop Conference Series Earth and Environmental Science, 2023. **1176**(1): p. 012046.

983. Sharma, A.K. and M. Srivastava, *A Grey-Dematel Approach to Prioritize Challenges for Decarbonization of Energy Sector.* 2023: p. 307-316.

984. Zhang, R., S. Fujimori, and T. Hanaoka, *The Contribution of Transport Policies to the Mitigation Potential and Cost of 2 °C and 1.5 °C Goals.* Environmental Research Letters, 2018. **13**(5): p. 054008.

985. Bompard, E., et al., *World Decarbonization Through Global Electricity Interconnections.* Energies, 2018. **11**(7): p. 1746.

986. Sahoo, B., D.K. Behera, and D.B. Rahut, *Decarbonization: Examining the Role of Environmental Innovation Versus Renewable Energy Use.* Environmental Science and Pollution Research, 2022. **29**(32): p. 48704-48719.

987. Ala, G., et al., *Electric Mobility in Portugal: Current Situation and Forecasts for Fuel Cell Vehicles.* Energies, 2021. **14**(23): p. 7945.

988. Korohod, A., *Decarbonization as One of the Ways to Solve the Problems of Globalization.* 2023. **17**(1): p. 39-47.

989. Aina, O., *Decarbonization, Opportunities and Dilemma for Developing Countries.* American Journal of Computing and Engineering, 2024. **7**(1): p. 1-12.

990. Gajdzik, B., et al., *Renewable Energy Share in European Industry: Analysis and Extrapolation of Trends in EU Countries.* Energies, 2024. **17**(11): p. 2476.

991. Gunawan, J., et al., *Decarbonisation: A Case Study of Malaysia.* Foresight and Sti Governance, 2023. **17**(4): p. 32-44.

992. Abbass, K., et al., *A Review of the Global Climate Change Impacts, Adaptation, and Sustainable Mitigation Measures.* Environmental Science and Pollution Research, 2022. **29**(28): p. 42539-42559.

993. Zamri, M.F.M.A., et al., *An Overview of Palm Oil Biomass for Power Generation Sector Decarbonization in Malaysia: Progress, Challenges, and Prospects.* Wiley Interdisciplinary Reviews Energy and Environment, 2022. **11**(4).

994. Bednar, J., et al., *Operationalizing the Net-Negative Carbon Economy.* Nature, 2021. **596**(7872): p. 377-383.

995. Foggia, G.D., *Energy Efficiency Measures in Buildings for Achieving Sustainable Development Goals.* Heliyon, 2018. **4**(11): p. e00953.

996. Cheekatamarla, P., *Hydrogen and the Global Energy Transition—Path to Sustainability and Adoption Across All Economic Sectors.* Energies, 2024. **17**(4): p. 807.

997. Ramsebner, J., et al., *The Sector Coupling Concept: A Critical Review.* Wiley Interdisciplinary Reviews Energy and Environment, 2021. **10**(4).

998. Best, R. and P.J. Burke, *Adoption of Solar and Wind Energy: The Roles of Carbon Pricing and Aggregate Policy Support.* Energy Policy, 2018. **118**: p. 404-417.

999. Su, Z., J. Huang, and L. Wu, *Economic Study of Wind and Solar Power Generation With Energy Storage Configuration.* Highlights in Science Engineering and Technology, 2024. **112**: p. 445-453.

1000. Guan, X., et al., *Hydride-Based Solid Oxide Fuel Cell–Battery Hybrid Electrochemical System.* Energy Technology, 2016. **5**(4): p. 616-622.

1001. Sun, W. and G. Harrison, *Wind-Solar Complementarity and Effective Use of Distribution Network Capacity.* Applied Energy, 2019. **247**: p. 89-101.

1002. Risco-Bravo, A., et al., *Modeling the Energy Transition: Multiobjective Optimization of Green Hydrogen Deployment Strategies.* E3s Web of Conferences, 2024. **545**: p. 01003.

1003. Dixit, A.C., et al., *Green Hydrogen for Karnataka: Regional Solutions for a Clean Energy Future.* E3s Web of Conferences, 2023. **455**: p. 02020.

1004. Tafarte, P., et al., *The Potential of Flexible Power Generation From Biomass: A Case Study for a German Region.* 2015: p. 141-159.

1005. Zhang, Y., et al., *Simultaneously Efficient Solar Light Harvesting and Charge Transfer of Hollow Octahedral Cu2S/CdS P–n Heterostructures for Remarkable Photocatalytic Hydrogen Generation.* Transactions of Tianjin University, 2021.

1006. Jou, H.L., et al., *Operation Strategy for Grid-Tied DC-coupling Power Converter Interface Integrating Wind/Solar/Battery.* Iop Conference Series Earth and Environmental Science, 2017. **93**: p. 012062.

1007. Pietzcker, R., et al., *System Integration of Wind and Solar Power in Integrated Assessment Models: A Cross-Model Evaluation of New Approaches.* Energy Economics, 2017. **64**: p. 583-599.

1008.   Nejabatkhah, F. and Y. Li, *Overview of Power Management Strategies of Hybrid AC/DC Microgrid.* Ieee Transactions on Power Electronics, 2015. **30**(12): p. 7072-7089.

1009.   Boretti, A., *Energy Storage Needs for an Australian National Electricity Market Grid Without Combustion Fuels.* Energy Storage, 2019. **2**(1).

1010.   Wehrle, N., *The Cost of Renewable Electricity and Energy Storage in Germany.* European Journal of Business Science and Technology, 2022. **8**(1): p. 19-41.

Index

D

E

Z

www.ingramcontent.com/pod-product-compliance
Lightning Source LLC
Chambersburg PA
CBHW080226270326
41926CB00020B/4164